NELSON

Science 9

Concepts and Connections

Program Consultants

Ted Gibb
Formerly of Thames Valley District School Board

Barry LeDrew
Formerly of Newfoundland Department of Education

Authors

Ted Gibb
Formerly of Thames Valley District School Board

Barry LeDrew
Formerly of Newfoundland Department of Education

Donna Osborne
Ottawa-Carleton District School Board

John Patterson
Algoma District School Board

Jill Roberts
Formerly of Ottawa-Carleton District School Board

Meredith White-McMahon
St. James-Assiniboia School Division

THOMSON
NELSON

Australia Canada Mexico Singapore Spain United Kingdom United States

Nelson Science 9
Concepts and Connections

Program Consultants
Ted Gibb
Barry LeDrew

Authors
Ted Gibb
Barry LeDrew
Donna Osborne
John Patterson
Jill Roberts
Meredith White-McMahon

Contributing Authors
Donald Plumb
Bob Ritter
Edward James
Alan J. Hirsch

Director of Publishing
David Steele

Publisher
Kevin Martindale

Program Manager
Tony Rodrigues

Developmental Editor
Lee Geller

Editorial Assistant
Matthew Roberts

Senior Managing Editor
Nicola Balfour

Senior Production Editor
Joanne Close

Copy Editors
Claudia Kutchukian
Dawn Hunter

Production Coordinator
Julie Preston

Creative Director
Angela Cluer

Art Management
Suzanne Peden

Design and Composition Team
Marnie Benedict
Angela Cluer
Nelson Gonzalez
Johanna Liburd
Peter Papayanakis
Amber Passalidis
Suzanne Peden

Illustrators
Andrew Breithaupt
Steven Corrigan
Deborah Crowle
Frank Netter
Myra Rudakewich
Bart Vallecoccia
Jane Whitney

Cover Design
Peter Papayanakis

Cover Image
NASA/Science Photo Library

Photo Research and Permissions
Christina Beamish
Vicki Gould
Cindy Howard

Printer
Transcontinental Printing Inc.

Printed and bound in Canada
1 2 3 4 05 04 03 02

For more information contact Nelson, 1120 Birchmount Road, Toronto, Ontario, M1K 5G4.
Or you can visit our Internet site at http://www.nelson.com

National Library of Canada Cataloguing in Publication Data

Gibb, Ted
Science 9: concepts and connections/Teb Gibb, Barry LeDrew.

Includes index.
ISBN 0-17-612179-X

1. Science—Textbooks.
I. LeDrew, Barry II. Title: Science 9: Concepts and Connections.

Q161.2.N45 2002 500
C2002–904392–1

Reviewers

Pearl Bradd
SciTech Exploratorium of Windsor

Paul Doig
Kawartha Pine Ridge District School Board, ON

Walter Hamilton
Simcoe County District School Board, ON

Shawna Hopkins
District School Board of Niagara, ON

Ann Jackson
Catholic District School Board of Eastern Ontario, ON

Herb Koller
retired York Region District School Board, ON

Ted Laxton
Wellington Catholic District School Board, ON

Michael McArdle
Dufferin-Peel Catholic District School Board, ON

Allan Mills
Simcoe County District School Board, ON

Jaime Perkins
Durham District School Board, ON

Brian Schroder
Dufferin-Peel Catholic District School Board, ON

John Sherk
Toronto District School Board, ON

Bruce Spurrell
Avalon East School board, NF

Robert Tkach
Peel District School Board, ON

Tanya Worobec Halton
District School Board, ON

ACCURACY REVIEWERS

Prof. Carey Bissonnette
University of Waterloo

Prof. Marko Horbatsch
York University

Prof. John Percy
University of Toronto

Mira C. Puri, Ph.D.
Mount Sinai Hospital, Toronto

SAFETY REVIEWER

Stella Heenan
STAO Safety Coordinator

Contents

Unit 1: Exploring Matter 3

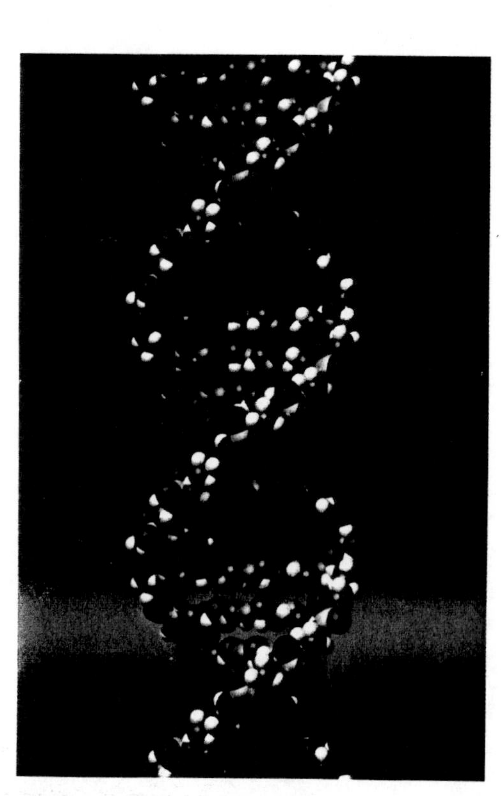

Unit 2: Reproduction: Processes and Applications 63

Unit 3: Electrical Applications 121

Unit 4: Space Exploration 185

Exploring Matter

What do your clothes, the food you eat, the products you use, and the places where you live, work, and learn have in common? They are all matter. Understanding how matter is composed, and how it changes, will help you to make good decisions about whether to use new products (for example, car care products) and how to safely dispose of them when they are no longer useful.

Unit 1 Overview

Overall Expectations

In this unit, you will be able to

- describe the structure of common elements and their grouping in the periodic table
- investigate the properties of common elements and compounds, and relate their properties to their location in the periodic table
- understand the importance, production, use, and hazards of common elements and compounds

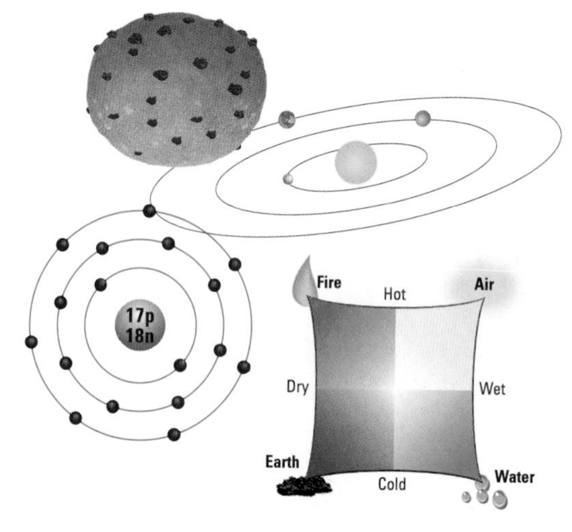

Matter and Change

All forms of matter have physical and chemical properties. These properties determine how matter is used. Combining samples of matter can result in new substances and combinations of properties. New and useful forms of matter are discovered this way.

Specific Expectations

In this unit, you will be able to

- investigate physical and chemical changes in matter
- distinguish between metals and nonmetals and identify their characteristic properties
- formulate questions about and safely plan and conduct investigations
- determine how the properties of substances influence how they are used, including associated risks
- identify and describe careers that require knowledge of the properties of matter

Models for Atoms

The tiniest building block of matter is the atom, a particle so small that it cannot be seen. Throughout history, the properties of matter have been explained by ever-changing models of the atom.

Specific Expectations

In this unit, you will be able to

- describe compounds and elements in terms of molecules and atoms
- describe an element as a pure substance
- recognize compounds as pure substances
- describe how elements are located, mined, and processed in Canada, along with related health and safety issues
- identify each of the three fundamental particles and its charge, location, and relative mass in a simple atomic model
- construct molecular models
- identify and write symbols/formulas for common elements and compounds

H																	He
Li	Be											B	C	N	O	F	Ne
Na	Mg											Al	Si	P	S	Cl	Ar
K	Ca	Sc	Ti	V	Cr	Mn	Fe	Co	Ni	Cu	Zn	Ga	Ge	As	Se	Br	Kr
Rb	Sr	Y	Zr	Nb	Mo	Tc	Ru	Rh	Pd	Ag	Cd	In	Sn	Sb	Te	I	Xe
Cs	Ba	La	Hf	Ta	W	Re	Os	Ir	Pt	Au	Hg	Tl	Pb	Bi	Po	At	Rn
Fr	Ra	Lr															

The Periodic Table

The positions of elements in the periodic table can be used to predict and explain the properties of elements and the compounds they will form.

Specific Expectations

In this unit, you will be able to

- identify general features of the periodic table
- identify the uses of elements in everyday life
- understand the relationship between the properties of the elements and their position in the periodic table
- select and integrate information from various sources
- investigate the chemical properties of representative families of elements

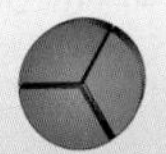

Challenge

Exploring Matter

As you learn about elements and compounds, you will learn how to explain the behaviour of matter. You can demonstrate your learning by completing a Challenge. For more information on the Challenges, see the next page.

1 Marketing Matter

Create a proposal that promotes a new material to a manufacturing company. Make sure to

- identify any raw materials that were used to make this new material
- describe the process that was used to make the new material
- identify any risks associated with the use of this new material

2 Time Capsule

Choose an artifact for a time capsule. The artifact should be

- a sample of a modern material
- a sample of a material whose properties are easily recognized
- represented by a model to explain the structure of the material and its properties

3 A Famous Scientist

Choose a famous scientist who discovered one of the elements, and for that person, write

- a journal entry from when he/she was in high school
- an obituary

Record your ideas for your Challenge as you progress through this unit, and when you see

Unit 1 Challenge

Exploring Matter

In the process of discovering new compounds, scientists must determine whether the process can be copied in a large-scale manufacturing setting. If it can be done, does this manufacturing process result in a negative impact on the environment? With the discovery of new compounds comes the added task of finding new uses for them. Once identified, do the intended uses bring with them any risks for the user?

Each of these Challenges allows you to examine how common elements and compounds are used in everyday life.

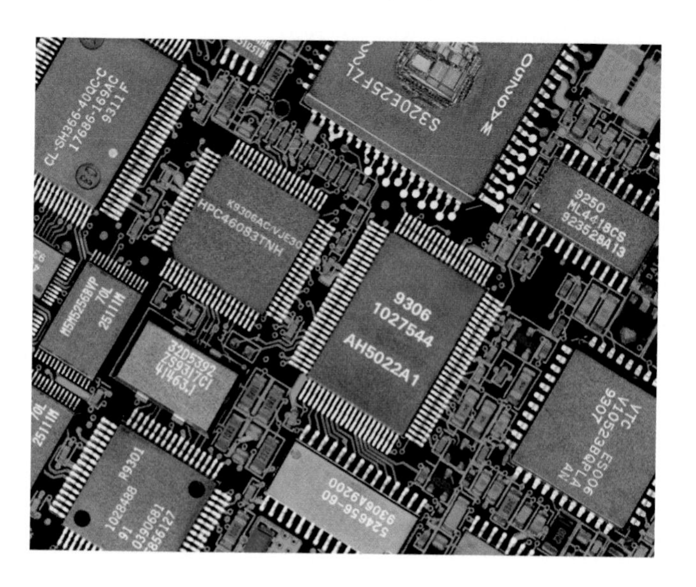

1 Marketing Matter

Research new substances that have been created during the past 10 years. Choose one of these substances, and prepare a marketing proposal that promotes this new material and its uses.

Create an electronic or print display that includes

- the identity of any raw materials that were used to make this new material
- a flow chart that describes the process used to make the new material
- information about any risks (to workers in the manufacturing process, to users, and to the environment) associated with the use of this new material

2 Time Capsule

Imagine that you have just opened a time capsule prepared by students in 1879 when the Corning Glass Company developed the first glass bulb for Thomas Edison's electric light. What do the artifacts tell you about the glass bulb—how it was made, intended uses other than for electric light, and so on.

Prepare artifacts for a time capsule today for students in the next century. The artifacts should include

- representative samples of materials discovered in the early twenty-first century
- materials whose properties can be easily observed.
- a model to explain the structure of the material and its related properties

3 A Famous Scientist

Imagine that you grew up with, went to school with, and lived next door to a famous scientist who discovered a new element that was very useful for society.

Choose a famous scientist and, for that person, create an electronic or poster display that includes

- a journal entry written when he/she was in high school
- an obituary

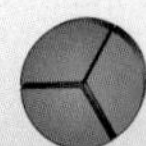

Assessment

Your completed Challenge may be assessed according to the following:

Process

- Understand the specific challenge.
- Develop a plan.
- Choose and safely use appropriate tools, equipment, materials, and computer software.
- Analyze the results.

Communication

- Prepare an appropriate presentation of the task.
- Use correct terms, symbols, and SI units.
- Incorporate information technology.

Product

- Meet established criteria.
- Show understanding of the science concepts, principles, laws, and theories.
- Show effective use of materials.
- Consider legal and ethical issues.
- Address the identified situation/problem.

Unit 1

Getting Started

Matter

What is matter? Matter is both the material of the soccer ball (**Figure 1**) and the air that is inside it. Matter exists in three forms or states: solid, liquid, and gas. When each state of matter is described, the descriptions are called properties. How many properties can you name for solids? liquids? gases?

Figure 1

A soccer ball is composed of matter in two forms.

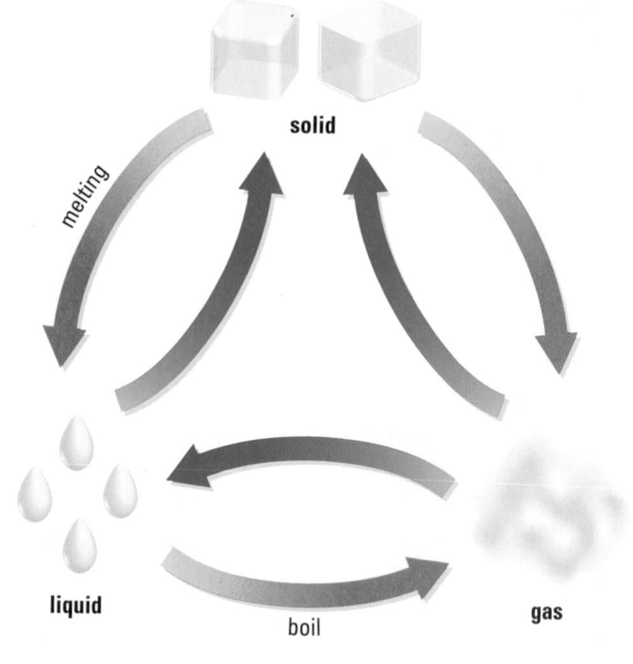

Figure 2

Changes in the states of matter

Each state of matter can also be changed into one of the other states. Solid water (ice) can melt to become liquid water. Liquid water can also be made to boil. As water boils, the liquid changes into a gas. Can you name the other changes in matter that are illustrated in **Figure 2**?

Have you ever carefully watched what happens when you cook a hot dog over an open fire (**Figure 3**)? As the hot dog is heated, it swells until the skin bursts. A boiling liquid runs out of the break in the skin. The juices inside the hot dog have been brought to a boil by the heat of the fire. But what other changes are occurring? The hot dog is changing colour, darkening on the outside as heating continues. If you are not careful to pull it away from the flame, it may even turn black on the outside.

There is a chemical change in the makeup of the hot dog in **Figure 3**. What other signs of chemical change can you name? The new black substance that forms on the outside of the hot dog is carbon. Carbon is an element. How is it different from other elements? How many other elements are present in the hot dog?

Figure 3

If you cook a hot dog too long, carbon forms on the outside of the hot dog.

The elements that make up matter are organized in the periodic table (**Figure 4**). The illustration of the carbon atom represents a model of what scientists believe to be the structure of the carbon atom. This model is based on evidence from chemical changes involving carbon.

C = 6 protons
6 neutrons
6 electrons

H																	He
Li	Be											B	C	N	O	F	Ne
Na	Mg											Al	Si	P	S	Cl	Ar
K	Ca	Sc	Ti	V	Cr	Mn	Fe	Co	Ni	Cu	Zn	Ga	Ge	As	Se	Br	Kr
Rb	Sr	Y	Zr	Nb	Mo	Tc	Ru	Rh	Pd	Ag	Cd	In	Sn	Sb	Te	I	Xe
Cs	Ba	La	Hf	Ta	W	Re	Os	Ir	Pt	Au	Hg	Tl	Pb	Bi	Po	At	Rn
Fr	Ra	Lr															

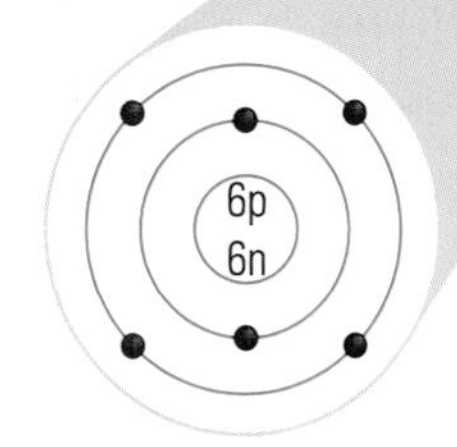

Figure 4

What DO YOU ALREADY *Know?*

1. What are some of the different ways that matter can be organized or classified?
2. With a partner, or in a small group, list as many changes in matter as you can. Use a different action word for each change (for example, a hot dog *cooks*, snow *melts*).
   - **(a)** In which changes do you think a new substance is formed?
   - **(b)** In which changes do you think matter is added to the air?
3. Models are used in science to help you see and understand an idea or concept.
   - **(a)** What scientific models are you already familiar with?
   - **(b)** Choose one scientific model. What does it explain?

Try This Activity Fill a Balloon Using Chemistry

- Examine the samples of baking soda and vinegar provided.
  (a) Describe each substance.
- Put two spoonfuls of vinegar into a glass.
- Add one spoonful of baking soda to the vinegar.
  (b) What happened when the two substances were mixed?
- You will be given a test tube and a balloon. Use the same amounts of each substance to blow up the balloon.

(c) Draw a diagram and explain how you (E3) blew up the balloon without using your breath.

(d) How much did your balloon expand compared to other groups' balloons?

(e) Using the same quantities, what change(s) would you make to get maximum expansion?

1.1 Activity

Safety and Science

Matter includes both helpful and harmful substances. A gas such as oxygen is necessary for survival. On the other hand, a gas such as carbon monoxide from car exhaust is poisonous if inhaled. How do we know whether a substance is safe to use?

Warning symbols are placed on containers of hazardous materials. **Table 1** and **Table 2** show these symbols. In this activity, you will learn about these warning symbols, and you will identify potential hazards in your science class. You will also learn how to work with hazardous products and set rules for working safely.

Part 1: Safety Scavenger Hunt

Materials

- samples of laboratory or workplace chemicals with WHMIS labels
- samples of household products with HHPS labels

Procedure

1 Draw a map of your science lab. Be sure to include the location of all lab benches and other furniture.

Table 1 Hazardous Household Products Symbols (HHPS)

The warning symbols on household products were developed to indicate exactly why and to what degree a product is dangerous.

	poisonous	flammable	explosive	corrosive

danger 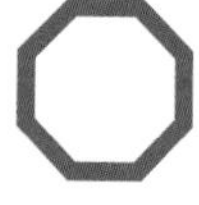				
warning 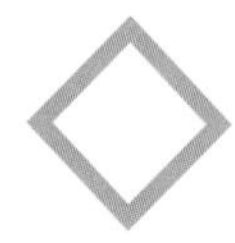				
caution 				

Table 2 Workplace Hazardous Materials Information System (WHMIS) Symbols

WHMIS symbols were developed to standardize the labelling of dangerous materials used in all workplaces, including schools. Pay careful attention to any warning symbols on the materials you use, and handle them appropriately.

Symbol	Meaning	Symbol	Meaning
	compressed gas		dangerously reactive material
	flammable and combustible material		biohazardous infectious material
  	oxidizing material	 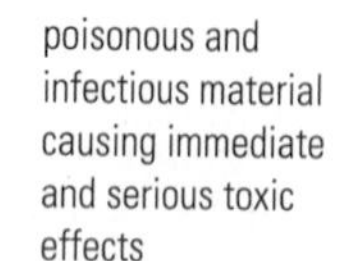	poisonous and infectious material causing immediate and serious toxic effects
	corrosive material		poisonous and infectious material causing other toxic effects

2 Locate each of the following safety devices. Indicate, using the letter, the corresponding location on your map.
A. fire extinguisher
B. fire blanket
C. closest fire alarm
D. safety goggles
E. aprons
F. eyewash station
G. disposal container for broken glass
H. disposal container for chemicals
I. intercom or closest telephone

3 Examine the samples of chemicals and household products provided. For each sample, record in your notebook:
- the name of the product
- the active ingredient (from the label)
- the warning symbol and the nature of the hazard (A)

4 Refer to the Skills Handbook at the back of the textbook. (B) With your partner, discuss and write a safety rule, or direction, for each of the following situations as they apply to laboratory activity:
- eye protection
- clothing protection
- burns, cuts, or injuries
- allergies
- broken glass
- waste or spilled chemicals
- skin contact with a corrosive chemical
- chemicals splashed into the eyes
- damaged equipment
- heating chemicals
- smelling chemicals
- fooling around, paying too little attention
- food and drink
- out-of-school investigations
- living things

Part 2: Kitchen Chemistry

The kitchen in your home can also be considered a chemical laboratory. It contains a variety of chemicals in solid and liquid form. A cook would call them ingredients. To prepare food, a cook often mixes the ingredients, sometimes using heat, to change them into a delicious meal. Often during this process, many changes in matter can occur.

Baking soda and salt are two chemical substances, or ingredients, found in most kitchens. What role do they play in the preparation of a meal?

Starch is a chemical that is found in foods that come from plants. Is there a way to determine whether a particular food item contains starch?

You will find the answer to these questions as you do the following activity.

Some household chemicals may form dangerous products when mixed together. Check with your teacher before mixing any substances other than those directed in the procedure that follows.

Materials

- 3 plates
- masking tape
- marking pen
- table salt (sodium chloride)
- powdered starch
- baking soda (sodium bicarbonate)
- vinegar (acetic acid)
- iodine solution
- water
- 3 micro droppers

Wear goggles and gloves when using iodine solution.

Design

(C6) Design a table for your observations.

Procedure

1 Attach a small piece of masking tape to each of three plates. Label the plates A—salt, B—starch, and C—baking soda.

2 On plate A, place a teaspoon of table salt in a pile. Leaving as much space as possible among piles, add two more teaspoons of salt. You should now have three piles of salt on the plate.

3 Repeat step 2 with plate B using starch, and with plate C using baking soda (**Figure 1**).

4 Fill a micro dropper with vinegar. Place 3–5 drops of vinegar on one of the piles on each of the three plates (**Figure 2**).

(E1) **(a)** Record your observations.

5 Fill a second micro dropper with iodine solution. Place 3–5 drops on one of the remaining piles on each of the three plates.

(E1) **(b)** Record all observations.

6 Fill a third micro dropper with water. Place 3–5 drops of water on the remaining pile on each of the three plates.

(E1) **(c)** Record all observations.

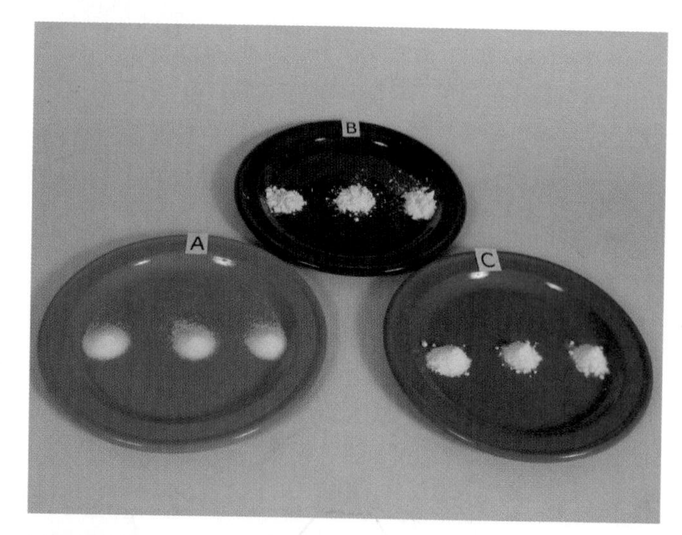

Figure 1

Figure 2

Analysis and Conclusion

(d) (F1) Which substances caused a change in the table salt? During which change, if any, was a new substance formed?

(e) (F1) What changes did you observe in the starch? Was a new substance formed during any of the changes?

(f) (F1) Consider the data you collected for the baking soda. Do any of your observations suggest that a new substance has formed? Explain.

Understanding Concepts

1. Why is it important to use a standard set of safety symbols when labelling substances
   - **(a)** in the home?
   - **(b)** in the science laboratory or the workplace?
2. Briefly describe what you would do if a corrosive chemical you were using came in contact with bare skin on your arm.
3. Do you think it is always safe to pour waste chemicals down the sink with lots of water? Why or why not?

Making Connections

4. **(a)** Why do body shop employees wear a facemask when painting a car (**Figure 3**)?

Figure 3

   (b) Why do hairdressers wear rubber gloves when colouring or chemically straightening hair?

   (c) Why do firefighters wear a breathing apparatus and protective clothing when entering a burning building (**Figure 4**)?

Figure 4

Exploring

5. Locate containers in your home that have HHPS symbols on the labels. (A)
   - **(a)** According to the symbol and information on the label, is each container and its contents stored safely? (Are flammable materials stored away from heat sources, for example?)
   - **(b)** Do any containers in your home have WHMIS symbols on the labels? If so, how do you suppose they got there?
   - **(c)** Prepare an action plan for safely disposing of the products.

Challenge

1 What types of hazards or safety issues should you consider when marketing a new substance?

2 If the material you chose for your artifact was found to be highly reactive with both water and oxygen, would it be a good choice to put into the time capsule? Why or why not?

Work the Web

Visit www.science.nelson.com and follow the links from *Science 9: Concepts and Connections,* 1.1, to find a Material Safety Data Sheet (MSDS). List the information that is on the sheet. What is the importance of this sheet? Why aren't only WHMIS labels used?

1.2

Physical and Chemical Properties

When you choose what to wear, what to eat, and what personal care products to buy, you are making your decisions based on the characteristics or properties of the items. Your decision on what to wear, for example, might be based on colour, style, or simply what is clean!

When it comes to the science of chemistry, the choice of substances for a particular purpose is also based on properties. Scientists have found it useful to group properties into two categories: physical properties and chemical properties.

Table 1 Physical Properties Observed with the Senses

Property	Describing the property	Sense used
colour	It is black, white, red,...	sight
texture	It is fine, coarse, smooth, gritty.	touch
odour	It is odourless, spicy, sharp, burnt	smell
lustre	It is shiny, dull,...	sight
clarity	It is clear, cloudy, opaque (thick).	sight
taste	It is sweet, sour, salty, bitter,...	taste
state	It is a solid, liquid, or gas.	sight

Physical Properties

When you observe any substance—whether you see it, touch it, hear it, smell it, or taste it—you are using your senses to observe its physical properties. A **physical property** is a characteristic or description of a substance that may help to identify it. Typical physical properties might include those listed in **Table 1**.

Other physical properties can be observed if simple tests and measurements are used to help your senses. For example, the mass and volume of a block of wood can be measured. The ratio of mass to volume describes the **density** of the wood, usually expressed in grams per cubic centimetre (g/cm^3).

Hardness is a measure of the resistance of a solid to being scratched or dented (**Figure 1**).

Melting point is the temperature at which a solid melts. **Boiling point** is the temperature at which a liquid begins to boil (**Figure 2**).

Solubility is the ability of a substance to dissolve in a solvent such as water. You may remember that this physical property depends on temperature (**Figure 3**).

Viscosity is a measure of how easily a liquid flows: the thicker the liquid, the more viscous it is. Honey is an example of a viscous liquid (**Figure 4**).

Figure 1

Figure 2

Figure 3

Figure 4

Chemical Properties

Have you ever accidentally put a piece of dark clothing into the washing machine along with a load of white clothing? You added bleach to make the white clothes whiter, so your dark clothing got "bleached" to a lighter colour at the same time.

When bleach interacts to lighten coloured fabric, this is called a chemical property. A **chemical property** describes the behaviour of a substance as it becomes a new substance.

Combustibility is a chemical property that describes the ability of a substance to burn. To burn, a substance requires oxygen from the air.

When a flame is brought close to a mixture of gasoline fumes and air, the mixture will ignite and burn. That is why there are "no smoking" signs around gas pumps (**Figure 5**).

If a substance is **flammable**, it will burn when exposed to a flame. What materials, other than gasoline, can you think of that are flammable?

Some substances will become new substances when they interact with light. **Light sensitivity** is a chemical property. The next time you go into a pharmacy, check to see how many products are stored in dark containers (**Figure 6**). One of these products is hydrogen peroxide. It is sometimes used as a disinfectant, protecting wounds from infection. When exposed to light, it changes into water and oxygen gas, and is no longer useful.

Figure 5

Figure 6

Challenge

1 List the physical and chemical properties of the new material that you are marketing.
2 Check the physical and chemical properties of the time capsule material. Will it resist air and moisture?

Understanding Concepts

1. What property is described by each of the following statements?
   - (a) A steel blade can scratch glass.
   - (b) Water boils at 100°C.
   - (c) Alcohol is flammable.
2. Give one physical and one chemical property of
   - (a) a marshmallow.
   - (b) a hot dog.
   - (c) an egg.

Making Connections

3. The top of a glass container of water breaks, and some pieces of glass fall into the container. What physical property of glass would help you to separate the broken glass from the water?

Exploring

Ⓝ Ⓞ 4. Find out about the disaster of the airship Hindenburg.
   - (a) What gas was used to inflate the airship? What property of this gas contributed to its disaster?
   - (b) What gas is used to inflate airships today? What property of this gas makes it safer?

Physical and Chemical Changes

Some of the most important uses for a substance are those that result from change. We experience many such changes every day. Applying heat to cook an egg, burning gasoline in a car, freezing water to make ice cubes, and mixing oil and vinegar for a salad dressing are just a few examples. Recognizing and recording what was done to make the change occur is an important first step. Discovering a use for the substance afterward requires observing its new properties.

Physical Change

In a **physical change**, the substance involved remains the same. It may change form or state, however. All changes of state (**Figure 1**) are physical changes. How many changes of state might you observe in the candle wax when the candle in **Figure 2** is burning? If you answered two, you are correct. There are actually two more, which are very difficult to observe. Look carefully at **Figure 1**. Can you predict what they might be?

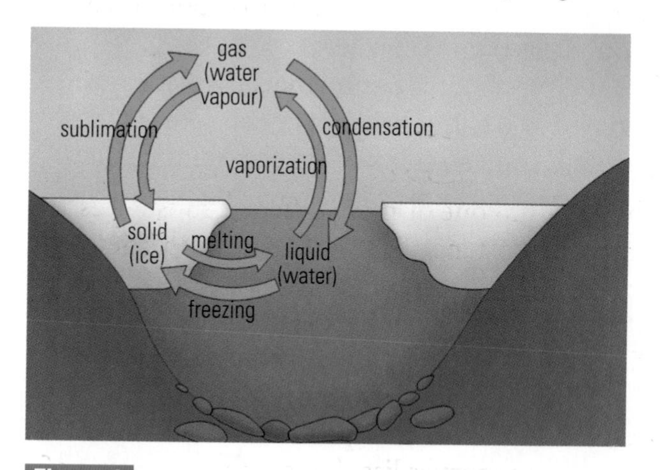

Figure 1

Changes of state

Try This Activity: Observing a Burning Candle

(E1) You can observe a lot about how a substance (E2) changes simply by observing a burning candle (**Figure 2**). Your goal is to make as many observations about the properties of the candle as you can in the time period specified by your teacher. You may want to divide your time in the following way:

> Be sure to tie back long hair when working around an open flame.

- Use your senses and any measuring instruments you have to describe as many properties as you can before the candle is lit.
- Light the candle. Again, use your senses and any measuring instruments to describe any new properties that you observe.
- Carefully blow out the candle. Describe any new properties that you observe.

Figure 2

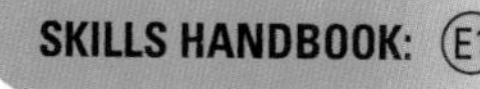
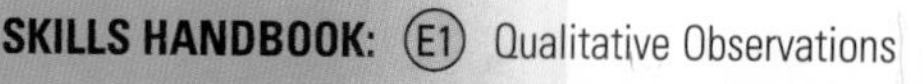
SKILLS HANDBOOK: (E1) Qualitative Observations (E2) Quantitative Observations

In addition to changes of state, there are other physical changes. Dissolving is a physical change. Have you ever thought about what happens when you dissolve a spoonful of sugar in a cup of tea or coffee? The sugar particles spread out, but they are still there as sugar particles.

You can reverse the dissolving process by evaporating the tea or coffee and collecting the sugar again. Most physical changes can be reversed. Can you think of one that cannot be easily reversed?

Chemical Changes

In a **chemical change**, the original substance is changed into one or more different substances. The new substances have different properties from the original substance.

Think about what happens as the candle in **Figure 2** continues to burn. The candle becomes shorter as it continues to burn. Where did the wax go? To answer this question, review the chemical properties listed in section 1.2. Some of the wax reacted with oxygen in the air to produce water vapour, carbon dioxide gas, heat, and light. The wax that seemed to disappear actually changed into new substances in a **combustion** reaction. The wax is combustible.

Most chemical changes are difficult to reverse. The new substances formed are unlikely to combine again to form the original substance. In addition to combustion, or burning, examples of chemical change include rusting and cooking.

When you observe a chemical change, often you cannot "see" the chemical change. Instead, you can observe only the results of the chemical change—the heat and light produced by a burning candle, for example. Heat and light are clues that a chemical change has happened.

Table 1 contains additional clues. But remember, they are just clues. They *suggest* that a new substance has formed. You must consider several clues to determine what type of change has taken place.

Table 1 Chemical Change Clues

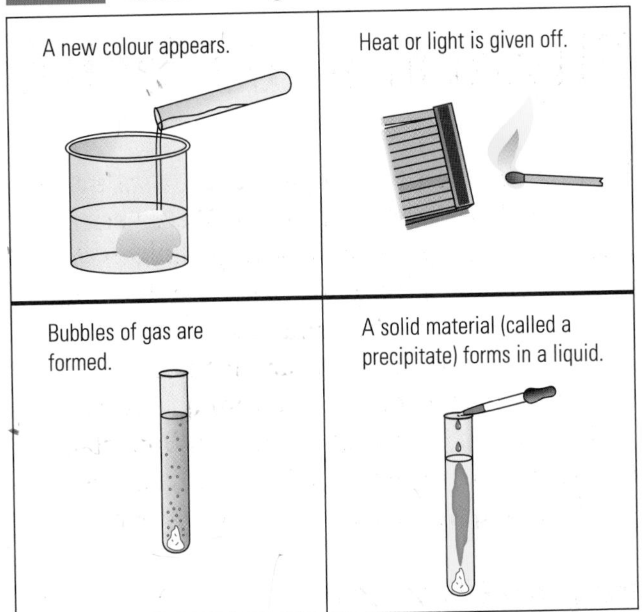

A new colour appears.	Heat or light is given off.
Bubbles of gas are formed.	A solid material (called a precipitate) forms in a liquid.

Understanding Concepts

1. Explain in your own words how a physical change and a chemical change are different.
2. Identify each of the following as a physical or chemical change. Explain your choice.
   (a) shattering glass
   (b) baking cookies in the oven
   (c) lights left on in a room
   (d) burning leaves in the fall

Making Connections

3. Why should you never operate a gas or charcoal barbecue inside your home?

Reflecting

4. Look at your observation table for Part 2: Kitchen Chemistry in section 1.1. Which combinations of kitchen chemicals produced physical changes, and which produced chemical changes? (F1)

Challenge

1, 3 What chemical changes were used to produce the new substance that you are marketing? Which scientist produced this new substance for the first time?

1, 2 Does the new substance that you have chosen undergo any chemical changes that might make it harmful to anyone who comes in contact with it?

1.4 Investigation

INQUIRY SKILLS
- ○ Questioning
- ○ Hypothesizing
- ○ Predicting
- ● Planning
- ● Conducting
- ● Recording
- ● Analyzing
- ● Concluding
- ● Communicating

Observing Changes

In the previous two sections you learned about the physical and chemical properties of substances, and clues for recognizing physical and chemical changes. In this investigation, you will observe changes in chemical substances and identify which are physical and which are chemical.

Question

How can you recognize a chemical change as different from a physical change?

Materials

- 2 plastic spoons
- baking soda (sodium bicarbonate)
- calcium chloride
- 25 mL of phenol red solution (50% phenol red with water)
- 1 large freezer bag
- plastic film canister

Design

(C6) **(a)** Design a table to record your observations.

Procedure

1 Put one spoonful of baking soda into one corner of the freezer bag.

(b) Describe the baking soda in your observation (E1) table.

2 Put two spoonfuls of calcium chloride in the opposite corner of the bag (**Figure 1**).

(c) Describe the calcium chloride in your (E1) observation table.

3 Measure approximately 25 mL of phenol red solution into a plastic film canister. Put the lid on the canister and carefully place it in the centre of the bag (**Figure 2**).

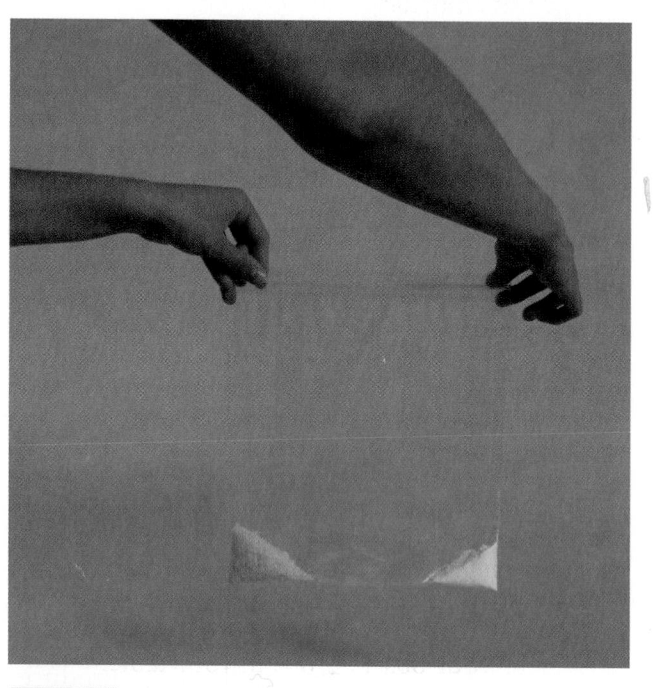

Figure 1

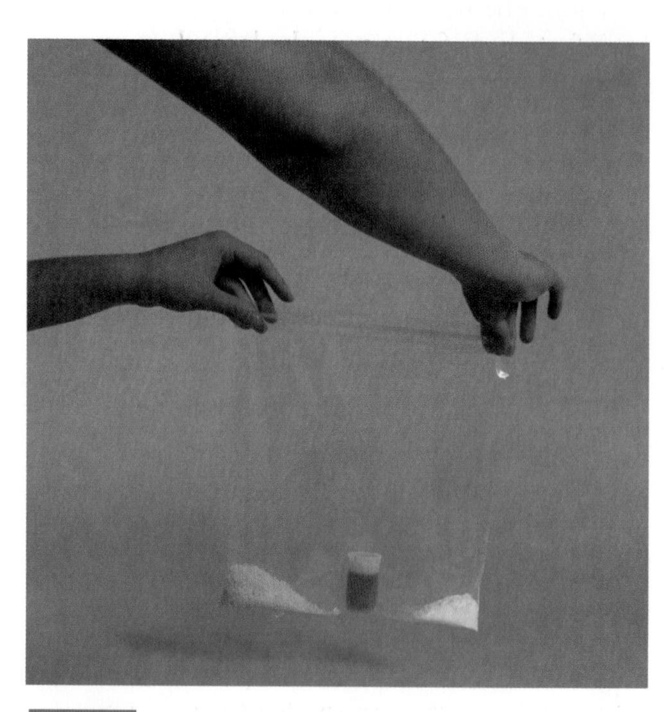

Figure 2

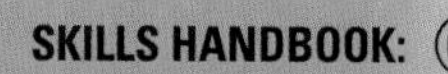
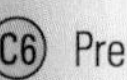

SKILLS HANDBOOK: (C6) Preparing Observation Tables (E1) Qualitative Observations

4 Seal the bag. From the outside of the bag, carefully remove the lid from the film canister. Pour half the contents onto the baking soda (**Figure 3**), and the other half onto the calcium chloride.

5 From the outside of the bag, use your hand to mix the phenol red and the baking soda in one corner of the bag.

(d) Describe any changes that occur as the two substances are mixed. Record your observations in your table. (E1)

6 Repeat step 5 with the phenol red and the calcium chloride in the other corner of the bag.

(e) Describe any changes when these two substances are mixed. Record your observations in your table. (E1)

7 Mix all the materials in the bag.

(f) Record your observations in your table. (E1)

8 Dispose of the bag as directed by your teacher.

Analysis and Conclusion

(g) What kind of change took place when you mixed the substances in step 5? Explain. (F1)

(h) What kind of change took place when you mixed the substances in step 6? Explain. (F1)

(i) What kind of change took place in step 7? Explain. (F1)

(j) At any point in the procedure, did you observe a physical change? If so, when? (F1)

(k) Use the evidence in your observation table to make a list of the clues that you can use to recognize a chemical change. (F1)

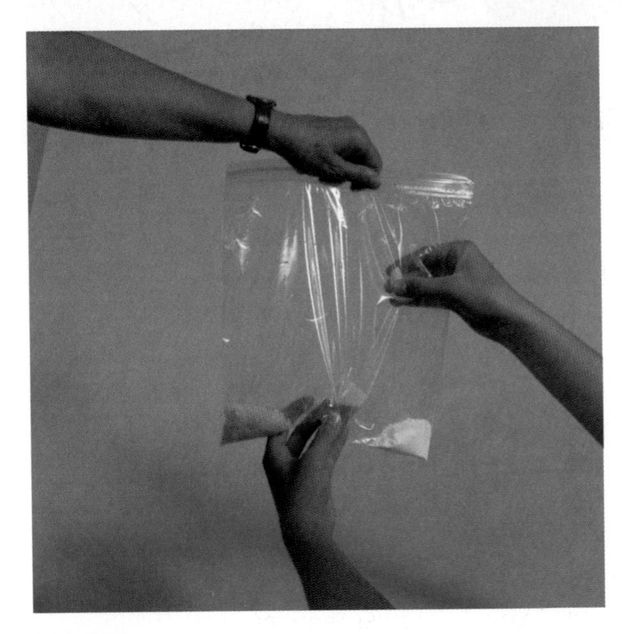

Figure 3

Understanding Concepts

1. Classify each of the following as a physical or a chemical change. Explain your decision.

(a) rotting garbage

(b) cutting up vegetables for a meal

(c) a silver necklace leaving a mark on your skin

(d) cooking an egg

(e) bleaching a stain out of your clothing

Making Connections

2. List three examples of physical changes and three examples of chemical changes that you have observed at home. For each example, give a reason for your decision.

Challenge

1 What kind of container was used in producing your new material? Do you think a plastic freezer bag would be suitable? Why or why not?

2 Would a plastic freezer bag be suitable to use as a time capsule? Why or why not?

Work the Web

Visit www.science.nelson.com and follow the links from *Nelson Science 9: Concepts and Connections*, 1.4, to find out what makes the best container for chemical reactions. From the information you find, choose the best container to use as a time capsule, and collect data to support your choice.

(E1) Qualitative Observations (F1) Interpreting Observation Tables

Everyday Chemical Changes

Have you ever thought about why cars rust? Or, more importantly, why only the metal parts rust while the plastic parts do not? You have learned that different substances have different physical and chemical properties that determine their uses.

One chemical property that has great economic importance is the slow chemical reaction of a metal with oxygen from the air, or **corrosion**. Can corroded metal be fixed, or must it be replaced? Can corrosion be prevented?

Rusting is a specific example of corrosion. Rusting involves the corrosion of iron (**Figure 1**). Iron reacts with oxygen from the air, water, and other chemical substances dissolved in the water. Rust, or iron oxide, is the product of this chemical change. Every year, millions of dollars of damage are caused to vehicles, building structures, and other iron products. Rust is particularly damaging because of one of its physical properties: rust is porous. It absorbs water almost like a sponge. The rust eventually flakes off, exposing fresh metal underneath to oxygen. This process continues until the rust has eaten its way right through the metal.

Aluminum, on the other hand, has a similar chemical property. It also reacts with oxygen in the air, but the aluminum oxide that forms is strong and unaffected by water. The oxide layer protects the aluminum from any further corrosion.

The corrosion of silver (**Figure 2**) results in a surface coating, or tarnish. The black layer can be removed by polishing the silver.

Figure 1

Rust damages the metal car body.

Figure 2

Silver slowly corrodes in the air.

Preventing Corrosion

There are several ways of preventing corrosion. Each involves protecting the metal surface from oxygen. Exposed metal surfaces, such as the bridge in **Figure 3**, and the outside of cars can be painted. As long as the painted surface is not broken or cracked, oxygen cannot get at the metal. For the same reason, the bottom and inside surfaces of cars can be sprayed with oil to protect them.

Figure 3

Some bridges are so big that painters take years to finish the whole structure. Then they have to start over again!

Another way to prevent corrosion is to use materials that do not react chemically with oxygen in air, or with water. Car bumpers and panels may get bumps or scratches. Plastics are being used for these parts because even when they are scratched, they never corrode. They always stay strong and flexible.

Try This Activity Preventing Corrosion

You work for a small corrosion laboratory. Your company has a contract to investigate ways to prevent corrosion of aluminum, magnesium, and steel metals. Your team is assigned to investigate and compare the effectiveness of two methods of preventing corrosion: painting and oiling. You will be given these materials:

- sample strips of aluminum
- magnesium and steel nails
- paint brushes
- 6 containers
- salt
- water
- paint
- motor oil

- (C3) Design a test, including safety procedures, to compare the effectiveness of paint and oil in protecting metals from corrosion.
- (C6) When you have your teacher's approval for your design, make a data table to record your observations.
- Conduct your test using the necessary materials.

(a) (E1) Describe the appearance of the materials before you start.

(b) (E1) Describe any changes in the materials during the test.

(c) Was your experimental design a fair test? Explain.

(d) (F4) Was paint or oil more effective in preventing corrosion? Use your observations to explain your answer.

(e) (G) What changes would you make to your experimental design? With your teacher's permission, try them.

(f) (R) Present your results in a report to be distributed to companies that manufacture aluminum, magnesium, and steel.

Understanding Concepts

1. What is corrosion?
2. How is iron oxide formed?
3. Describe three ways to protect a metal from corrosion.
4. **(a)** Which parts of a car corrode the most? Why?
   (b) How can car owners help to reduce the effects of corrosion?

Making Connections

5. List objects in your home that can corrode. What steps can you take to protect these items from corrosion?

Challenge

2 How can you make sure that the artifacts or materials that you choose for your time capsule will not undergo corrosion?

Work the Web

Visit www.science.nelson.com and follow the links from *Science 9: Concepts and Connections,* 1.5, to find additional ways of preventing rust and corrosion.

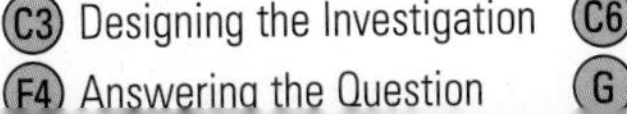

(C3) Designing the Investigation (F4) Answering the Question (C6) Preparing Observation Tables (G) Concluding (E1) Qualitative Observations (R) Writing for Specific Audiences

DECISION-MAKING SKILLS

- ● Define the Issue
- ○ Identify Alternatives
- ● Research
- ● Analyze the Issue
- ● Defend a Decision
- ○ Evaluate

"Xtreme" Chemical Changes

When you see a colourful fireworks display, you probably don't think about chemistry. But it is chemistry that gives us the special effects of bursts of colour, flashes, and sound (**Table 1**).

Table 1 Some Chemicals Used for Special Effects

Materials	Special effects
magnesium metal	white flame
sodium oxalate	yellow flame
barium chlorate	green flame
cesium (II) sulphate	blue flame
strontium carbonate	red flame
iron filings and charcoal	gold sparks
potassium benzoate	whistle effect
potassium nitrate and sulfur	white smoke
potassium perchlorate, sulfur, and aluminum	flash and bang

The creators of fireworks displays are called pyrotechnics technicians. They are specialists in controlled explosions and must keep chemistry in mind at all times.

Each firework explosion is a carefully controlled series of chemical changes that occur at just the right times. These chemical changes produce large amounts of heat in short periods of time.

Chemistry and Fireworks

A typical firework contains a fuel, a source of oxygen (called an oxidizer), a fuse (a source of heat to start the reaction), and a colour producer. Suppose the technicians have the job of making a firework that will rise 50 m and then produce a red burst of fire, followed by a loud bang and a flash. The technician would have to make, by hand, three different explosive mixtures: one to lift the firework shell into the air, and one for each of the special effects (**Figure 1**).

The first and the most dangerous step is mixing the ingredients. The oxidizer is the main component, making up anywhere from 34% to 68% of the material in the firework. Typical oxidizers are potassium nitrate (KNO_3), potassium chlorate ($KClO_3$), and ammonium perchlorate (NH_4ClO_4). When the oxidizer in the fuel reacts with sulfur or aluminum, it creates great amounts of heat, as well as a "bang" and flashes of light. The bang comes from the rapidly expanding gases that are being produced.

Each mixture also contains binders such as paraffin oil and red gum. Binders act as a fuel and also hold the mixture together. Magnesium metal, or any of the metallic salts in **Table 1**, is added to

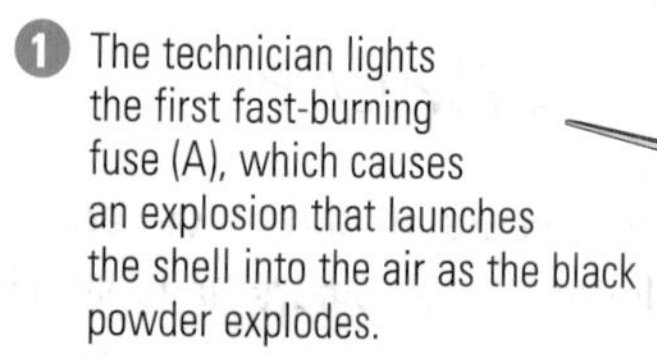

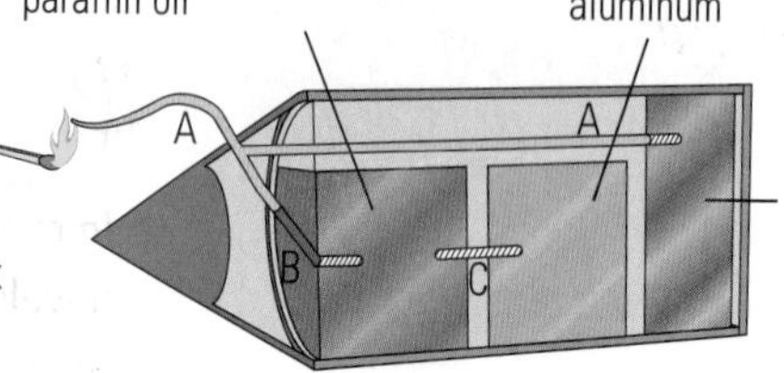

2 Fuse A also lights a slow-burning fuse (B) which ignites a mixture that produces a red burst, when the shell is high in the air.

3 The red explosion lights another slow-burning fuse (C) which ignites a mixture that produces a final flash and loud bang.

Figure 1
A fireworks shell

produce a specific colour. It is very important to understand how each chemical will react with others. Pyrotechnic technicians must be careful in choosing ingredients so that the oxidizer will not react with the metallic salt while it is in storage. If they were to react, there could be a dangerous explosion. Finally, the technician wraps each mixture in a cardboard package and links the packages together with fuses.

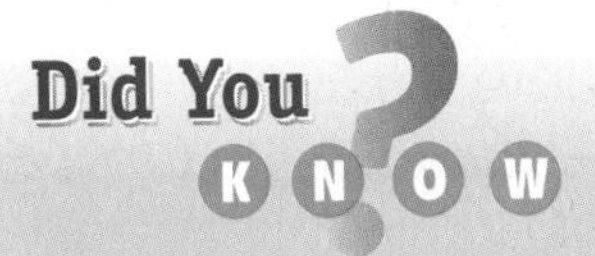

Watching fireworks on Canada Day is an annual tradition for many Canadians. However, many people do not realize how dangerous fireworks can be, particularly when children handle them. Thousands of injuries occur every year as a result of improperly using fireworks. Data released from the Canadian Hospitals Injury Reporting and Prevention Program (CHIRPP) report suggests that following proper safety procedures would prevent most of these injuries ("Injuries Associated with Fireworks," Health Canada, 1999).

Work the Web

Go to www.nelson.science.com and follow the links from *Science 9: Concepts and Connections,* 1.6. Find out which salts are used to make each of the following colours: red, orange, gold, yellow, electric white, green, blue, purple, and silver. What do pyrotechnic technicians need to get pure colours?

Understanding Concepts

1. Explain how fireworks give off bursts of light and sound.
2. What does an oxidizer do?
3. Why are oxidizers so dangerous?

Challenge

1 In recent years, an indoor or "cold" firework has been developed. New colours have also been developed as the research in colour technology advances. Could you market either of these new "products"?

2 Could you create a 3-D, cross-sectional model of a "cold" firework that would show the elements of design that differ from a regular firework?

3 Many elements are used to create fireworks. Could the scientist who discovered one of these elements be your famous scientist?

Debate Should fireworks be banned?

Statement

No one should be allowed to buy or use fireworks at any time within city boundaries.

THE PROPOSAL
No one should be allowed to buy or use fireworks at any time within city boundaries.

Laurel Bishop
Julie F. Fletcher
P. Tolean
John Hornstein
Olive Travers
Mary Tong
Anthony Petwell

Point

- Fireworks are dangerous mixtures of chemicals. When ignited, they can explode in unpredictable ways. People have been terribly injured through the unsafe use of fireworks.
- Fireworks displays pollute the environment. The reactants involved produce nitrogen dioxide and sulfur dioxide, both of which are poisonous gases and produce acid precipitation. Fireworks also create noise pollution.
- Fireworks are very expensive and last only a few seconds. We would be better off using the money to celebrate special occasions in other ways.

Counterpoint

- Fireworks are a traditional way of celebrating a special occasion in some cultures. The move to prevent fireworks could be seen as discriminating against those groups.
- People who want to use fireworks will continue to do so outside the city. In wooded areas or farmlands far from emergency services, the risk of fire or accident would be greater.
- Fireworks displays that mark special events promote tourism and bring economic benefits to the community.

What Do You Think?

- A group will be assigned to each side of the debate. You will argue for or against the statement.
- (I) In your group, discuss the statement and the points and counterpoints. Write down additional points and counterpoints that your group considers.
- (J) (N) (O) Search newspapers, a library periodical index, a CD-ROM directory, and the Internet for information on the safe use of fireworks.
- (L) Elect your spokesperson(s) and prepare to defend your group's position in a class debate.

SKILLS HANDBOOK: (I) Identifying Alternatives/Positions (J) Researching the Issue (N) General Research (O) Internet Reseach (L) Defending a Decision

Hair Colourist

Mary Talbot

If you have the right personality and an interest in the science of hair, then Mary Talbot, a colourist who teaches at Winnipeg Technical College, recommends a career as a hair colourist.

The hair colouring business begins and ends with the clients. "You may have your own idea of what auburn means, but the client may have something else in mind—and you have to figure out exactly what colour they want, and then how to make it!" Students must have an understanding of colour theory, chemical substances, and the changes they undergo.

Colouring hair is a complex chemical process. Once the hair is coloured, the change is permanent until new hair grows in or the coloured hair is cut off. Lightening the natural colour of a client's hair requires a two-step process. First, the natural colour pigment in the hair must be removed. The "lifting" of the original pigment is done using chemicals called peroxides, or with ammonia. If the client's hair becomes as white as paper, the addition of the artificial colour pigment is straightforward. Often though, bleaching leaves a client's hair yellow or orange. The colourist must now mix and match colours to get the desired colour.

The chemical substances used in the hair colouring process must be stored, used, and disposed of safely. These chemicals use the WHMIS labelling system. As well, a file of MSDS (Material Safety Data Sheet) with more detailed information on these chemicals must be maintained. Students at the college are tested on their knowledge of the chemicals used in the business. They must score 100% on these safety tests to complete program requirements!

Clients spend a lot of time with their colourists, so communication and listening skills are important. Talbot's advice: "Study long and hard at school, and apprentice in a good salon."

Making Connections

1. Find out whether there are any courses for hairstylists at your school or at a nearby school. What are the requirements, if any, to get into the course? How long is the training?
2. Are there any advantages to taking an apprenticeship program? Explain.

Reflecting

3. A customer asks you to colour his dark brown hair with "dark blond" tips. How would you get him to tell you exactly what he wanted before you begin?

Work the Web

There are two basic types of hair colour: temporary and permanent. Visit www.science.nelson.com and follow the links from *Science 9: Concepts and Connections*, 1.7, to find out how they differ chemically and how each affects hair.

Models of Matter

Up to this point, you have observed matter and its behaviour. You have observed both physical and chemical properties, and how these have changed during a physical and chemical change. You have seen a number of ways in which matter is used, based on these properties.

Now comes the hard part: how do you explain the physical evidence that you have observed in earlier sections of this unit? Science has always used models to explain behaviour. A model for matter that you may remember is the particle model.

The Particle Model

When you first learned about the particle model you may have been asked to imagine taking a small sample of a chemical substance and breaking it up into smaller and smaller pieces. You would eventually come to the smallest possible particle. This smallest particle was called a "building block." The basic principles of the particle model are illustrated and explained in **Table 1**.

Table 1 A Particle Model for Matter

Principle	Illustration
1. All matter is made up of tiny particles.	
2. All particles of one substance are the same. Different substances are made of different particles.	
3. The particles are always moving. The more energy the particles have, the faster they move.	hot cold
4. There are attractive forces between particles. These forces are stronger when the particles are closer together.	particles far apart—force weak particles close together—force strong

Pure Substances, Mixtures, and the Particle Model

You have learned that matter can be classified into three categories: elements, compounds, and mixtures.

A particle model for a pure substance shows that it contains only one kind of particle. A mixture, on the other hand, contains at least two kinds of particles (**Figure 1**).

Each kind of particle that is observed in a mixture, then, must represent a pure substance. A mixture consists of two or more pure substances.

When you make pizza, for example, you first spread tomato sauce over the dough. The dough can be considered a pure substance, and the tomato sauce can be considered a pure substance. So can the mushrooms, pepperoni, and other ingredients you scatter over the cheese and

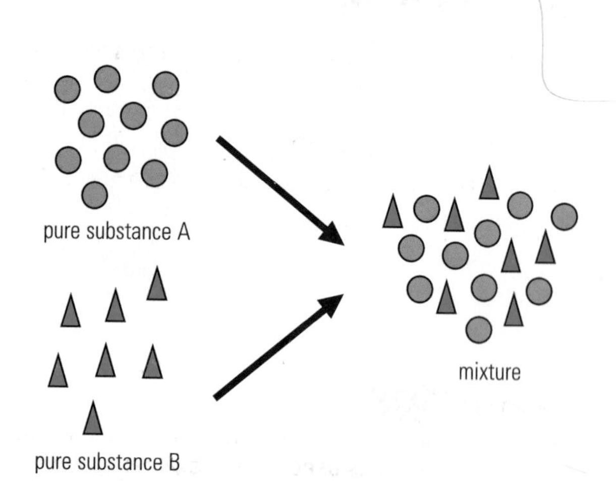

Figure 1
Most substances you will come into contact with are mixtures. Mixtures contain at least two pure substances.

tomato sauce. Each part can be clearly seen before and after baking the pizza. Pizza is an example of a mixture (**Figure 2**).

Elements, Compounds, and the Particle Model

Many properties of matter can be explained by using the particle model. But what are these particles?

Observations of pure substances as they were discovered led scientists to believe that these particles, or building blocks, were elements and compounds. On the classification of matter chart in section 1.9, pure substances were subdivided into elements and compounds. Today there is evidence that elements and compounds can be broken down into smaller particles. These particles are called **atoms** and **molecules**.

Figure 2

A mixture contains several kinds of particles.

Understanding Concepts

1. Use a diagram to illustrate and explain the difference between a pure substance and a mixture.
2. Give two examples of molecules that are made from the same kind of atom. Which compounds do they represent?

Making Connections

3. Copy and complete **Table 1** by filling in the middle column. Choose from
   (i) pure substance—element
   (ii) pure substance—compound
   (iii) mixture

Table 1

Name of substance	Type of substance	Description
table salt		white crystalline solid
orange juice		mixture of juice and pulp
copper		reddish-brown wire
iron		coarse black powder
salad dressing		oil and vinegar

4. Now that you know that atoms and molecules are the "particles" in the particle model, do you think atoms can be broken down into smaller particles? Explain.

Challenge

1, 2 The particle model is used to help explain matter. How could you use this model in the Challenge you have chosen?

3 The scientists credited with exploring the atom include Dalton, Thomson, Rutherford, and Bohr. How was their world different from the world we live in today?

Classifying Matter

Tens of thousands of different chemical substances make up the Earth. Some of these substances appear to be quite similar, while others are very different from one another. Have you ever thought about the clothes people wear, for example? Some are made from cotton, which comes from the cotton plant. Others are made from wool, which comes from an animal.

Some clothes are also made from substances that are **synthetic**—invented and produced by people. Nylon, polyester, and Gore-Tex are all synthetic materials used to make clothing. Which of the materials in **Figure 1** are natural, and which are synthetic?

Organizing chemical substances into two groups, those that are natural and those that are synthetic, is the first step in classifying matter.

Figure 1

Try This Activity Classification

Think about the last time you went shopping. How did the stores you visited organize their merchandise for sale? If you visited a music store, how were the CDs or tapes organized? If you visited a clothing store, how were the articles of clothing organized?

For this activity, pick one of the following places and describe how the items are organized there. Your description may be a paragraph or a labelled sketch.

- the place where you keep your clothes (for example, a chest of drawers, a closet)
- the place where food is kept in your home (for example, cupboards)
- your favourite music store
- your favourite clothing store

(a) For the place that you chose to investigate, (E1) name the categories into which items were sorted.

(b) Did students who investigated the other choices observe the same categories? Why or why not?

Pure Substances

All substances, whether natural or synthetic, are made of the same building blocks—atoms or molecules. A substance that contains only one of these building blocks is called a **pure substance**. Pure substances that consist only of atoms are called **elements**. The element carbon, for example, which makes up the soft, dark core of a pencil, consists of atoms of carbon. Oxygen is an element that consists of pairs of atoms joined together as diatomic (two-atom) molecules (**Figure 2**).

Pure substances that consist of molecules that are not diatomic are called **compounds**. Each of the atoms that make up the molecules of a compound are from different elements. The air around the pencil in **Figure 2**, for example, contains a compound called carbon dioxide. The molecules that make up this compound consist of one carbon atom linked to two oxygen atoms. The air will also likely contain some water vapour, which is composed of single oxygen atoms, each linked to two hydrogen atoms.

The number of possible chemical compounds is almost endless. Atoms of the elements carbon, hydrogen, and oxygen combine in unique ways (**Figure 3**) that produce more compounds than those of all the other elements combined!

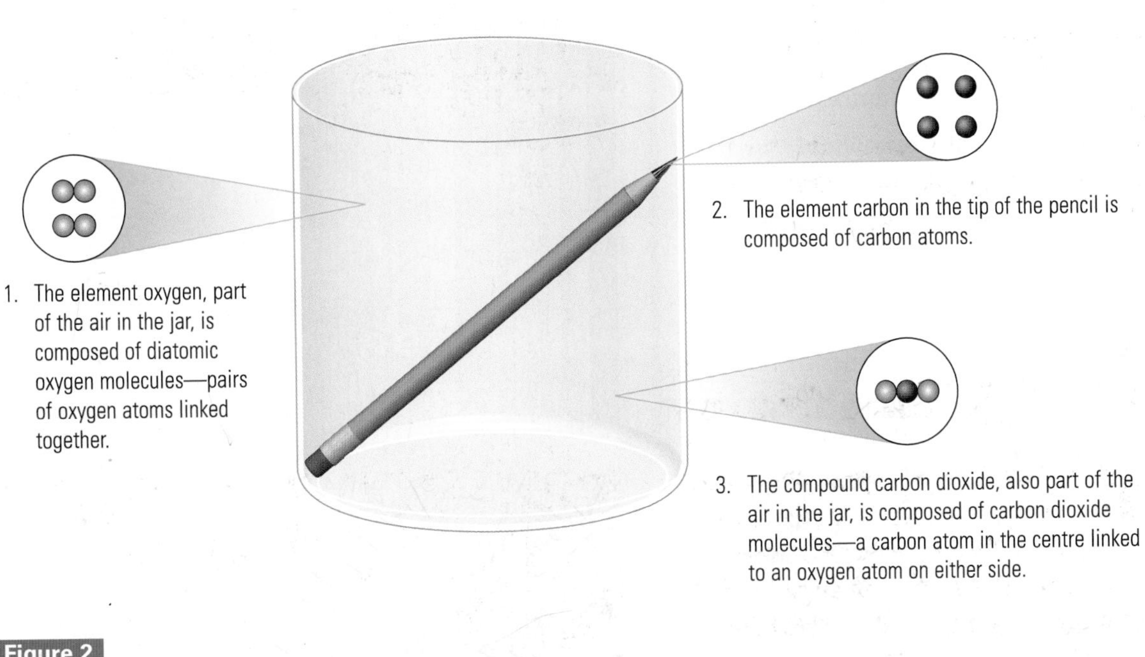

Figure 2

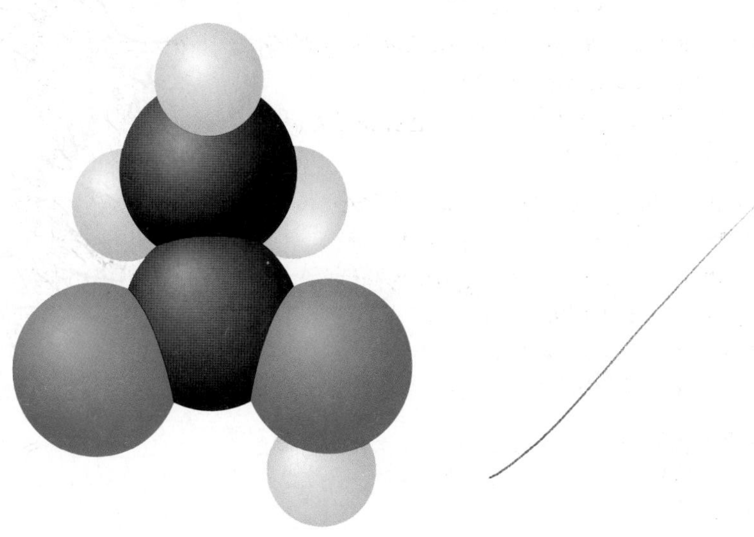

Figure 3

Mixtures

Metal ores are examples of mixtures. A **mixture** generally consists of two or more pure substances. Almost all the natural substances found on or in the Earth are mixtures. So are most human-made and manufactured products. Mixtures can be any combination of solids, liquids, and gases. For example, soft drinks (**Figure 4**) are mixtures of liquid water, solid sugar, and carbon dioxide gas.

Figure 4

Heterogeneous and Homogeneous Mixtures

In a mixture like granola, you can clearly see separate pieces. Each spoonful of granola looks different (**Figure 5**). The composition and properties of one spoonful may differ from another—they are not pure substances. This type of mixture is called a **heterogeneous mixture.** Heterogeneous means "different kinds." In these mixtures two or more substances can be seen or felt.

In other mixtures, the particles of the pure substances mix together so completely that the mixture looks and feels like it is made of only one substance. This is called a **homogeneous mixture.** Steel, composed of iron and carbon, is a homogenous mixture. No matter where you cut a steel bar, it always looks the same. However, the amount of carbon and iron used may change from one steel bar to another, depending on the intended use of the steel. For this reason, homogeneous mixtures are not pure substances.

Just as you organize your clothing and just as merchants organize products for sale in their stores, science organizes matter. Observe this classification of matter in **Figure 6**.

Figure 5

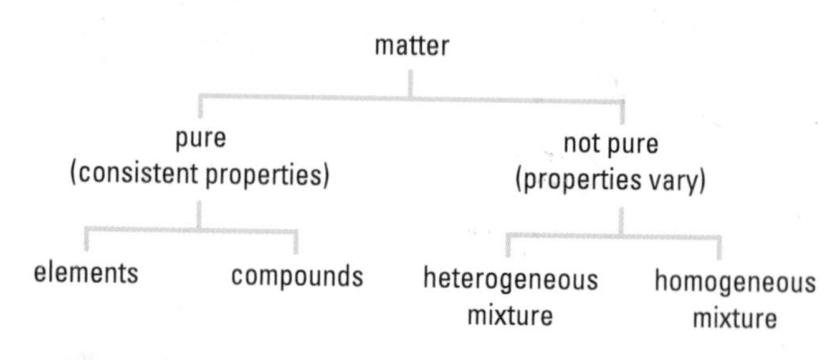

Figure 6

Classification of matter

Understanding Concepts

1. **(a)** What is meant by the term "natural" when describing a substance? Give two examples.

   (b) What is meant by the term "synthetic substance"? Give two examples.

2. How are elements and compounds

   (a) the same?

   (b) different?

3. **(a)** What is a pure substance? Give an example.

   (b) What is a mixture? Give an example.

4. Explain the difference between heterogeneous and homogeneous mixtures.

5. Give an example of a homogeneous mixture that is

   (a) a solid

   (b) a liquid

   (c) a gas

Making Connections

6. The company you work for has asked your team to come up with a new mixture that can be made using two or more substances from the following list. Invent a use for your mixture.

Substance	Useful property
A	sticks to plastic
B	is bright blue
C	boils at 20°C
D	smells like bananas
E	is elastic
F	glows in the dark
G	conducts electricity
H	bends without breaking
I	repels insects

Work the Web

Joseph Priestly, who lived in the 1700s, is famous for his discovery of the element oxygen. Did you know that Priestly also discovered soda water? You can read about Priestly at www.nelson.science.com. Follow the links from *Science 9: Concepts and Connections,* 1.9.

Challenge

1 Most chemical substances exist in either the solid, liquid, or gas states. A fourth state, plasma, has been observed. Which state does your material exist as?

3 Joseph Priestly is recognized for the discovery of the element oxygen. Check out other chemical elements. Were they always recognized as elements from the day they were first observed in nature, or were they "discovered" as well? For those elements that were discovered, find out who is credited with the discovery. Was that person recognized for any other scientific work?

Mineral Extraction and Refining in Canada

Now that you have learned that all matter exists as elements, compounds, solutions, or mixtures, look around at the different materials you see. How many of these materials are elements? Very few elements occur naturally in pure form. Elements such as gold, silver, and copper exist in pure form in nature. Most of the other elements, however, combine easily with oxygen, sulfur, or other elements to form compounds. Some compounds are called **minerals** (**Table 1**). For example, most iron mines produce the compound iron oxide or magnetite and not the element iron.

Table 1 Types of Minerals

Element	Mineral name	Mineral formula	Mineral sample
silver, gold, platinum	silver, gold, platinum	Ag, Au, Pt	
calcium	limestone	$CaCO_3$	
aluminum	bauxite	Al_2O_3	
lead	galena	PbS	
mercury	cinnabar	HgS	
iron	magnetite	Fe_3O_4	
copper	malachite	$CuCO_3$	

Minerals such as iron oxide are rarely found in pure form in the ground. They are often mixed with many less useful compounds in rock formations called **ore**. The minerals first need to be separated from the ore. Once this is done, the elements need to be separated from the mineral.

Mining and Metallurgy

The technology of separating a mineral from its ore is called **metallurgy** (**Figure 1**). It requires knowledge of both the physical (magnetism, for example) and chemical (what compounds it reacts with) properties of the minerals and elements.

To obtain the pure metal in the form of an element, two steps are required:

1. The mineral must be separated from the ore.
2. The metal must be separated from the mineral.

The ore must first be broken away from the rock face using explosives. Large ore carriers remove the ore from the base of the rock face and transport it to a crusher. It is now easier to separate the important mineral from waste material. Magnets, for example, can be used to separate magnetic elements such as iron from the waste.

Producing the element from its mineral involves chemical changes. Some of these chemical changes produce oxides of carbon and sulfur, both of which contribute to acid rain. The industry is working to remove these gases in a process called "scrubbing," so that they will not be released into the atmosphere.

Challenge

2 What element(s) would you include in your time capsule? Why?

3 Are there any famous scientists that played a significant role in developing safe mining and metallurgy practices in Canada?

Figure 1

Iron metal is produced at temperatures of over 1000°C in a blast furnace.

Understanding the Issue

1. (a) Use **Table 1** to write the mineral or chemical formula for each of the following:
   - (i) bauxite
   - (ii) cinnabar
   - (iii) galena

   (b) Identify the desired element in each mineral.
2. Name three elements that occur as pure substances in nature.
3. List four steps that are required to separate an element from the ore in which it is found.

Exploring

4. (N) (O) Research the methods used to obtain one of the following elements in Canada: nickel, copper, aluminum, iron, gold. Prepare a Bristol board presentation of the information you were able to find.
5. Many communities collect aluminum beverage cans at the curbside for recycling.

   (a) (N) (O) Is recycling a good alternative to using raw aluminum?

   (b) (N) (O) How does the cost of recycling compare with the cost of extracting the aluminum from the ground?

   (c) (N) (O) Which process results in fewer environmental problems?

   (d) (K) (L) Use the answers to (a)–(c) to decide whether recycling is a reasonable alternative for producing aluminum beverage cans.

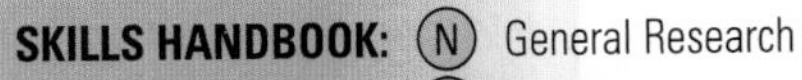
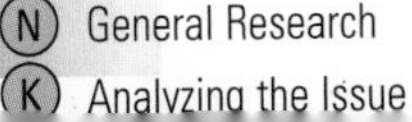

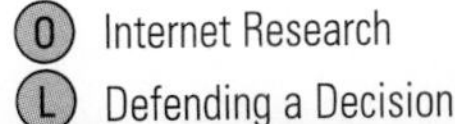

SKILLS HANDBOOK: (N) General Research (O) Internet Research (K) Analyzing the Issue (L) Defending a Decision

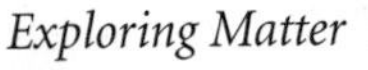

INQUIRY SKILLS
- ○ Questioning
- ○ Hypothesizing
- ● Predicting
- ○ Planning
- ● Conducting
- ● Recording
- ● Analyzing
- ● Concluding
- ● Communicating

Classifying Elements

In the previous section, you learned that matter can be organized into four categories: elements, compounds, solutions, and mixtures. Can any of these categories be further subdivided? In this investigation, you will examine whether a number of elements can be grouped according to the properties they have in common.

Question

Can elements be grouped together by properties which they share?

Materials

- apron
- safety goggles
- small pieces of paper or baking cups
- samples of some of the following elements: aluminum, iron, nickel, tin, lead, copper, magnesium, silicon, zinc, carbon
- magnet
- electrical conductivity apparatus

Prediction

Look at the list of elements in the materials list.

(a) Predict which elements should be classified (C2) together. Explain your groupings.

Procedure

1 Draw a data table in your notebook like (C6) **Table 1**.

Table 1

Element	Properties					
	Colour	Lustre	Hardness	Density	Magnetic	Other

2 Put on your apron and safety goggles. Obtain a sample of each element that you will observe. Place it on a labelled piece of paper or into a labelled baking cup (**Figure 1**).

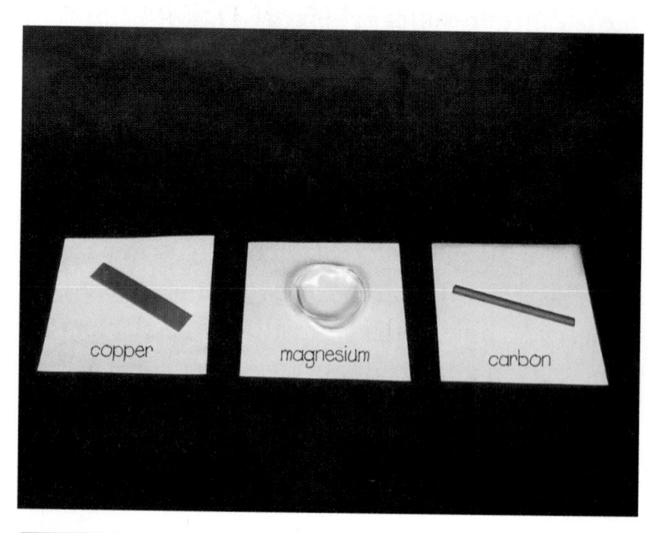

Figure 1

3 Examine each element and note its colour and lustre (shininess).

(E1) **(b)** Record your observations.

4 Try to bend or break each sample. Substances that shatter or crumble can be described as brittle.

(E1) **(c)** Record your observations.

5 Pick up each of the elements. Which elements seem heavy or light for their size? (This is how you rate their densities.)

(E1) **(d)** Record your observations.

6 Use a magnet (**Figure 2**) to determine which elements are magnetic and which are nonmagnetic.

(E1) **(e)** Record your observations.

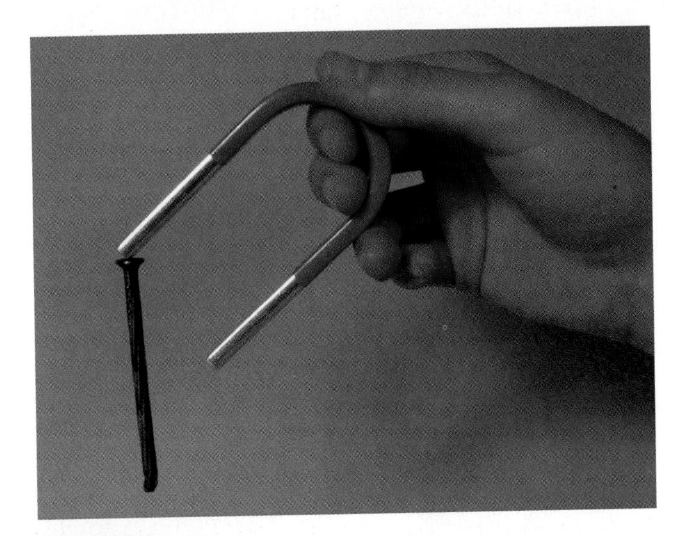

Figure 2

7 Assemble the electrical conductivity apparatus (**Figure 3**). Touch the leads to opposite ends of each sample. If the lamp glows, the element is a conductor. If it doesn't, the element is an insulator.

E1 **(f)** Record your observations.

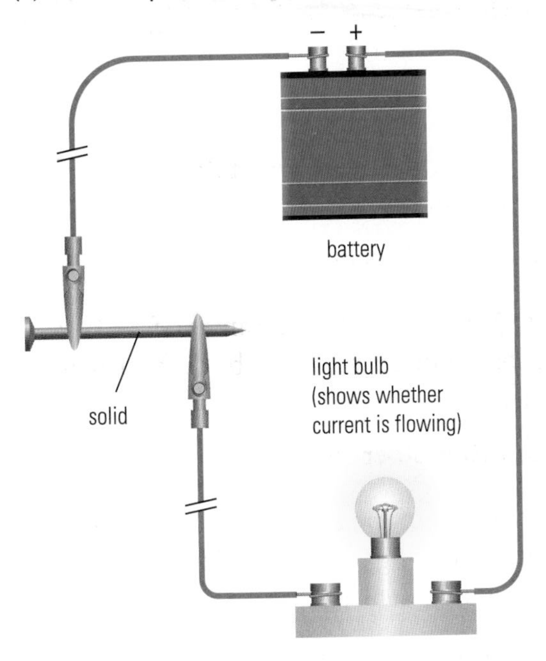

Figure 3

8 Return the samples of the elements to your teacher. Clean up your work area, and wash your hands.

Analysis and Conclusion

(g) Based on your observations, can you group F4 the sample elements into categories according to their properties? If you can, how many categories can you use?

(h) For each category you chose in answer to (g), G list the elements that you would place in each category. How are the elements in one category the same or different from the elements in the other categories?

Making Connections

1. How do the properties of the following elements determine their use?
   - **(a)** Copper and aluminum were both used in electrical wiring at one time.
   - **(b)** Carbon rods are used in some batteries.
   - **(c)** Steel (iron) cans can be separated easily from aluminum cans.
2. You may have noticed that when you put certain solid objects against your skin they feel cold.
   - **(a)** Why does this happen?
   - **(b)** What physical property explains this observation?
   - **(c)** Describe how an element with this property might be used.

Work the Web

Research other ways of classifying elements. Go to www.nelson.science.com and follow the links from *Science 9: Concepts and Connections*, 1.11. How do other classification systems of elements that you find compare with the groupings you have determined in this investigation?

Challenge

1 Why is your new material unlikely to be an element? Explain.
2 Are any of the material samples in your time capsule elements? If so, name them.
3 Try to find out who is credited with the discovery of as many of the elements in this investigation as you can.

Putting Metals to Work

You have already learned that many of the chemical substances that you encounter in everyday life have uses that are based on their physical and chemical properties. Metals, for example, have been used for thousands of years to make tools, weapons, and jewellery. Today, many different mixtures of metals, called **alloys**, are used for everything from airplane parts to braces for teeth. In each case, the metal is chosen because of its properties.

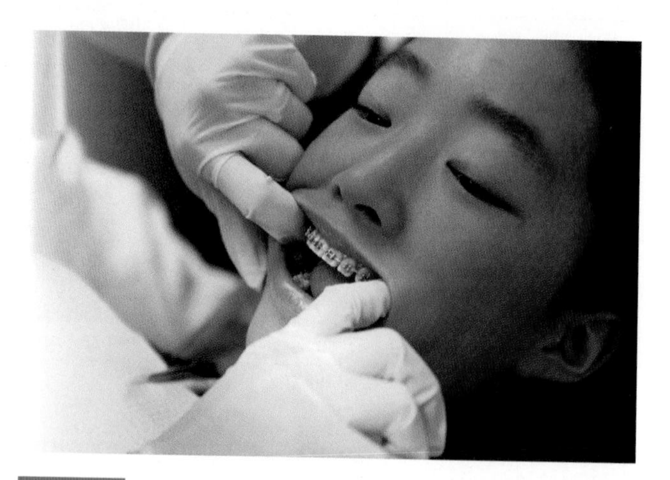

Figure 1

Braces are made from unreactive metals.

Metals and Alloys

The metal alloys used in the braces in **Figure 1** must have specific chemical properties:

- They must not react with saliva.
- They must not react with chemicals in food.

They must also have specific physical properties:

- They must be relatively stiff, but easy to bend into the shape of the teeth.
- They must have enough "spring" to push or pull individual teeth into position.

The metals used in braces are not the only ones that must be unreactive. The lack of reactivity of gold and silver, for example, make them valuable for jewellery. These metals do not leave marks on the skin, nor do they become dull when exposed to air or moisture. Other metals, such as platinum, as well as mixtures of metals are used to make jewellery (**Figure 2**).

The uses of many other metals are listed in **Table 1**.

Figure 2

Jewellery that appears to be gold or silver is actually made from a mixture of these and other metals.

Heavy Metals

Another group of chemicals that are required in small amounts in healthy plants and animals are **heavy metals**, such as mercury and lead. In large quantities, these very dense metals can cause damage to plants and animals. The damage done by lead and mercury in humans has been well documented. Both metals affect the nervous system and the brain. Today, lead-free paint products and gasoline have reduced human exposure to lead. Pulp mills and industrial plants that discharge mercury into freshwater systems are under pressure to reduce the amount of mercury they release (**Figure 3**).

Table 1 **Metals and Their Uses**

Metal	Selected properties	Typical uses
tungsten	very high melting point	light-bulb filaments
chromium	resists corrosion	chrome plating
iron	forms strong alloys	structural steel
copper	good conductor of electricity	electrical wire
nickel	resists corrosion	coins
lead	resistant to acid, soft	batteries
zinc	forms protective coating	galvanized containers
tin	resists corrosion	coating for steel cars
mercury	conductor	home thermostat switches
magnesium	light and strong	car wheels, luggage
aluminum	good conductor of heat	pots and pans

Figure 3

Mercury compounds are discharged by pulp mills and industrial plants.

In the past, mercury was used in felt-making for top hats, so many hatters went mad with mercury poisoning, hence the expression "Mad as a hatter."

Work the Web

Research other alloys to find out what metals have been mixed together to make them. Go to www.nelson.science.com and follow the links from *Science 9: Concepts and Connections, 1.12.* Describe one use for each alloy that you find. What properties does the alloy have that make it suitable for this use?

Understanding Concepts

1. Name two properties that would be required of a metal used for braces for teeth.
2. Name a metal that is
   (a) a good conductor of heat
   (b) used to make jewellery
   (c) no longer a part of gasoline
   (d) no longer in many paint products
   (e) used to make tire rims
3. Identify three chemicals that are needed in large amounts by plants.

Making Connections

4. If you were to eat mercury-contaminated fish, what would likely be the
   (a) short-term effect?
   (b) long-term effect?

Challenge

1 Is the substance you have chosen to market an element or a compound? What properties make it extremely useful?

3 You have learned about some new elements in this section. Try to find out how and when they were discovered, and by whom.

Atoms—The Inside Story

Can matter be continually divided into smaller and smaller pieces? This question had puzzled scientists for hundreds and hundreds of years. As they tried to find an answer to that question, models of the atom slowly developed. Many of these models were strange and definitely wrong but they all helped scientists develop the model of the atom that we use today.

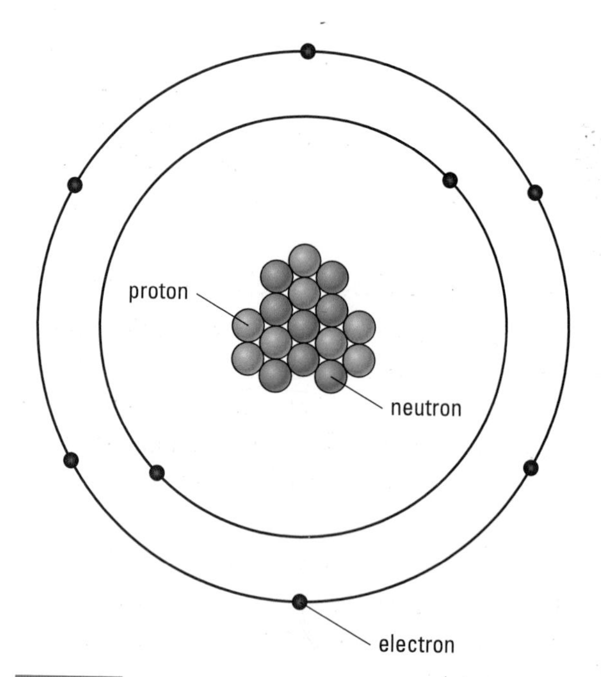

Figure 1
An oxygen model

Types of Subatomic Particles

We now know that the atom is made up of three different types of particles called subatomic particles:

- **Protons** are positively charged particles located in the nucleus, or core, of the atom. Each proton has a mass of 1.
- **Neutrons** are neutral particles also located in the nucleus. They also have a mass of 1.
- **Electrons** are negatively charged particles. They have almost no mass at all—1/2000 of the mass of a proton or neutron. They move rapidly in the space around the nucleus.

All atoms have this basic structure, but not all atoms are alike.

Table 1

Element	Number of protons	Total positive charge	Number of electrons	Total negative charge	Net charge of atom
hydrogen	1	1+	1	1–	0
oxygen	8	8+	8	8–	0
copper	29	29+	29	29–	0

Important Numbers and Atoms

The number of protons in the nucleus, called the **atomic number**, determines the identity of an atom. If you know the atomic number of an atom, you know how many protons and how many electrons that atom contains because each atom must have an equal number of protons and electrons. The element oxygen has an atomic number of 8 (**Figure 1**). This tells us that an atom of oxygen has 8 protons in its nucleus and 8 electrons moving around the nucleus. The number of positive charges equals the number of negative charges, so the overall charge of the atom is zero (**Table 1**).

Another important number is the **mass number**, which represents the sum of the protons and the neutrons in an atom. Therefore, if you know the atomic number (the number of protons) and the atomic mass (the number of protons plus the number of neutrons), you can easily calculate the number of neutrons:

$$\text{mass number} - \text{atomic number} = \text{number of neutrons}$$

Let's use oxygen as an example. Oxygen has a mass number of 16 and an atomic number of 8. $16 - 8 = 8$, so oxygen has 8 neutrons.

Scientists show the numbers of subatomic particles using an internationally recognized system that allows anyone to communicate information about the atom. This is called **standard atomic notation**. In this notation, we write the chemical symbol of the atom and place the atomic number to the lower left and the mass number to the upper left. For example, the atomic notation of chlorine is

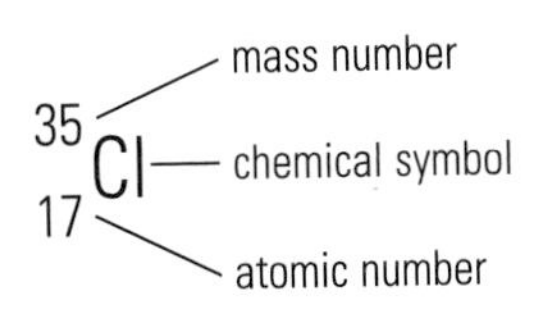

This tells us that chlorine has 17 protons and 35 – 17 = 18 neutrons. Since the atom is neutral, it also tells us that the number of electrons is 17.

The Bohr Model of the Atom

One of the scientists who helped to develop the model of the atom that we use today was a Danish physicist named Niels Bohr (**Figure 2**). He suggested that there was a regular pattern to the position and motion of electrons. He believed that:

- Electrons move in definite orbits around the nucleus, much like planets orbit the Sun.
- These orbits are located at certain distances from the nucleus.
- Electrons cannot exist between these orbits, but they can move up and down from one orbit to another.
- The maximum number of electrons in the first three orbits is 2, 8, and 8.
- Electrons are more stable when they are closer to the nucleus.

In **Bohr-Rutherford diagrams**, a circle is drawn in the centre to represent the nucleus of the atom. The numbers of protons and neutrons are written in this circle. Electrons are shown in circular orbits around the nucleus. Let's look at a diagram of chlorine (**Figure 3**).

The atomic number of chlorine is 17 and the mass number is 35. There are 17 protons and 18 neutrons in the nucleus. Chlorine has 17 electrons: 2 in the first orbit, 8 in the second orbit, and 7 in the third orbit.

Challenge

2 How could you create a three-dimensional Bohr model for your time capsule?

3 Many scientists were involved in developing the present-day atomic model. Could one of these be your "famous scientist"?

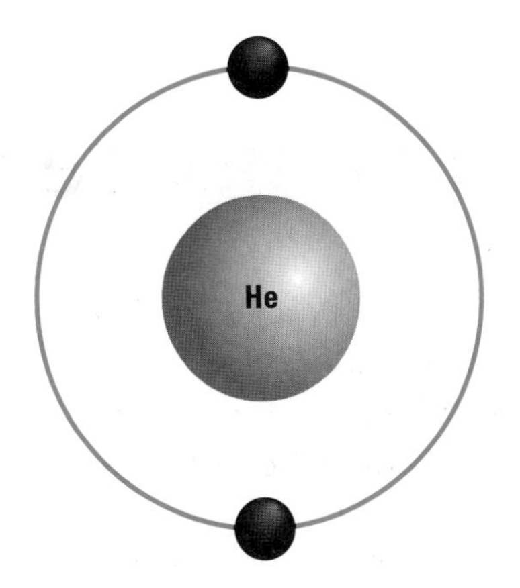

Figure 2

In Bohr's model of the atom, electrons travel around the nucleus in nearly circular orbits, much like planets around the Sun.

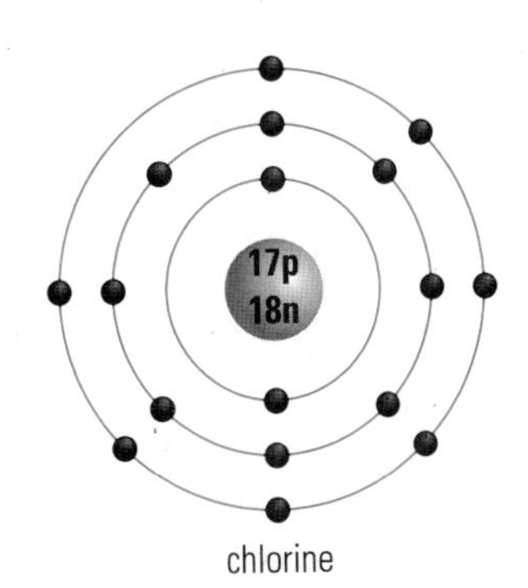

Figure 3

Chlorine

Understanding Concepts

1. In your notebook, draw and complete the following table:

Particle	Proton	Neutron	Electron
mass			
charge			
location in atom			

2. Write the standard atomic notation for
   (a) an atom of nitrogen with 7 protons and 7 neutrons.
   (b) an atom of sulfur with 16 protons and 16 neutrons.
3. Draw Bohr-Rutherford diagrams for
   (a) oxygen (O): 8 protons, 8 neutrons.
   (b) aluminum (Al): 13 protons, 14 neutrons.
   (c) sodium (Na): 11 protons, 12 neutrons.

Chemical Symbols and Formulas

In the first half of this unit, you learned that an element is a pure substance that is made up of identical atoms and has definite properties. Scientists needed a way to represent these elements that people all over the world would understand.

In 1817, a Swedish scientist named Jakob Berzelius developed a system of representing elements with symbols that was soon accepted and used around the world. All countries now use the same chemical symbols to represent elements and compounds, even when their language makes the names different. For example, the symbol Fe represents the element that people call *iron* in Canada, *fer* in France, and *fier* in Romania. A scientist in any country can identify the contents of a bottle by reading the symbols in the formula on the label (**Figure 1**).

Berzelius's system worked because it provided symbols for all the known elements of the time. It also set up a system for naming any new elements that had not yet been discovered.

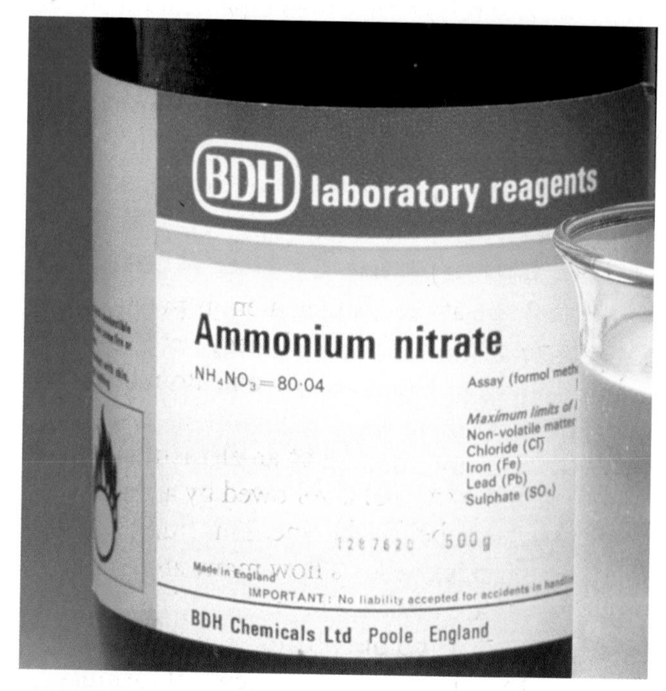

Figure 1

Scientists rely on symbols and formulas to help them keep track of chemicals.

Chemical Symbols

Today, each **chemical symbol**, an abbreviation of the name of the element, consists of one or two letters. For example, the symbol for oxygen is O, and the symbol for carbon is C. When the first letter has already been used as the symbol for another element, the first two letters are used. For example, the symbol for calcium is Ca, and the symbol for cobalt is Co. Notice that when two letters are used, the first is always a capital letter while the second is not.

The names for symbols and their elements come from many sources. The symbol for silver is Ag because it comes from the Latin name for silver, *argentium*. Hydrogen (H) comes from the Greek term *hydor gene*, which means water producer. Mercury's symbol (Hg), comes from the Latin word *hydragyrum*, which means liquid silver. Some elements and the origin of their symbols are listed in **Table 1**.

Table 1 Some Chemical Symbols of Elements

Element	Symbol	Origin of name
aluminum	Al	Latin: *alumen*, a bitter salt
fluorine	F	Latin: *fluor*, a flow
carbon	C	Latin: *carbo*, charcoal
potassium	K	Latin: *kalium*
sodium	Na	Latin: *natrium*
tin	Sn	Latin: *stannum*
helium	He	Greek: *helios*, Sun
chlorine	Cl	Greek: *chloros*, pale green
neon	Ne	Greek: *neos*, new
uranium	U	named after planet Uranus
magnesium	Mg	named after a district in western Turkey

Chemical Formulas

Just as single symbols are used to represent elements, chemical symbols can be put together to represent chemical substances. Combinations of chemical symbols are called **chemical formulas** (**Table 2**). The chemical formula indicates which elements are present and how many atoms of each element are found in that substance (**Figure 2**). For example, table salt is a compound made up of the elements sodium (Na) and chlorine (Cl). The chemical formula for table salt (sodium chloride) is NaCl.

If only one atom of an element is present in a compound, no number is included. NaCl is made up of one atom of sodium and one atom of chlorine.

If more than one atom of an element is in a compound, the symbol is followed by a small number written below the line. This number, called a **subscript**, tells us how many atoms of that element are present. Water, H_2O, is a compound made up of two atoms of hydrogen and one atom of oxygen. The chemical formula for ammonia is NH_3. What elements make up this compound? How many atoms of each element are found in ammonia? If you answered one atom of nitrogen and three atoms of hydrogen, you are correct.

Some formulas contain more than two elements. The formula for baking soda (sodium hydrogen carbonate) is $NaHCO_3$. It contains one atom of sodium, one atom of hydrogen, one atom of carbon, and three atoms of oxygen. The formula for acetic acid (vinegar) is $C_2H_4O_2$. Can you list the elements and how many atoms of each element are found in each of these formulas?

Challenge

1 What chemical symbols and formulas will you need to identify the elements and compounds found in your substance or material?

Work the Web

Learn about the elements and their symbols the interactive way. Visit www.science.nelson.com and follow the links from *Science 9: Concepts and Connections,* 1.14.

Table 2 Some Examples of Chemical Formulas

Name of Substance	Formula
sodium bicarbonate (baking soda)	$NaHCO_3$
calcium carbonate (chalk)	$CaCO_3$
sodium nitrate (fertilizer)	$NaNO_3$
calcium phosphate (fertilizer)	$Ca_3(PO_4)_2$
sodium chloride (salt)	NaCl
acetylsalicylic acid (ASA or aspirin)	$C_9H_8O_4$
acetic acid (vinegar)	$C_2H_4O_2$

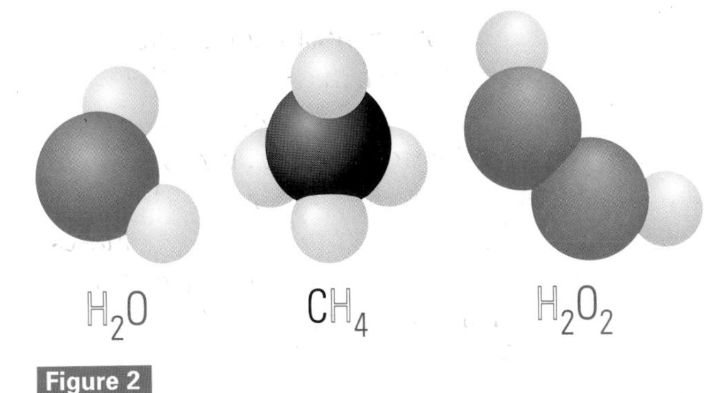

Figure 2

Understanding Concepts

1. Why are symbols useful in describing chemicals?
2. Use the periodic table at the back of this text to find the symbols for the following elements: lithium, silicon, argon, copper, phosphorus, and gold.
3. What two things does a chemical formula tell us about a compound?
4. Write a chemical formula for the following:
   - **(a)** a molecule of carbon dioxide that is made up of one atom of carbon and two atoms of oxygen.
   - **(b)** a molecule of aspirin that is made up of nine atoms of carbon, eight atoms of hydrogen, and four atoms of oxygen.
   - **(c)** a molecule of sugar (glucose) that is made up of 6 atoms of carbon, 12 atoms of hydrogen, and 6 atoms of oxygen.

Making Connections

(N) (O) 5. Research a common use for each of the following pure substances:
   - **(a)** helium gas
   - **(b)** acetone
   - **(c)** tartaric acid

Compounds and Molecules

Atoms are seldom found alone in nature. They have a tendency to combine with the atoms of other elements. But when they do, how do we know how many atoms of each element are needed to form a compound? Why is table salt (sodium chloride) NaCl? Why isn't it Na_2Cl or $NaCl_2$? The key is in knowing how the atoms combine.

Combining Capacity

After years of experimenting, scientists found that each element was able to make a specific number of connections with other elements. They called these connections the element's **combining capacity**. Scientists gave a number value to the combining capacity of each metal and nonmetal to explain the compounds they form. The scientists also came up with rules to follow. **Table 1** lists the rules for how some elements combine.

For example, both sodium and chlorine were assigned a combining capacity of one. Each element needs to make one connection, so when sodium and chlorine combine their formula is NaCl.

Aluminum, however, has a combining capacity of three. It needs to make three connections. Each chlorine atom needs to make only one connection. When aluminum combines with chlorine, for every atom of aluminum three atoms of chlorine are needed. Therefore, the chemical formula for aluminum chloride is $AlCl_3$.

Table 2 lists the combining capacities of some metals, and **Table 3** lists the combining capacities of some nonmetals.

Both sodium and bromine have a combining capacity of one. Sodium bromide has the formula NaBr, meaning it contains one atom of sodium for each atom of bromine (**Figure 1a**).

Table 1 How Elements Combine

Rule	
Rule 1:	Metals combine with nonmetals in many compounds.
Rule 2:	Write the name of the metal first and the nonmetal second.
Rule 3:	Change the ending of the nonmetal to "ide."
Rule 4:	Each atom has its own combining capacity.
Rule 5:	Atoms combine so that each can fill its combining capacity.

Table 2 Combining Capacities of Some Metals

Element	Symbol	Combining capacity
aluminum	Al	3
barium	Ba	2
calcium	Ca	2
magnesium	Mg	2
potassium	K	1
silver	Ag	1
sodium	Na	1
zinc	Zn	2

Table 3 Combining Capacities of Some Nonmetals

Element	Symbol	Combining capacity	Combined name
bromine	Br	1	bromide
chlorine	Cl	1	chloride
fluorine	F	1	fluoride
iodine	I	1	iodide
oxygen	O	2	oxide
sulfur	S	2	sulfide

Aluminum has a combining capacity of three, and oxygen has a combining capacity of two. Therefore, in the compound aluminum oxide (**Figure 1b**), two atoms of aluminum must combine with three oxygen atoms. The chemical formula of aluminum oxide is Al_2O_3.

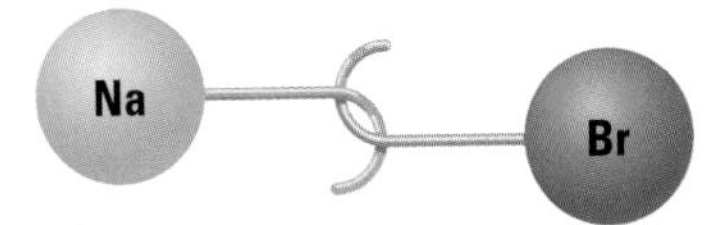

a sodium bromide

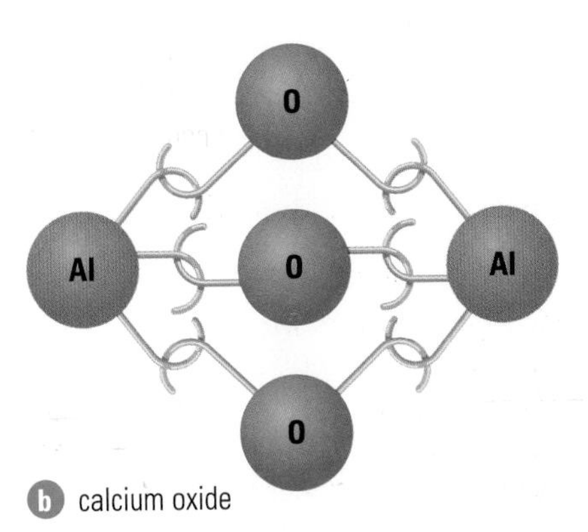

b calcium oxide

Figure 1

Some early models pictured atoms with hooks that could attach to the hooks of other atoms.

a nitrogen molecule

b oxygen molecule

Figure 2

Nitrogen (a) and oxygen (b) molecules in their natural state

When Like Atoms Combine

Sometimes chemical formulas represent a molecule of an element instead of a compound. This happens when two or more atoms of the same element join together. A molecule that forms when two atoms of the same element join together is called a **diatomic molecule.** This is not a compound because it contains atoms of only one element. Seven nonmetal elements are found naturally as diatomic molecules of two identical elements: hydrogen (H_2), nitrogen (N_2) (**Figure 2a**), oxygen (O_2) (**Figure 2b**), fluorine (F_2), chlorine (Cl_2), bromine (Br_2), and iodine (I_2).

Understanding Concepts

1. What does the term "combining capacity" mean?
2. Elements can be classified as metals or nonmetals. Which elements change their names when they form compounds? Explain, using an example.
3. What are the names of the following compounds?
   - **(a)** $CaCl_2$, used in bleaching powder and for melting ice
   - **(b)** CaO, used in plaster and construction
   - **(c)** CuCl, used to make red glass
   - **(d)** AgCl, used in photography
4. Use the combining capacities shown in **Tables 2** and **3** to write chemical formulas for:
   - **(a)** sodium fluoride
   - **(b)** magnesium fluoride
   - **(c)** potassium bromide
   - **(d)** silver oxide
   - **(e)** aluminum sulfide
5. Draw "hook-and-ball" diagrams for the compounds in question 4.
6. Which of the following is not a compound? Why?
   - **(a)** CH_4
   - **(b)** H_2O
   - **(c)** NH_4
   - **(d)** N_2

Somebody who gets a lot of public attention is described as "being in the limelight." This expression refers to calcium oxide or lime, which was used in stage shows decades ago to produce a brilliant white light for footlights.

Work the Web

Learn more about the way that elements bond together to form compounds. Visit www.science.nelson.com and follow the links from *Science 9: Concepts and Connections,* 1.15.

Building Molecules

Try to imagine a molecule. Would it be easier if you had a model in front of you? Models are important tools in many lines of work.

Models can be built or drawn. Digital animators use computer-generated models to allow dinosaurs to roam the Earth in movies like the *Jurassic Park* series. Firefighters use models of burning buildings to help them understand how certain types of fires burn and to improve their skills in putting out these fires. Scientists use models of molecules to help them understand and predict how molecules will behave (**Figure 1**).

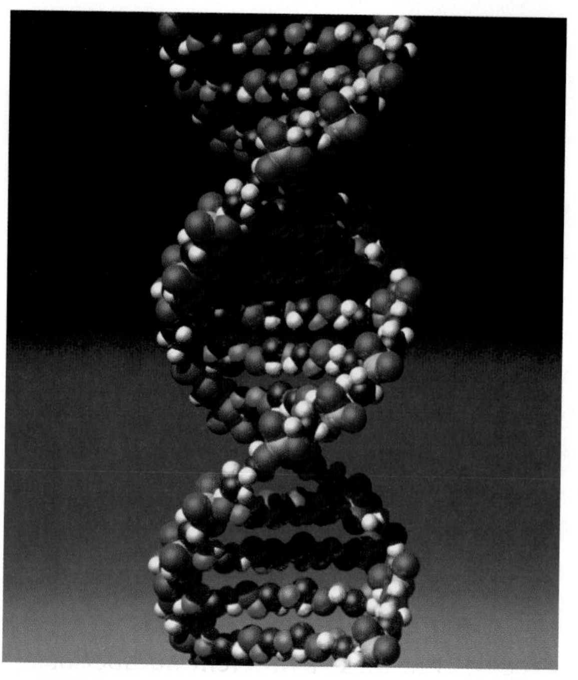

Figure 1

Francis Crick and James Watson discovered the structure of the DNA (deoxyribonucleic acid) molecule, which carries our genetic characteristics. They figured out the structure by working with molecular models.

The structure of molecules refers to how they are made up as well as how they are organized. For example, aspirin and sucrose (table sugar) are both made up of the elements carbon, hydrogen, and oxygen, but you wouldn't use aspirin to sweeten your cereal or take a spoonful of sugar for your headache.

In the models of molecules that scientists build (**Figure 2a**) the atoms are held together by connections called **bonds**. The connections represent electrons that "glue" the atoms together. The molecules can also be represented by drawings on paper called **structural diagrams** (**Figure 2b**). In these diagrams, each atom is represented by its chemical symbol, and each bond or connection is represented by a straight line drawn between the symbols.

Figure 2

The ball-and-stick model and the structural diagram for rubbing alcohol

Each kind of atom generally forms a set number of connections (bonds). For example, each hydrogen atom forms only one bond. Each oxygen atom forms two bonds: either two single bonds with two other atoms such as water (H_2O) (**Figure 3a**), or one double bond with another atom such as a diatomic molecule of oxygen (O_2) (**Figure 3b**).

In this activity, you will build models of some common molecules and see how bonds can link atoms in molecules.

a

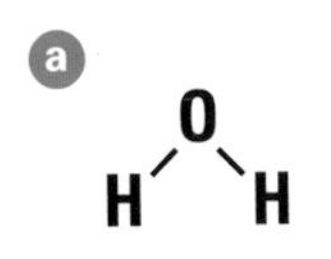

b

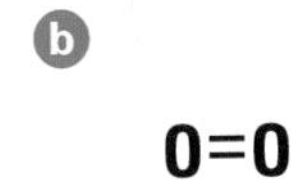

Figure 3

Structural diagrams of water and oxygen

Materials

- 54 large marshmallows (per group)
- toothpicks
- food colouring (red, blue, green, yellow)

Procedure

1 Prepare the marshmallow atoms by applying food colouring to the marshmallows as follows:
N (nitrogen)—red (7)
H (hydrogen)—blue (31)
O (oxygen)—green (5)
C (carbon)—yellow (11)
Let the marshmallows dry for two hours.

2 Each atom can make a different number of connections. The number of connections is summarized in **Table 1**.

Table 1

Atom	Number of connections per atom
hydrogen	1
oxygen	2
nitrogen	3
carbon	4

3 Make a model of hydrogen gas by connecting two atoms of hydrogen with a connector (toothpick).

(a) Draw a structural diagram of the molecule. Write the name and formula, H_2, beside your diagram. (E3)

4 Make a model of oxygen gas. Join two atoms of oxygen with two connectors to represent the two connections that each atom usually makes.

(b) Draw a structural diagram of the molecule. Write the name and formula, O_2, beside your diagram. (E3)

5 Make models of the following molecules: nitrogen (N_2), ammonia (NH_3), methane (CH_4), water (H_2O), ethane (C_2H_4), and carbon dioxide (CO_2). Make sure that each atom makes the right number of connections.

(c) Draw structural diagrams of the models. Write the formula beside each diagram. (E3)

6 If you have time, take three carbon and eight hydrogen atoms. See how many molecules you can make with some or all of the atoms. Remember how many connections each atom can have.

(d) Draw structural diagrams of the models. Write the formula beside each diagram. (E3)

Understanding Concepts

1. What is a model?
2. Why do scientists find making models useful?
3. Usually more bonds between two atoms make a stronger connection. Which of all the molecules you made probably has the strongest bond? Explain your reasoning.

Making Connections

4. How are the marshmallow models like real atoms?
5. How are the marshmallow models different from real atoms?

Work the Web

Find out more about molecular models by visiting www.science.nelson.com and following the links from *Science 9: Concepts and Connections,* 1.16. Draw and colour a molecule of your choice. Label or describe what you are seeing.

Challenge

2 You have seen several ways models are made. Could you create your model in one of these ways?

SKILLS HANDBOOK: (E3) Scientific Drawings

1.17 Investigation

INQUIRY SKILLS
- ○ Questioning
- ○ Hypothesizing
- ○ Predicting
- ● Conducting
- ○ Planning
- ● Recording
- ● Analyzing
- ● Concluding
- ● Communicating

Black Box Atoms

Imagine that you are standing in front of a soft drink vending machine (**Figure 1**). You put in a coin, press a button, and a can falls down into the tray at the bottom. How does the machine work? You cannot see inside it, so you have to create a model that could explain the workings of the machine.

One possibility is that there is a very small person working inside. When the coin appears, the person checks to see which button was pushed, searches for the right can, and puts it in the tray.

A second possibility might involve a mechanical system with various electronic sensors, levers, motors, and slots that operate when the coin is inserted to release the can.

How could you test these two hypotheses? You could pull out the electric plug in the back. If no can appears when the power is off, maybe the second hypothesis is correct. Or maybe the person inside refuses to work in the dark, so the first hypothesis is still a possibility.

The best model you can create is the one that allows you to predict how the vending machine will behave in as many situations as you can imagine. You may come up with a model that works perfectly for all the evidence that you have. But someone else might try using the vending machine in a completely new situation. If the can does not fall, and the model cannot explain it, then the model needs to be adjusted.

Like a scientist exploring models of matter, all of your testing, thinking, and experimenting are based on the fact that you can't see what is going on inside the vending machine. Atoms cannot be seen, so we can only see how matter behaves in certain circumstances. The model of matter changes when that model is tested in new situations and produces different results.

In this investigation, you will try to guess what is inside a sealed box.

Figure 1

Materials

- sealed "black box" (such as a shoebox), numbered and taped shut, containing an unknown object or objects (1 per group)
- empty sealed box (1 per class)
- ruler
- balance
- magnet

Question

How can we use a model to explain what we cannot see?

Prediction

(C2) **(a)** Write a prediction for this investigation.

Procedure

1 Your teacher will give you a "black box" with an object or objects inside. Do not damage or open the box.

2 Measure the outside dimensions of the box.

(E1) **(b)** Record your observations in a data table.

3 Use a magnet to determine whether the object(s) have any magnetic properties.

(E1) **(c)** Record your observations.

4 Determine the mass of the empty box. Determine the mass of your "black box."

(E1) **(d)** Record your observations.

The difference between the two masses is the mass of the object(s) in your "black box."

(E1) **(e)** Record your observations.

5 Carefully tilt, shake, and move the "black box." Does the object slide (is it flat)? Does the object roll (is it round)? Does it bump into something else in the box (is there more than one object)? Does it bounce? If so, how hard does it bounce? Does it flip?

(E1) **(f)** Record your observations.

6 Examine your observations and invent new movements for the box that will help you determine the size, shape, and other physical properties of the object(s).

(E1) **(g)** Record each new movement and the observations you make.

7 With your group, discuss a model for the object(s) in the box.

(h) Write a description (or model) of what you think the object(s) is/are. For example, describing an object as "a 15-cm long metal object, branched into four small projections at one end" is better than describing it as a "fork."

(i) Make a drawing of what you think is in your (E3) "black box." Draw the object(s) to show their relative size.

Analysis and Conclusion

After completing your description and drawing, open the box and look at the object(s).

(j) Write a description of the object(s).

(k) Make a drawing of the contents of the box.

(l) How does your first prediction and drawing (G) compare with the actual contents of the box? Was your prediction correct? Explain why or why not, based on your observations.

(m) Describe how you can develop a model of an (F4) object without directly observing the object.

Making Connections

1. Think about a gumball machine.

(a) What experiments could you do to find out how the gumball machine operates?

(b) Draw a model of how you think the machine operates.

Exploring

2. (E1) (E3) Your teacher has buried an object in a soft ball of modelling clay. Take the ball, and using a probe provided by your teacher, carefully insert it into the ball. Make a systematic series of probings. Record your observations and make a model drawing to describe what is inside the clay "atom."

Reflecting

3. Think about your group's success with determining the identity of the object(s) in the "black box." Is there any reason why you successfully determined some characteristics but not others? Explain.

1.18

Organizing the Elements

By the mid-1800s, about 60 elements had been discovered. Scientists had a lot of information about these elements, but it wasn't organized and therefore wasn't very useful. The scientists had tried to arrange the elements alphabetically, but found that each time they discovered another element, the whole list had to be changed. Organizing them by colour didn't work because too many elements looked the same. Taste didn't work because many elements were poisonous. The scientists found that properties such as conductivity, malleability, and lustre allowed them to group the elements into metals and nonmetals (**Figure 1**).

This was a good start, but most elements were metals. All metals conducted electricity, were malleable, and looked shiny. A better system was needed.

Finally, scientists found a property that not only could be measured, but also was different for every element. **Atomic mass**—the average mass of one atom of an element—was used to arrange the elements.

Figure 1

Physical and chemical properties suggest that elements can be organized into metals and nonmetals.

Mendeleev and the First Periodic Table

A Russian chemist, Dmitri Mendeleev, came up with the best arrangement for the 64 elements then known. Mendeleev looked at the chemical and physical properties of the elements and found that, based on their properties, some elements were similar to others. Mendeleev (**Figure 2**) had an idea that there was a pattern or relationship among the elements, but he was not quite sure what it was.

He began with each element's atomic number and then arranged and rearranged the elements until he began to see regular patterns. He found that elements with similar properties fit into the same vertical columns. Some elements fit because of their mass, but their properties were not similar. Mendeleev moved these elements to a column with similar properties and ignored their

Figure 2

Dimitri Mendeleev's organization of the elements into a periodic table made the study of chemistry manageable.

mass. He even left spaces in his table for elements that had not yet been discovered. When he had finished with his arrangement, his table of elements showed a pattern that repeated based on the elements' properties. Anything that repeats according to the same pattern can be called **periodic** (for example, the days of the week, the months of the year, and even the migration of birds).

Mendeleev created a periodic law that stated:

> *If the elements are arranged according to their atomic mass, a pattern can be seen in which similar properties occur regularly.*

Today's Periodic Table

Mendeleev's **periodic table** was a good place to start, but scientists like Mendeleev didn't know anything about the atomic structure of the atom. By the twentieth century, the nuclear atom and its subatomic particles had been discovered. Scientists had realized that the key to the identity of an element was the number of protons in the nucleus—the atomic number—not the atomic mass. A **modern periodic law** was written and a new periodic table was created.

> *If the elements are arranged according to their atomic number, a pattern can be seen in which similar properties occur regularly.*

Elements in the modern periodic table are arranged by atomic number (**Figure 3**). With a few exceptions, atomic numbers are in the same order as atomic mass. The lightest element has the lowest atomic number, and the heaviest element has the greatest atomic mass.

The periodic table is divided into three main groups: metals, nonmetals, and metalloids. There are more metals than nonmetals, and the metals are on the left side of the table. The nonmetals are on the right side. The two groups are divided by a heavy, steplike line. Some elements have the properties of both metals and nonmetals. They are called **metalloids**. They are found on both sides of the step-wise line that divides the metals from the nonmetals. For example, silicon is a metalloid. It is shiny and silvery, but it is not malleable and is only a partial conductor of electricity.

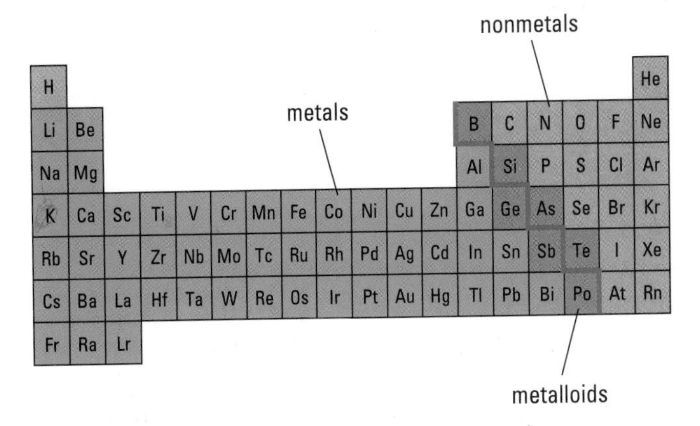

Figure 3

Location of the metals, nonmetals and the metalloids on the periodic table

Understanding Concepts

1. What are some of the properties that helped scientists organize elements into metals and nonmetals?
2. **(a)** What property of atoms did Mendeleev use to organize elements?
   (b) How did he use this property to organize them?
   (c) When did he ignore this property in building his table?
3. Explain why Mendeleev included spaces in his periodic table.
4. Why were properties such as colour and taste not used to arrange elements on the first periodic table?
5. Where on the modern periodic table do you find metals, nonmetals, and metalloids?

Challenge

3 Another scientist and his discoveries have been introduced in this section. Could he be your "famous scientist"?

Work the Web

Visit www.science.nelson.com and follow the links from *Science 9: Concepts and Connections,* 1.18. Read the biography of Dmitri Mendeleev, and make a list of his scientific contributions other than the periodic table.

Exploring the Modern Periodic Table

In section 1.18 you learned that the periodic table is divided into three main groups: metals, nonmetals, and metalloids. These three groups are quite large. Although members of each group share certain similarities, they also have noticeable differences. For that reason, elements in the periodic table are arranged in columns, or **families**, based on groups with similar properties and characteristics. These properties include melting and boiling points and the sizes of the atoms. The size of an atom is called the **atomic radius**, the distance from the nucleus to the "outer edge" of the atom.

You might also notice that each horizontal row, or **period**, has a pattern as well. Unlike the elements in a family, the elements in a period do not have similar properties. But as you move across a period from left to right, the first element is always an extremely reactive solid, and the last element is a nonreactive gas.

In this activity, you will work with a partner to find patterns or trends in the elements in the periodic table.

Procedure

1 Look at the periodic table at the back of this text. The key at the top indicates the state at room temperature (solid, liquid, or gas) by the colour of the element's symbol. If the element's symbol is black, it is a solid. If the symbol is blue, the element is a liquid. If the symbol is red, it is a gas.

(a) Which elements are gases at room temperature?

(b) Name two elements that are liquids at room temperature.

2 Metals have a green background. Nonmetals have an orange background.

(c) On which side of the table do you find the elements that are metals?

(d) On which side of the table do you find the elements that are nonmetals?

(e) Are most of the elements metals or nonmetals?

(f) What are the elements called that have a purple background?

(g) Why are they called that?

3 Find the names of the elements with the following chemical symbols: H, Al, Fe, Eu, Kr, Xe.

4 Find the chemical symbols for the following elements: helium, iodine, lead, plutonium, uranium, einsteinium.

5 Find the atomic numbers for the following elements: Lr, Cs, Pt, Ag, He, Si.

6 Find the atomic mass for the following elements: sodium, zinc, chlorine, lithium, bromine, argon.

7 Mendeleev used atomic masses to organize his periodic table. What two elements are "out of order" in the fifth row, according to atomic mass?

8 Look at the densities and melting points of the elements.

(h) Which element has the highest melting temperature? What is it?

(i) Which element has the lowest melting temperature? What is it?

(j) Which element has the greatest density? What is it?

(k) Which element has the lowest density? What is it?

9 Elements 1, 3, 11, and 19 are in the first column of the periodic table.

(l) Draw Bohr-Rutherford diagrams for these elements. (Remember: The order of filling in the first three orbits is 2, 8, 8.)

(m) How many electrons are in the outer orbit of each of these elements?

(n) How many electrons do you think there are in the outer orbit of the elements Rb and Cs?

10 Elements 9 and 17 are in the second-last column of the periodic table.

(o) Draw Bohr-Rutherford diagrams for these elements.

(p) How many electrons are in the outer orbit of each of these elements?

(q) How many electrons do you think are in the outer orbit of the elements Br and I?

11 Look at elements 3 to 10.

(r) Draw Bohr-Rutherford diagrams for these elements.

(s) Describe the general pattern (trend) that you observe across a row of the periodic table. (Hint: Look at the outer orbit of the electrons.)

12 Elements in the periodic table have been arranged in columns or groups according to their properties. Name four elements that have properties similar to lithium.

13 Helium is a gas that will not burn. Name three other gaseous elements that probably will not burn either.

Challenge

1 What group of elements does your substance or material belong to? What does this tell you about its properties?

2 Does your model show specific relationships or properties that give clues to the chemical or physical behaviour of your matter?

Understanding Concepts

1. At room temperature, in what state are most elements?
2. Using the words "increase" or "decrease," describe how the following properties generally change as you go across the rows (from left to right) of the periodic table:
   - **(a)** atomic number
   - **(b)** melting temperature
   - **(c)** atomic radius
3. Using the words "increase" or "decrease," describe how the following properties generally change as you go down the columns (groups) of the periodic table:
   - **(a)** density
   - **(b)** melting temperature
   - **(c)** atomic radius

Making Connections

4. Elements in the same group have similar properties. Think of two examples in everyday life where similar substances could be substituted for each other. What other factors would you consider before making the substitutions?

Work the Web

Go to www.nelson.science.com and follow the links from *Science 9: Concepts and Connections*, 1.19. Choose two elements and write a short report comparing their properties and uses.

Elemental Magic

The big game is tomorrow. You are the team's leading goal scorer and you are shopping for a new hockey stick. Years ago, wood was the only choice in hockey sticks, but today's players can choose from wood, aluminum, and composites or combinations of carbon, glass, and aramid fibres. The salesperson tells you about the fibres that make up composite hockey sticks. She explains that these fibres have properties that allow them to be shaped, positioned, and angled as required. This gives players a stick that improves the feel for puck handling and shooting. The sticks are well balanced, extremely light, and can be reproduced so that each one feels the same.

You have many choices of materials because scientists use the "magic" of chemistry to make thousands of materials. These "magicians" make new products by using their knowledge of the periodic table.

Understanding the arrangement of the elements in the periodic table makes it possible to predict new ways of combining atoms into molecules (**Figure 1**).

Over the centuries, people have used four general types of materials to make everything they need: metals, polymers, ceramics, and composites (**Table 1**). The importance of these substances has changed over time. For example, in 5000 B.C.E., metals were not very important, but by 1960 they had become the most important materials in the world.

Table 1 Percentage Use of Different Materials in History

Material	5000 B.C.E.	1800	1960	2000
metals	5	30	80	40
polymers	40	35	10	25
ceramics	40	30	8	20
composites	15	5	2	15
total	100%	100%	100%	100%

Metals

Metals make up most of the elements in the periodic table. In 5000 B.C.E., gold and copper were used for jewellery and containers but not much else because they were so hard to produce. Over time, people learned how to mix metallic elements and make alloys such as bronze and steel. These alloys could be used for weapons, utensils, and even buildings. By the twentieth century, alloys were used in everything from tractors (**Figure 2**) to kitchen utensils. Today, superalloys can be made heat-resistant for use in jet engine parts and corrosion-resistant for chimney linings in power plants.

(a) What types of materials would likely have been used in 5000 B.C.E.?

(b) What types of metals were being used in 1960?

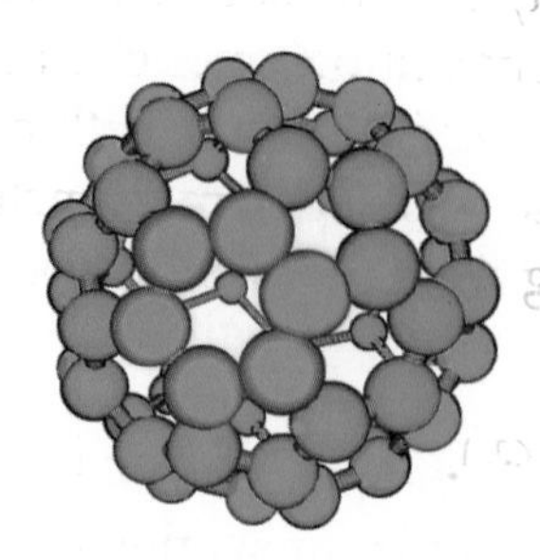

Figure 1
Materials scientists are always looking for new ways of assembling atoms into molecules. "Bucky-balls" are made from carbon atoms arranged in a sphere.

Figure 2

Polymers

Polymers are materials made up of long chains of molecules joined together. The most important elements in these substances are carbon, hydrogen, and oxygen (**Figure 3**). The word polymer comes from the Greek words *poly*, which means many, and *meros*, which means parts. The individual molecules that make up polymers are called **monomers**. The type of monomers and the length and shape of the polymer chain determine the properties of the polymer.

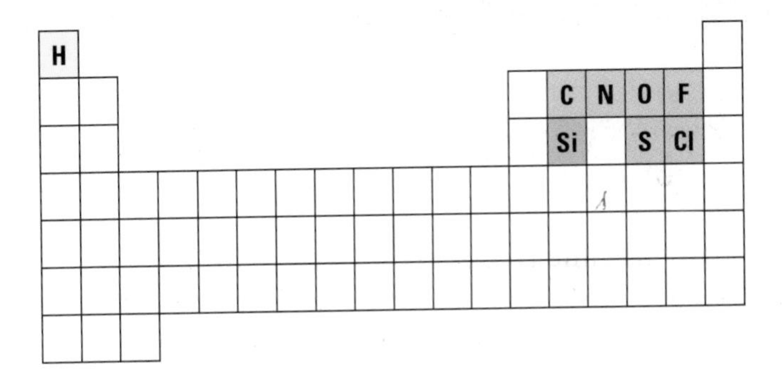

Figure 3

The elements in polymeric materials

(c) What is the relationship between a monomer and a polymer?

(d) What determines the properties of a polymer?

In 5000 B.C.E., the only polymers that existed were natural polymers found in animal and plant fibres. Wool, leather, wood, linen, and cotton were natural polymers used for clothing, shelter, tools and boats.

Over the years, scientists discovered and developed glues, rubbers, and other synthetic polymers. Synthetic polymers are used to make fabrics such as nylon, rayon, orlon, and dacron. A major breakthrough in polymer chemistry was the invention of plastics in the twentieth century (**Figure 4**). These plastic polymers included nylon, nonstick coating for cookware, and polystyrene for drink cups.

Figure 4

All plastics are examples of polymers.

(e) Give three examples of natural polymers.

(f) Give three examples of synthetic polymers.

Since the first synthetic polymer was made in the early 1900s, polymerization has come a long way. **Polymerization** is the process of chemically bonding monomers to form polymers. Early polymers consisted of less than 200 monomers. Today's polymers can be made up of thousands of monomers. They can be linked together in chains, loops, and spirals.

(g) What is polymerization?

Some new polymers are tough, light, and heat-resistant enough to be used in jet aircraft. Polymers are also used as substitutes for human arteries and bones. Polymers are replacing glass, metal, and cardboard for food containers. Polymer materials are also being used for furniture, rugs, draperies, and wall coverings. Try to imagine a world without synthetic polymers: no CDs, no food storage containers, and no plastic toys.

Ceramics and Glass

Ceramics and glass are materials that come from minerals and rocks. The elements that are most important in these compounds include silicon, carbon, and oxygen, but many elements can be used, as you can see in **Figure 5**.

(h) Carbon fibre, a ceramic, is used in sports equipment such as tennis racquets. What property does it have that makes it suitable for this use?

In 5000 B.C.E., ceramics used were stone for weapons, flint for tools, and pottery for containers. Glass was invented about 2000 years ago. Today, tough new ceramics are being developed with very special properties (**Figure 6**).

H																	
Li	Be											B	C	N	O		
Na	Mg											Al	Si	P			
K	Ca	Sc	Ti	V	Cr	Mn	Fe	Co	Ni	Cu	Zn	Ga	Ge				
Rb	Sr	Y	Zr	Nb	Mo	Tc	Ru	Rh	Pd	Ag	Cd	In	Sn	Sb			
Cs	Ba	La	Hf	Ta	W	Re	Os	Ir	Pt	Au	Hg	Tl	Pb	Bi			
Fr	Ra	Lr															

Figure 5

The elements used in ceramic materials

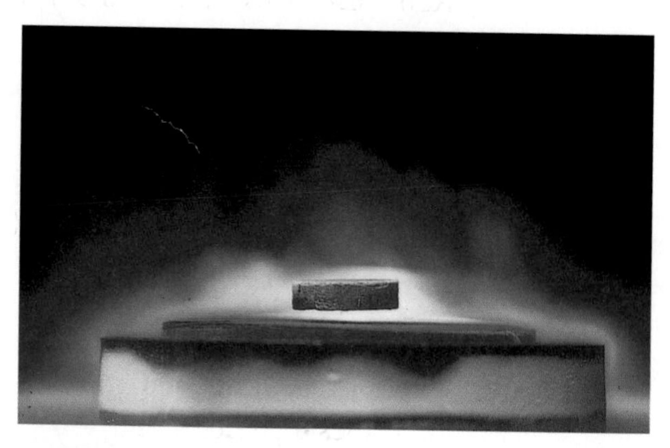

Figure 6

In the future, superconducting electromagnets may power high-speed trains.

Composites

Composites are materials that are formed by mixing two materials together. The first composite was probably a straw-clay combination used in building bricks. When scientists make composites, they try to combine the best properties of the polymers, ceramics, or metals that they put together. For example, fibreglass is a composite of a polymer and tiny glass fibres that can be used to repair a car or make a boat hull, the body or frame of the boat.

(i) Paper is a composite. What substances might it contain?

Material Science and Hockey Sticks

Back in the sporting-goods store, you have to make a decision about what hockey stick you are going to buy. You choose a stick made of an advanced thermoplastic composite, a combination of resin, nylon (a semicrystalline polymer), and a very strong carbon used in airplanes and space shuttles. According to the salesperson who helped you pick out the stick, this composite is tougher and less likely to break than other composites—only the best for the team's leading goal scorer.

Even for something as simple as a hockey stick, materials science offers choices in all the categories you have just learned about. Depending on whether you have dreams of playing professional hockey or just playing street hockey with your friends, the "materials magicians" have given you a wide choice of materials.

Challenge

1 Identify three products that are valuable to you. What are they made of? Include this information in your display. Which substance would you choose to market?

2 Scientists are always looking for new ways of assembling atoms into molecules. Can you include models of any of the substances mentioned in this section in your display?

Work the Web

Visit www.science.nelson.com and follow the links from *Science 9: Concepts and Connections,* 1.20. Find out about molecules called "Bucky-balls." What are they? What makes them so important? What could they replace in computers? Draw a diagram of this molecule. What other information can you add?

Try This Activity The Superball Polymer Goo Challenge

It is "Take Your Kid to Work Day" and you have joined your parent at the toy factory. You will work with the plastics team and have been assigned the task of finding the right recipe for a new superball. You can make your own synthetic polymer by dissolving 5 g of borax in 5 mL of water. Add 5 mL of white glue to 5 mL of water. Add the borax solution to the glue solution. Add one drop of green food colouring to the mixture and mix well.

Wear goggles and gloves when handling borax.

When the borax is dissolved in water, it forms borate ions. When this solution is added to white glue, the borate ions form bonds that join the borax molecules together. This mixture has some properties of a solid and some properties of a liquid. As the reaction continues, you will be able to pick up the polymer and work it in the palm of your hand. Instant goo! You don't need to wear gloves—the ball will stick to them if you do.

The polymer goo recipe above has very little bounce and tends to lose its original shape when left to sit. However, when mixed in the proper proportion, that polymer goo will produce a ball that is very bouncy. It is your job to experiment with the ingredients to make the bounciest polymer ball possible.

Experiment with the amounts of borax, water, and glue. Change only one ingredient in each trial. Do not exceed 6 mL of glue. (Hint: You don't need much borax.) You can also experiment with the order in which you combine the ingredients.

- In your group, decide on the amount of each ingredient you will use. (C4)
- Record your test amounts and the order in which you will combine them. (E2)
- Make a prediction about how the change will affect the bounciness. (C2)
- Make your polymer goo based on that recipe.
- Drop your superball from a height of 60 cm.
- Record the height of the resulting bounce. (E2)
- Repeat this process until you think you have created the bounciest superball.
- Test your superball against the superballs created by other groups in your class to see who created the bounciest formula.

Airborne cleaning powder can irritate your eyes, skin, and respiratory tract. Avoid direct contact with this powder.

Understanding Concepts

1. **(a)** Draw a bar graph to summarize the information in **Table 1**.
   (b) Examine your graph. What types of materials were most important in (i) 5000 B.C.E., (ii) 1800, (iii) 1960, and (iv) 2000?
2. Give five examples of each type of material that you would use in a typical day.
   (a) polymers
   (b) ceramics
   (c) metals
   (d) composites

Making Connections

3. The development of almost indestructible polymers is not entirely good news. What drawbacks do they have?
4. Polymers are used to make artificial body parts such as artificial knee joints and artificial hearts. What important properties do you think polymers used in artificial body parts must have?

SKILLS HANDBOOK: (C4) Choosing Materials (E2) Quantitative Observations (C2) Predicting Hypothesizing

Unit 1 Summary

Key Expectations

Throughout this unit, you have had opportunities to do the following:

- Describe an element as a pure substance made up of one type of particle or atom with its own distinct properties (1.8, 1.9, 1.13, 1.14)
- Recognize compounds as pure substances that may be broken down into elements by chemical means (1.9, 1.14)
- Describe compounds and elements in terms of molecules and atoms (1.15, 1.16)
- Identify each of the three fundamental particles (neutron, proton, and electron), and its charge, location, and relative mass in a simple atomic model (1.13, 1.19)
- Identify general features of the periodic table (1.18, 1.19)
- Demonstrate an understanding of the relationship between the properties of the elements and their position on the periodic table (1.18, 1.19)
- Identify and write symbols/formulas for common elements and compounds (1.14, 1.15)
- Describe, using their observations, the evidence for chemical changes (1.1, 1.3, 1.14, 1.7)
- Distinguish between metals and nonmetals, and identify their characteristic properties (118, 1.19)
- Demonstrate knowledge of laboratory safety, and disposal procedures while conducting investigations (1.1, 1.4, 1.11, 1.17)
- Determine how the properties of substances influence their use (1.6, 1.7, 1.12, 1.20)
- Formulate scientific questions about a problem or issue involving the properties of substances (1.1, 1.11, 1.17)
- Demonstrate the skills required to plan and conduct an inquiry into the properties of substances, using apparatus and materials safely, accurately, and effectively (1.4, 1.11, 1.17)
- Select and integrate information from various sources, including electronic and print resources, community resources, and personally collected data, to answer the questions chosen (1.1, 1.11, 1.17, 1.19)
- Organize, record, and analyze the information gathered (1.1, 1.4, 1.11, 1.18, 1.19)
- Communicate scientific ideas, procedures, results, and conclusions using appropriate language and formats (1.1, 1.2, 1.4, 1.16, 1.17, 1.19)
- Investigate, by laboratory experiment or classroom demonstration, the chemical properties of representative families of the elements (1.4, 1.11)
- Investigate the properties of changes in substances, and classify them as physical or chemical based on experiments (1.3, 1.4)
- Construct models of simple molecules (1.15, 1.16)
- Identify the uses of elements in everyday life (1.1, 1.5, 1.7, 1.12, 1.20)
- Describe the methods used to obtain elements in Canada, and outline local environmental concerns and health and safety issues related to the ways in which they are mined and processed (1.10)
- Explain how a knowledge of the physical and chemical properties of elements enables people to determine the potential uses of the elements and assess the associated risks (1.2, 1.5, 1.6, 1.20)
- Identify and describe careers that require knowledge of the physical and chemical properties of elements and compounds (1.6, 1.7)

Key Terms

alloy
atom
atomic mass
atomic number
atomic radius
Bohr-Rutherford diagram
boiling point
bond
chemical change
chemical formula
chemical property
chemical symbol
combining capacity
combustibility

combustion
composite
compound
corrosion
density
diatomic molecule
electron
element
family
flammable
hardness
heavy metal
heterogeneous mixture
homogenous mixture
light sensitivity
mass number
melting point
metalloid
metallurgy
mineral
mixture
modern periodic law
molecule
monomer
neutron
ore
period
periodic
periodic table
physical change
physical property
polymer
polymerization
proton
pure substance
rusting
solubility
standard atomic notation
structural diagram
subscript
synthetic
viscosity

What HAVE YOU *Learned?*

Revisit your answers to the What Do You Already Know? questions on page 9 in Getting Started.

- Have any of your answers changed?
- What new questions do you have?

Unit Concept Map

Use the concept map to review the major concepts in Unit 1. This map can help you begin to organize the information that you have studied. You may copy the map, and then add more links to your map. Also, you may add more information in each box.

A concept map can be used to review a large topic on a general level, or it can be used to examine a very specific topic in detail. Select one concept from this unit that you need to study more, and make a detailed concept map for it.

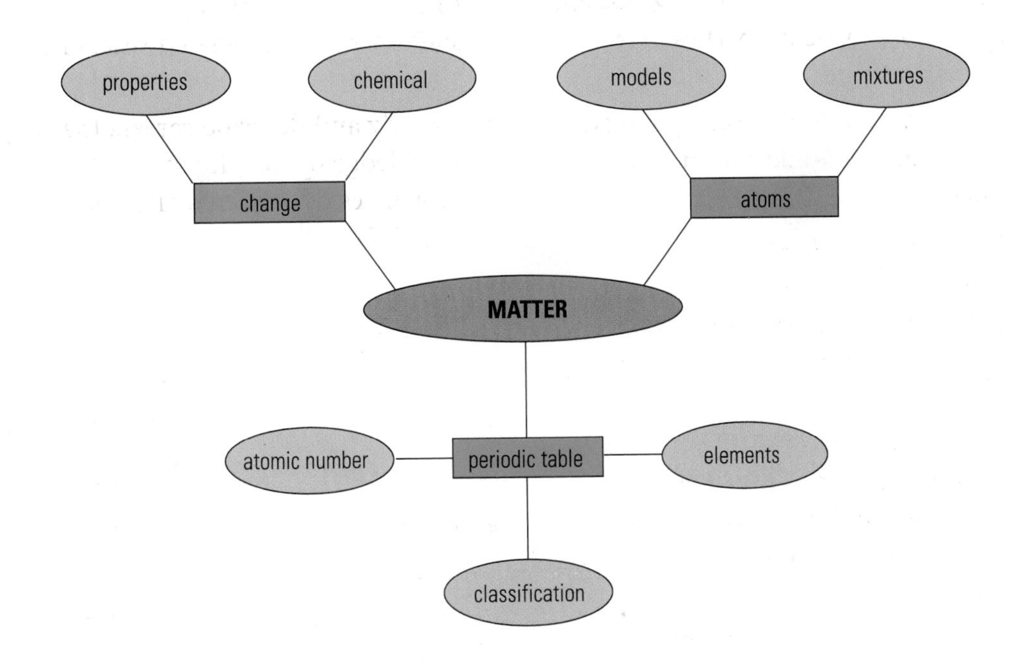

Unit 1 Review

Understanding Concepts

1. In your notebook, write the word(s) needed to complete each statement below.
   (a) In a(n) ________ change, a new substance is produced.
   (b) A(n) ________ is a mixture of metals.
   (c) A solid produced when two solutions are mixed together is a(n) ________.
   (d) A(n) ________ is a sample of matter containing only one type of atom.
   (e) ________ are shiny, malleable, and conduct electricity.
   (f) The ________ is the number of protons in the atom.
   (g) An electrically charged atom is a(n) ________.
   (h) The ________ is the core of the atom, containing most of its mass.
   (i) The size of an atom is described by its atomic ________.
   (j) ________ materials are made by humans, rather than naturally.
   (k) Elements can be arranged in a(n) ________ table.

2. Indicate whether each of the statements is true or false. If you think the statement is false, rewrite it to make it true.
   (a) Combustion is the chemical reaction between a fuel and hydrogen.
   (b) Colour and hardness are examples of chemical properties.
   (c) The measure of how easily a liquid flows is called viscosity.
   (d) A molecule is a combination of atoms.
   (e) The chemical symbol for calcium is Cal.
   (f) The modern periodic table organizes elements by atomic mass.
   (g) A Bohr diagram shows electrons in orbits around the nucleus.
   (h) A neutron is positive and located in the nucleus.

3. (a) What is the difference between a physical property and a chemical property?
   (b) Give an example of one physical property and one chemical property for each of the following: wood, gasoline, and baking soda.

4. For each of the following, replace the description with one or two words:
   (a) the measure of resistance of a solid to being scratched or dented.
   (b) the chemical property that describes the ability of a substance to burn
   (c) a change in which a new substance is produced
   (d) a change in which no new substance is produced
   (e) able to dissolve in a solvent
   (f) substances invented by people

5. The sentences below contain mistakes or are incomplete. Write complete, correct versions.
   (a) A physical change produces a new substance.
   (b) The formation of frost is a chemical change.
   (c) A new colour indicates a physical change.
   (d) The ability to react with an acid is an example of a physical property.
   (e) It is safe to taste some substances in the lab.
   (f) A chemical change is a change of state or form.
   (g) Corrosion is the reaction of a metal with nitrogen in the air.
   (h) Goggles may be taken off if you have finished your experiment.
   (i) Elements are made up of compounds.
   (j) Nonmetals are shiny and good conductors.
   (k) Protons are negative particles in orbits around the nucleus.

(l) The mass number is the number of neutrons.

(m) A Bohr diagram shows protons in orbit.

(n) Mendeleev's table organized elements by atomic number.

(o) Elements in the same period have similar properties.

6. List five things you would consider before deciding whether a change is chemical or physical.

7. Indicate whether each of the following is a physical or a chemical change.

   (a) water freezing on a pond

   (b) soap removing grease from hands

   (c) a light bulb glowing

   (d) a cake baking

   (e) wood burning

   (f) kitchen scraps composting

   (g) a paper clip bending

   (h) dynamite exploding

8. Describe two compounds that contain atoms of the same elements, but in different proportions.

9. State the types of atoms and the numbers of each type that are present in the following molecules: copper phosphate (Cu_3PO_4) and sodium nitrate ($NaNO_3$).

10. Describe the similarities and/or differences between each pair of terms.

   (a) physical property, chemical property

   (b) combustion, corrosion

   (c) element, compound

   (d) atom, molecule

   (e) metal, nonmetal

   (f) mineral, ore

   (g) natural material, synthetic material

11. In a Bohr-Rutherford model of the atom,

   (a) Where are the protons found?

   (b) Where are the neutrons found?

   (c) Where are the electrons found?

   (d) Which particles make up most of the mass of the atom?

   (e) Which particles take up most of the space in the atom?

12. State Mendeleev's periodic law.

13. How does Mendeleev's periodic law differ from the modern periodic law?

14. (a) In the periodic table, where are the metals found?

   (b) Where are the nonmetals found?

Applying Skills

15. Name four materials or pieces of equipment that you used in your investigations to ensure lab safety. Explain the function of each.

16. A yellow solid is heated and is observed to change to a brown liquid. Explain whether the change is chemical or physical.

17. A white solid is heated and is observed to change to a liquid at 65°C. When the liquid is cooled, it becomes a white solid again at 65°C. Is the change chemical or physical? Explain.

18. Examine the models in **Figure 1**.

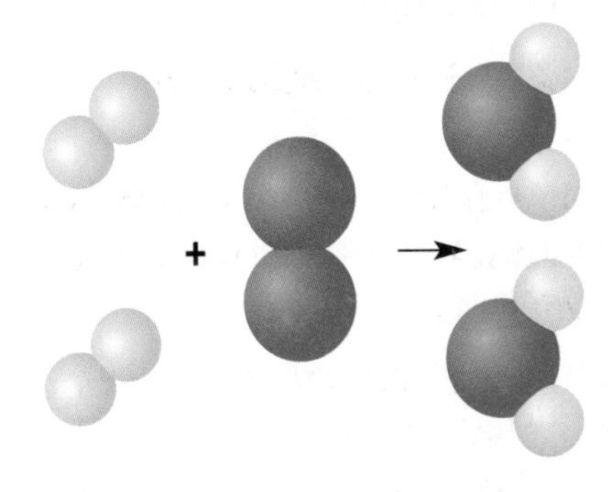

Figure 1

   (a) What substances could the drawings represent?

   (b) Write a word equation for the reaction.

19. Copy **Table 1** into your notebook. Fill in the blanks with the missing numbers.

Table 1

Element	Symbol	Atomic number	Mass number	No. of protons	No. of electrons	No. of neutrons
beryllium	Be	4	9	?	?	?
carbon	C	6	?	?	?	8
silicon	S	?	?	?	14	14
potassium	K	?	?	19	?	20

20. Write the formula, name, and structural diagram for the compound formed by each of the following combinations of elements:

(a) potassium and chlorine

(b) calcium and oxygen

(c) aluminum and sulfur

21. Draw Bohr-Rutherford diagrams for each of the atoms in question 20.

22. Identify the numbers of protons and neutrons in each of the atoms in **Figure 2** by interpreting their standard atomic notation.

(a) $^{40}_{19}K$ (b) $^{28}_{13}Al$ (c) $^{14}_{6}C$

Figure 2

23. Match the description on the left with the term on the right. Use each term only once.

Description	Term
A smallest particle of an element	**1** element
B substance containing only one type of atom	**2** synthetic
C connection between atoms	**3** proton
D particle made up of two or more atoms	**4** composite
E number of protons	**5** molecule
F positive subatomic particle	**6** polymer
G sum of protons and neutrons	**7** atomic number
H uncharged subatomic particle	**8** mass number
I very long molecule	**9** neutron
J material formed by mixing two or more materials	**10** atom
K produced by people	**11** bond

24. List the four different types of materials that people have used over the centuries and give one ancient and one modern example of each.

25. Classify each of the following as a metal, polymer, ceramic, or composite.

(a) concrete reinforced with steel bars

(b) a pottery coffee mug

(c) a bronze statue

(d) a polyethylene drink bottle

(e) a shirt that is 40% cotton and 60% nylon

26. Suppose someone tells you that a green object contains copper. You are not convinced because you have seen copper wires and jewellery that are reddish-brown. Is it possible that this green substance does contain copper? Explain.

Making Connections

27. The four Hazardous Household Product Symbols indicate products that are poisonous, flammable, explosive, and corrosive. Which labels would be on containers of

(a) an aerosol insect spray?

(b) a drain cleaner?

(c) an ant powder?

(d) furniture polish?

28. Think of a group of objects (e.g., leaves, food, animals, children's toys, or drug store products) and a way to categorize them so that, if a new one were to be discovered, it could be included in your system. Create a computer or poster display of your organization system, and explain why you chose to categorize the objects in this way.

29. Elements have been named for many reasons. Using library resources, or the Internet, research the following:
   (a) Germanium, lutetium, and polonium were named to honour the geographic origin of their discoverers. Who were the discoverers, and where did they come from?
   (b) Which heavenly bodies were the following named after: mercury, uranium, neptunium, plutonium, tellurium, selenium, palladium, cerium? (Some are very easy, while others are not so obvious.)
   (c) Some elements were named to honour people. Which people were honoured by the following: gadolinium, curium, einsteinium, fermium, mendelevium, lawrencium, nobelium, seaborgium? Write out their full names, and write a couple of sentences about each person.
   (d) Some elements were named after places. What places are the following named after: europium, hafnium, americium, berkelium, californium?
30. Carbon dioxide ejected from a fire extinguisher is so cold that it changes to snow.
   (a) Is this a chemical or a physical change?
   (b) The carbon dioxide snow, when applied to a burning object, is said to smother the flame. What kind of chemical change is the carbon dioxide snow preventing? How does the carbon dioxide stop the fire?
31. Corrosion is the reaction of metals with oxygen in a chemical change.
   (a) What kinds of changes occur in other substances over time? For example, what happens to plastic products left out in below-zero temperatures or in intense sunlight for long periods of time?
   (b) In what instances could the use of plastics practically replace the use of metals?
32. You have learned that models are modified as scientists gather new evidence. Has this happened with the atomic model? Explain.
33. Elements are the basic building blocks of all the substances in the world. Elements are made up of atoms. Think about the structure of the atom.
   (a) Which part of the atom is involved in the chemical reactions that form these substances? Give reasons for your answer.
   (b) Identify one substance that is produced by industry, and describe its potential uses and associated risks.
34. Customs officials investigating a crate shipped from Central America wanted to know what was in the crate before allowing it into Canada. All the labels were in Spanish, but the following chemical symbols were printed on the crate: $NaHCO_3$, $NaNO_3$, $Ca_3(PO_4)_2$. Would you recommend that the officials allow the crate to continue, or should they call the shipping company for more information? Explain your answer.
35. Design a poster to represent an element from the periodic table. Include the following information: atomic mass, atomic number, the origin of the symbol, properties of the element, where it can be found, can it be produced, and what is it commonly used for? You may include any other information that you feel is important or relevant. Be sure to include a neat and colourful drawing of its Bohr model (including the number of protons and neutrons).
36. In general, the atomic mass of elements increases as the atomic number increases. Find three pairs of elements in the periodic table that are exceptions to this generalization. Why do you think these exceptions occur?

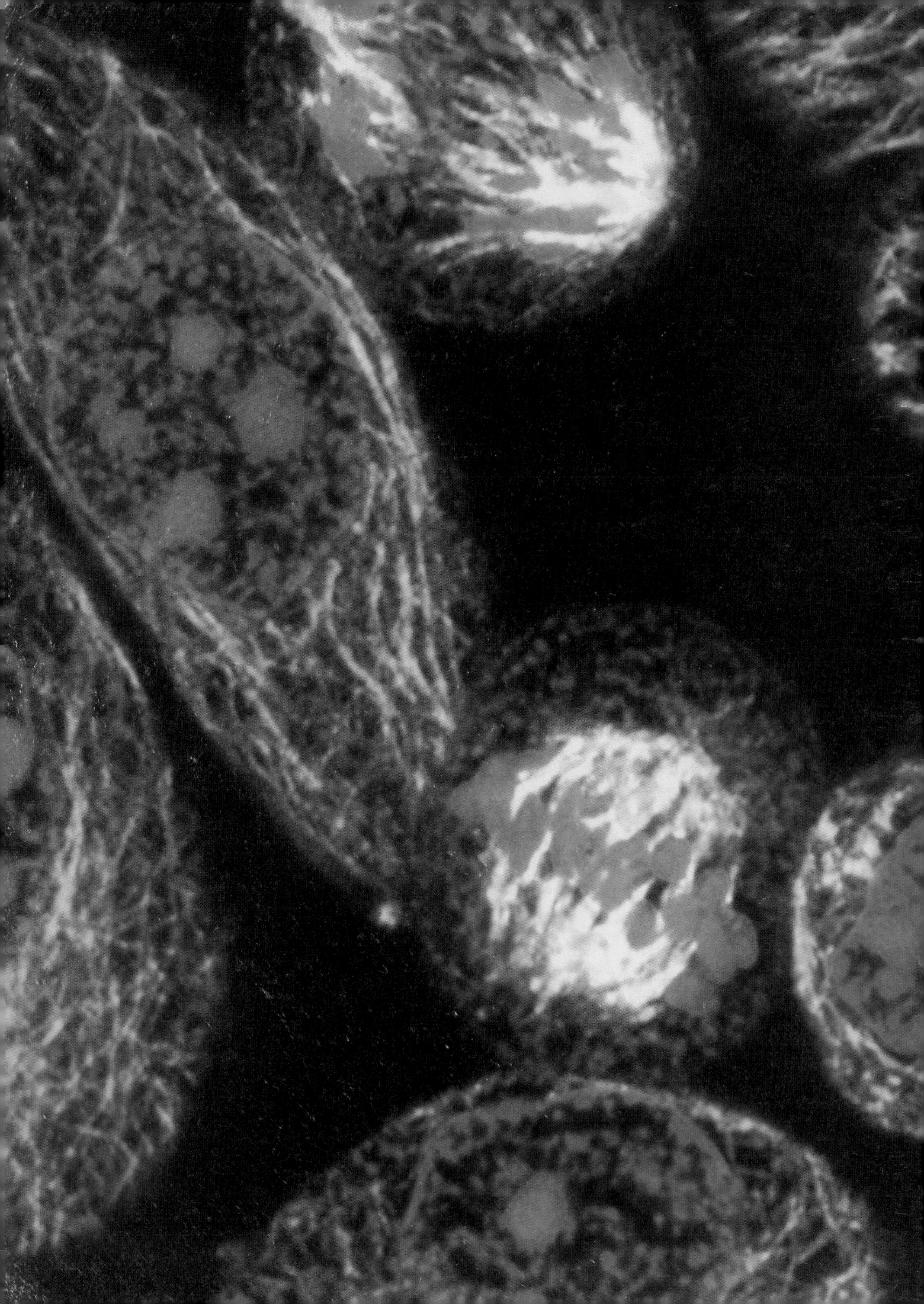

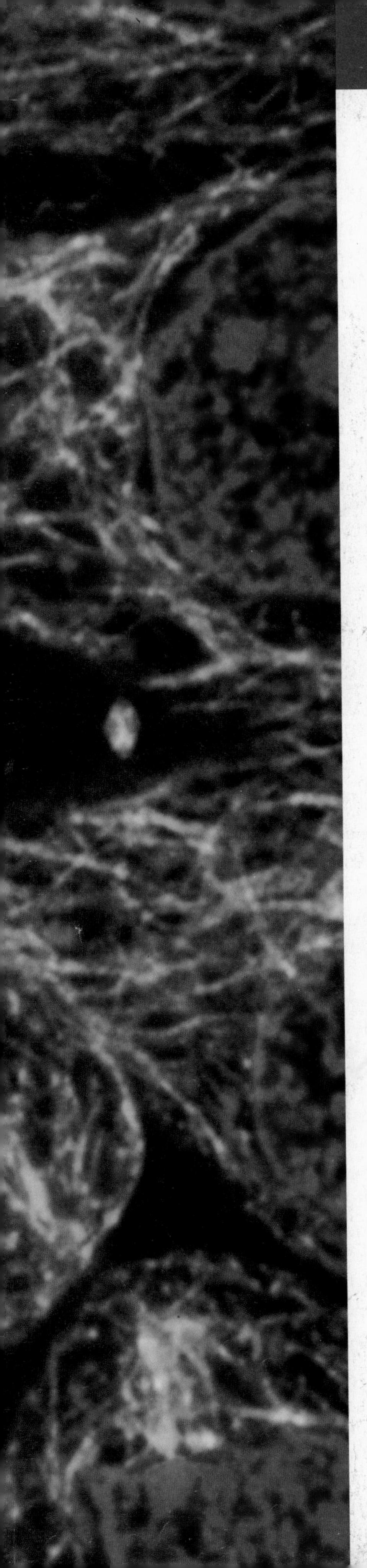

Reproduction: Processes and Applications

How does a living thing reproduce? Do all living things reproduce in the same way? What are the possible risks and benefits of using scientific knowledge to create clones of living things? This unit will help you understand what is known about the development and reproduction of living things, including humans, and to address some of the issues related to growth, development, and reproduction.

Unit 2 Overview

Overall Expectations

In this unit, you will be able to

- understand the importance of cell division in growth and reproduction
- investigate reproductive strategies
- analyze issues related to reproductive technologies

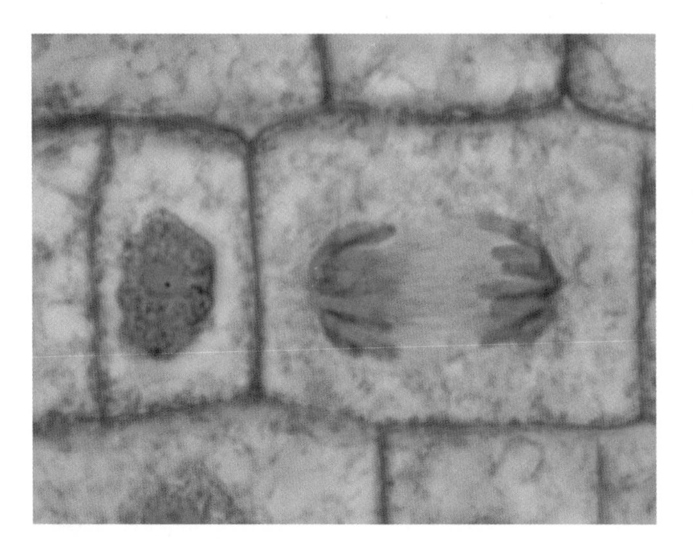

Cell Division in Growth and Reproduction

Every living organism must grow and reproduce. Neither growth nor reproduction can occur without cell division. Cell division and all of the functions of cells are determined by the genetic information contained in the nucleus of each cell.

Specific Expectations

In this unit, you will be able to

- describe the processes and explain the importance of cell division
- explain how cell division determines the growth of organisms
- use a microscope to examine cells and the stages of cell division
- explain the importance of DNA replication to the survival of an organism
- explain how changes to the DNA material can affect an organism
- analyze the factors that affect the risk of developing cancer

Reproduction

There are two main types (sexual and asexual) and many different methods of reproduction. Sexual reproduction creates variety among and within species. Asexual reproduction does not require two organisms to reproduce.

Specific Expectations

In this unit, you will be able to

- describe the differences between sexual and asexual reproduction and indicate the advantages and disadvantages of each
- identify and describe the advantages of different sexual reproductive strategies, such as conjugation, hermaphroditic reproduction, and separate sexes
- identify reproductive structures within flowers and humans
- explain why sex cells have half as many chromosomes as other cells

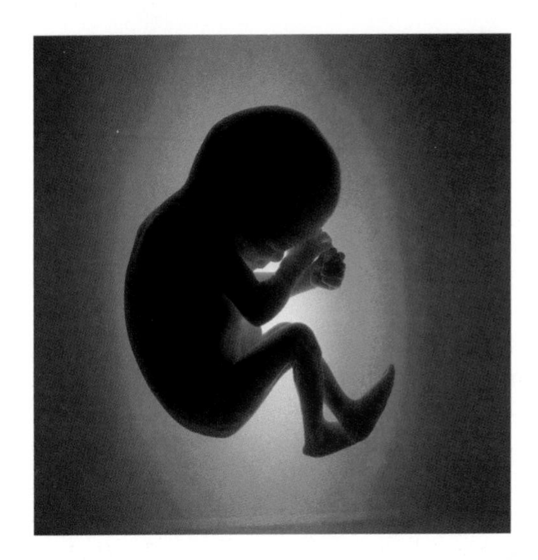

Reproductive Strategies and Technologies

Different organisms use different strategies to ensure survival of their offspring. Science has helped us understand the biology of reproduction. Technology has helped us solve some reproductive problems.

Specific Expectations

In this unit, you will be able to

- identify and describe the advantages of different reproductive strategies, such as spores, seeds, eggs, and development within the uterus
- explore seed formation and germination of plant embryos in a laboratory setting
- examine the events of zygote formation, embryo growth, fetal development within the uterus, and birth of humans
- examine and evaluate the implications of various reproductive technologies designed to help people who want to have children
- explain how materials pass between the mother and the unborn child, and examine the implications of drug use by the mother during pregnancy

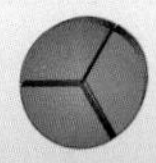

Challenge

Reproduction and Reproductive Technology

As you learn about cell division and reproduction, you will be able to explain how this knowledge is important to humans. You will also be able to demonstrate your learning by completing a Challenge. For more information on the Challenges, see the following page.

1 Reproductive Technology

Choose a reproductive technology and prepare a report in pamphlet form that will inform the general public about the issues and the advantages and disadvantages of the technology.

2 Plant Experiment

Plan and conduct an experiment to investigate factors that affect the growth and development of plants, and report the results of your investigation.

3 Analyzing Cancer Risk

Design an instrument or rating scale that will help people analyze their cancer risk, and make recommendations that will help people lower their risk of cancer.

Record your ideas for your Challenge as you progress through this unit, and when you see

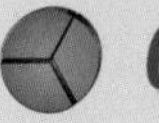

Challenge

Reproduction and Reproductive Technology

Scientists are making advances in their understanding of cell division and reproduction, both in animals and plants. These advances are happening rapidly, so it is difficult for most ordinary citizens to be aware of them.

Each of these Challenges allows you to research an area of science that interests you so that you will understand it better and will be able to make informed decisions, if necessary.

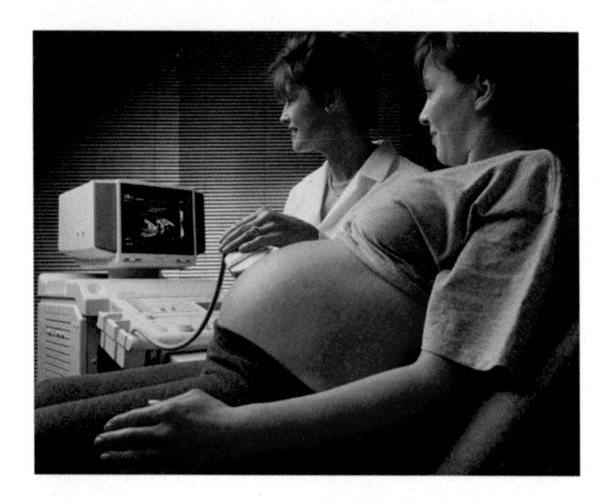

1 Reproductive Technology

Throughout this unit, you will be introduced to a number of common reproductive technologies. Choose one of these, or another technology, and prepare a pamphlet to inform your classmates and the general public about it. Before developing your pamphlet, find out what people already know about the technology.

In researching and reporting on your chosen reproductive technology, you should

- create and conduct a brief survey to determine what people already know about the technology
- describe the science behind the technology
- identify and explain any issues associated with the technology
- prepare a pamphlet that will educate the general public about the advantages and disadvantages of the technology

2 Plant Experiment

Through experiments, attempts have been made to improve the quality and yield of plants that reproduce sexually. Plan and conduct an experiment to illustrate plant reproduction. Using seeds from a fast-growing plant, investigate factors that might affect germination and growth of the plant. Determine how this plant can reproduce asexually. Report on the advantages and disadvantages of each method.

In designing and carrying out your experiment, you should

- choose a plant and plan the experiment, taking into account all necessary safety precautions, investigating at least two factors that might affect germination and growth
- after your teacher has approved your plan, conduct the experiment
- prepare a report in which you accurately describe the plants and the experimental procedure, summarize the results, and present your conclusions
- submit your plants to support your report

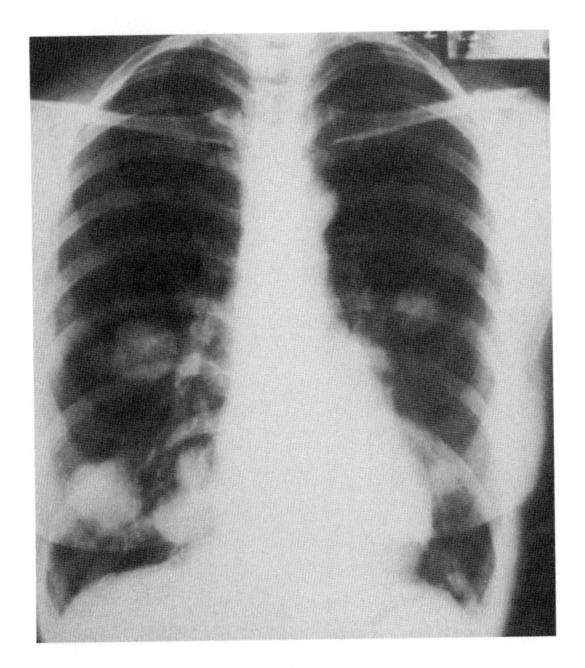

3 Analyzing Cancer Risk

Scientific research has established that certain factors in our lives increase the probability that we will develop cancer. Identify three such factors, and research how each factor affects people's cancer risk. Prepare a report that includes a rating scale to help individuals evaluate their lifestyle and determine their personal cancer risk.

In preparing to analyze factors that pose cancer risks, you should

- identify the factors that increase the risk of cancer
- summarize the scientific research about each of the factors
- create a questionnaire or survey that will allow individuals to assess their lifestyles and determine their cancer risk
- develop a rating scale based on the questionnaire or survey
- recommend strategies that will help individuals reduce their cancer risk

When preparing to use any test or carry out an experiment, have your teacher approve your plan before you begin.

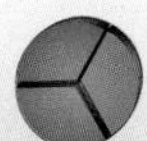

Assessment

Your completed Challenge will be assessed according to the following points:

Process

- Understand the specific challenge.
- Develop a plan.
- Choose and safely use appropriate tools, equipment, materials, and computer software.
- Analyze the results.

Communication

- Prepare an appropriate presentation of the task.
- Use correct terms, symbols, and SI units.
- Incorporate information technology.

Product

- Meet established criteria.
- Show understanding of the science concepts, principles, laws, and theories.
- Show effective use of materials.
- Consider legal and ethical issues.
- Address the identified situation/problem.

Unit 2 Getting Started

Cell Division, Growth, and Reproduction

Have you ever wondered how an organism grows? There are approximately 100 trillion cells in your body, all of which started from a single cell—a fertilized egg. How does one cell grow into an organism with trillions of cells? If they all come from one cell, do all the cells look alike? The obvious answer is no: if they did, organisms would simply be big blobs of cells.

The 35-m blue whale (**Figure 1**) is about 18 times longer than the average human. Does it have the same number of cells as humans, or do bigger organisms have bigger cells?

Some plants grow very quickly. Is cell division in plants similar to that in animals? Do all cells divide at the same rate?

Why do young organisms look like their parents? Some organisms are identical to their parents (**Figure 2**) while others only resemble them (**Figure 3**). It is obvious that certain characteristics are passed on from parents to offspring. Since an egg from the mother and a sperm from the father unite to create a new individual, these cells must contain the information that determines the offspring's characteristics.

There are several ways of creating plants that are identical to their parents. Can we use the same methods to create identical animals, including humans?

Figure 2

These plants are identical. The one on the right grew from a cutting of the one on the left.

Figure 1

The average blue whale is many times larger than the average human.

Figure 3

It is easy to see the resemblance between Kiefer Sutherland and his father, Donald Sutherland.

Only one egg cell and one sperm cell are needed to start the development of a new organism (**Figure 4**). In fact, as soon as the egg is fertilized, it forms a barrier to keep all other sperm out. Both plant and animal males produce millions of sperm cells. Why are so many produced if only a single sperm is needed?

How is the development of a human baby different from the development of the young of other organisms? Human babies develop inside their mother's body. The young of some other animals develop outside the mother's body. Yet some other organisms, like plants, develop without any contact with their parents. What are the advantages of each method? Can you identify any disadvantages?

A human baby develops for nine months inside its mother's body before being born. What happens during that nine-month period? How is the baby affected by the mother's lifestyle?

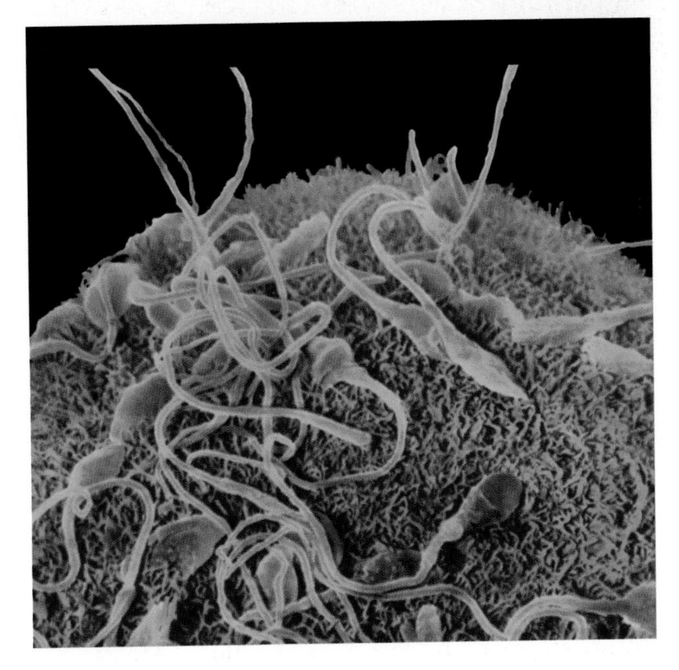

Figure 4

Why are all the sperm cells produced if only one can fertilize the egg?

What DO YOU ALREADY *Know?*

1. Do all cells look alike? Explain.
2. Explain the similarities and differences between plant and animal cells.
3. Why is it important for cells to divide rather than to simply grow larger?
4. Where in the cell is genetic information found? What is the purpose of this genetic information?
5. **(a)** Give two examples of different methods of reproduction in plants.
   (b) Do all animals reproduce in the same way? Explain.

Throughout this unit, note any changes in your ideas as you learn new concepts and develop your skills.

Try This Activity Cell Replacement

Use a permanent marker to put a spot of ink in the palm and on the back of your hand. Because the marker ink does not dissolve in water, the cells that absorb the ink are permanently stained.

Do not use the permanent marker if you suspect that you may be allergic to it. Try food colouring instead.

C2 (a) Predict from which area the stain will first disappear.

E1 Observe the stained areas daily and record your observations.

(b) Explain your observations.

The Importance of Cell Division

How are a radish leaf cell and a human skin cell alike (**Figure 1**)?

Plant and animal cells each have **cell membranes**. The membrane acts as a gatekeeper, controlling the movement of materials into and out of the cell. Cells are filled with **cytoplasm** containing specialized structures called **organelles**.

Near the centre of each cell is the **nucleus**. This organelle acts as the control centre, directing all cell activities, including cell division.

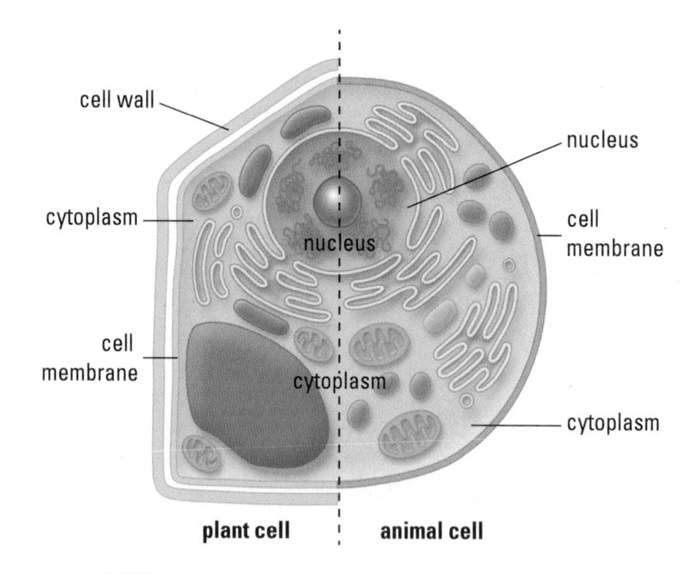

Figure 1
A comparison of plant and animal cell structures

Why Cells Divide

Healing and Tissue Repair

You do not go through life with the same cells you had at birth. Every second, millions of your body cells are injured or die and must be replaced (**Figure 2**). This replacement also occurs in plants.

Growth

All plants and animals begin life as a single cell. There are only two ways an organism can grow: the single cell gets bigger, or the single cell divides into more cells.

One of the most important jobs a cell must do is to exchange materials with its environment. Food comes in and wastes go out. These substances travel through the cytoplasm, and enter and exit the cell through the cell membrane.

When cells become very large, there is too much cytoplasm for the cell membrane to exchange materials efficiently. As well, it takes too long for messages to reach the nucleus. Cell division allows one big cell to become two small cells. In this way, an organism grows while still maintaining an efficient cell size.

Not all cells grow and divide at the same rate. In adult humans, blood cells divide at enormous rates, but brain cells rarely divide.

Figure 2
Normal activities remove millions of skin cells that must be replaced every day. It has been estimated that about 50% of the dust in a furnace filter consists of dead human skin cells.

Try This Activity Limits on Cell Size

How large can cells be? Why are humans made of millions of tiny cells rather than one large cell? In this activity, you will use a model to determine whether cell size affects the cell's ability to respond to the environment.

The "cells" you will examine are small cubes of jelly called agar. Liquids can easily move in and out of the agar as they do in real cells. The agar has been treated with an indicator that turns pink in a special solution. The pink colour indicates how well liquids pass in and out of the "cell."

Exercise caution when using a knife. Use tongs or wear rubber gloves. Wear safety goggles.

- Using tongs, put a large and a small agar cube into a beaker.
- Carefully pour the special solution into the beaker, covering both cubes.
- After exactly one minute, take the cubes out of the solution. Immediately pat them dry using a paper towel.
- Cut each cube in half using a knife.

Examining the agar cubes, answer the following:

(a) Describe the appearance of each cube.

(b) If the nucleus (control centre) is located in the middle of each cube, how quickly will each cube react to changes in the environment?

(c) If the coloured part of the cell receives food and gets rid of wastes, which cell would do a better job of carrying out these functions? Explain why.

(d) Why are large animals made up of many cells and not just one giant cell?

Reproduction of Organisms

Cells divide to create new organisms. A single-celled bacterium divides to form two identical bacteria. Cell division is also necessary to reproduce multicellular organisms.

In the human body, red blood cells live 120 days, white blood cells anywhere from 1 day to 10 years, and platelets, the cells that help your blood clot, only about 6 days.

Understanding Concepts

1. How are plant and animal cells alike?
2. What differences do you notice between plant and animal cells, from **Figure 1**?
3. Give three reasons cells divide.
4. (E3) Draw a large cell and a small cell. Label the nucleus, cytoplasm, and cell membrane. Show the directions in which food and wastes travel. Referring to your diagrams, explain why a small cell works better than a large cell.

Making Connections

5. In the early 1900s, doctors gave elderly patients blood transfusions from younger people. The doctors believed that the younger blood would give the elderly people more energy. Do older people actually have older blood? Support your answer.

Challenge

2 Explain why your plant isn't one giant cell. What functions must different plant cells perform?

SKILLS HANDBOOK: (E3) Scientific Drawings

Cell Division

The nucleus of a cell acts as the control centre. Inside the nucleus is the genetic material **deoxyribonucleic acid** (**DNA**). This genetic material is organized into threadlike structures called **chromosomes** (**Figure 1**). Each chromosome contains many different **genes**, small units of DNA that contain the information necessary for all cell functions and that determine the characteristics of an individual.

Cells divide for three reasons: healing and repair, growth, and reproduction. The process of cell division for healing and repair and growth is the same. Cell division for reproduction is a similar process and will be discussed in section 2.8. Whatever the reason for cell division, it is essential that a copy of the genetic material be passed on to each new cell.

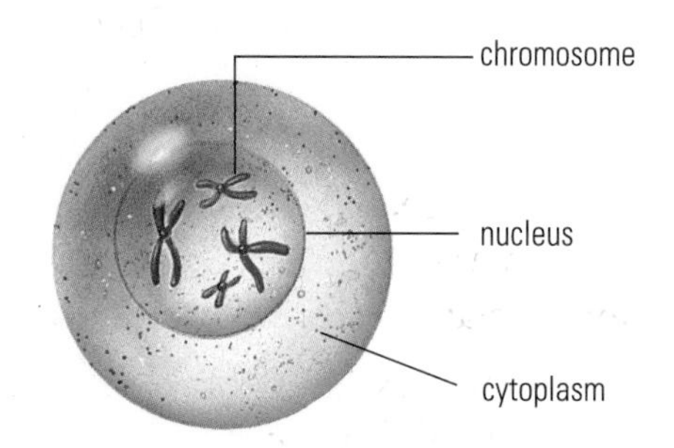

Figure 1

At certain times, the chromosomes can be clearly seen as threadlike structures. The individual genes cannot be seen.

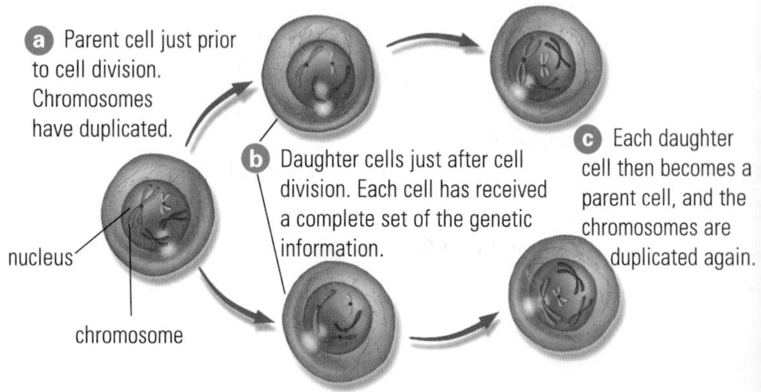

Figure 2

A parent cell produces two identical daughter cells.

The Cell Cycle

Cells are not continuously dividing. There is a period when they are carrying out their normal functions, such as using nutrients to produce energy. And there is a period when they go through the process of cell division. Both of these periods together are known as the **cell cycle**. For most cells, the cell division stage is only a small part of the cell cycle.

Mitosis and Cytokinesis

Cell division involves two processes: the division of the material in the nucleus and the sharing of the cytoplasm. During cell division, the duplicated chromosomes separate and move to the opposite ends of the cell. This process of dividing the nuclear material is called **mitosis**. Cell division continues with the separation of the cytoplasm and its contents into equal parts. This process is called **cytokinesis**.

As a result of cell division, the initial cell (called the parent cell) divides into two identical cells (called daughter cells), as shown in **Figure 2**.

The Phases of Cell Division

The period between cell divisions is called **interphase**. During this period, the cell grows and duplicates the chromosomes in the nucleus. This process ensures that the same genetic information is passed on to each of the two new daughter cells.

The process of mitosis has four phases (see **Figure 3**):

1. **Prophase**
   In prophase, the individual chromosomes, now made up of two identical strands, shorten and thicken. The membrane around the nucleus starts to dissolve.

2. **Metaphase**
   In metaphase, double-stranded chromosomes line up in the middle of the cell.

3. **Anaphase**
During anaphase, each chromosome splits. The two halves move to opposite ends of the cell. If this process is successful, each daughter cell will have a complete set of genetic information.

4. **Telophase**
During telophase, the chromosomes reach the opposite poles of the cell and a new nuclear membrane starts to form around each set.

Cytokinesis begins and the cytoplasm is divided into roughly equal parts. Following cytokinesis, the two new daughter cells are in interphase and begin to grow. They eventually duplicate the genetic material and prepare for further cell division.

Work the Web

To research cell division, visit www.science.nelson.com and follow the links from *Nelson Science 9: Concepts and Connections*, 2.2. Do all cells divide at the same rate?

Challenge

3 Cancer is a cell division problem. What could go wrong with the cell division process that might cause problems for the cell or for the organism?

Understanding Concepts

1. Describe the cell cycle. What happens during interphase?
2. Why is it necessary to duplicate the nuclear material?
3. List and describe the four phases of mitosis.
4. A normal human cell has 46 chromosomes. After the cell has undergone mitosis, how many chromosomes would you expect to find in each daughter cell? Explain.

Making Connections

5. X-rays and other forms of radiation can break chromosomes apart. Doctors and dentists ask women whether they are pregnant before taking X-rays. Why don't they want to X-ray pregnant women?

Reflecting

6. Sketch an outline of a human body. On the sketch, identify the areas of the body where you think cell division is most rapid. Why do you think cells from these areas divide most rapidly?

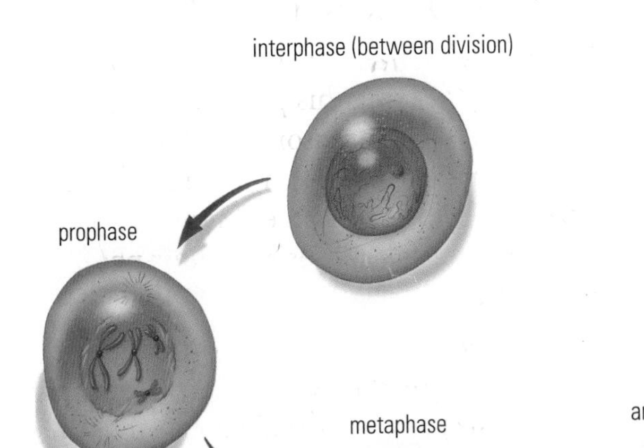

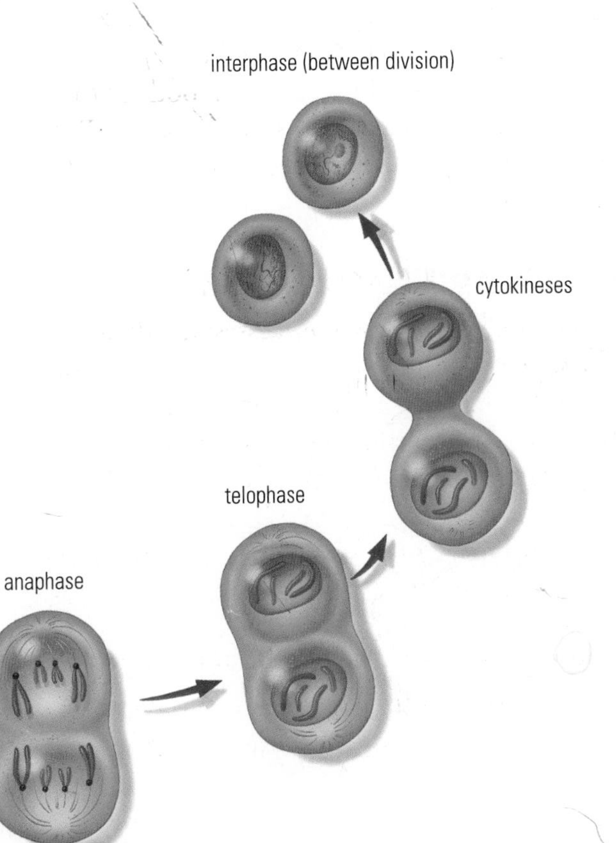

Figure 3
Animal cell division

2.3 Activity

Observing Cell Division

In the previous sections, you learned why and how cells divide. In this activity, you will have an opportunity to view and compare plant and animal cells during mitosis. You will examine prepared slides of the onion root tip and the whitefish embryo to identify cells that are dividing. Because prepared slides show a cell at one moment in time, you will not be able to watch the complete division of a single cell.

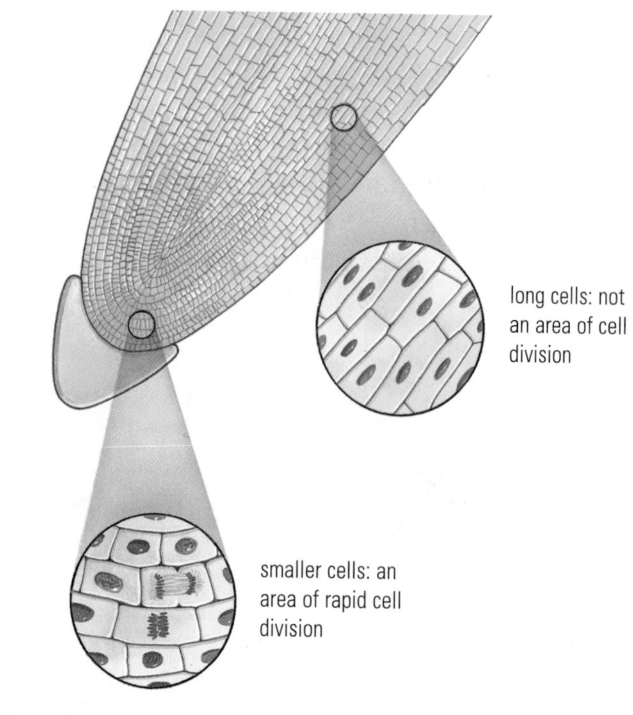

Figure 1
The onion root tip is an area of rapidly dividing cells.

Materials

- microscope
- lens paper
- prepared microscope slide of an onion root tip
- prepared microscope slide of a whitefish embryo

Procedure

(B) Refer to section D of the Skills Handbook for information on proper use of the microscope. Handle glass slides with care, and dispose of any broken slides as directed by your teacher.

1 Obtain an onion root tip slide and place it on the stage of your microscope.

2 Look at the slide under low-power magnification (the shortest objective lens). Focus using the coarse-adjustment knob. Find the cells near the root tip (**Figure 1**). This is the area of greatest cell division for the root.

3 Centre the root tip and then rotate the nosepiece to the medium-power objective lens. Focus the image using the fine-adjustment knob only. Identify a few dividing cells.

(a) How can you tell whether the cells are dividing?

4 Rotate the nosepiece to the high-power objective lens. Use only the fine-adjustment knob to focus the image. Locate and observe cells in each phase of mitosis. Use the photographs of cells dividing in **Figure 2** to help you. Don't worry if what you see does not look exactly like the photographs.

(b) Draw and label each of the phases that you (E3) see. Label chromosomes if they are visible. It is important to draw and label only the structures that you see under the microscope.

5 Return your microscope to the low-power objective lens and remove the onion root tip slide.

6 Place the slide of the whitefish embryo on the stage. (An embryo is an organism in the very early stages of its development.) Focus the slide using the coarse-adjustment knob.

SKILLS HANDBOOK: (B) Safety in the Laboratory (E3) Scientific Drawings

7 Repeat steps 3 and 4 for the whitefish cells.

(c) Draw and give a title to each of the phases that (E3) you see. Label the chromosomes if they are visible.

8 Compare your diagrams with those of other students. Help others locate phases or cell structures.

9 Return your microscope to the low-power objective lens. Remove the slide of the whitefish embryo. Put away your microscope and return the slides to your teacher.

(a)

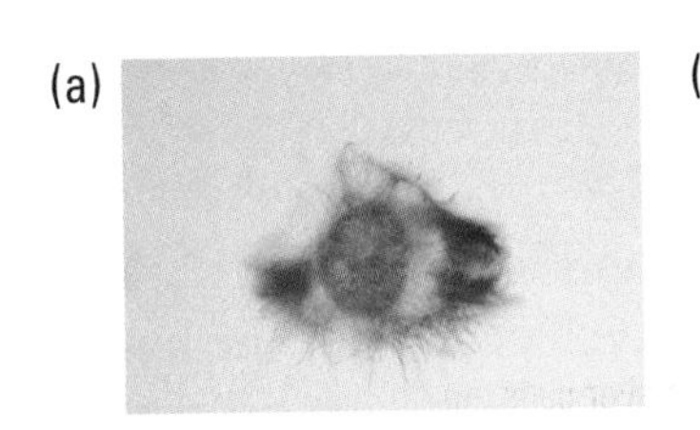

(b)

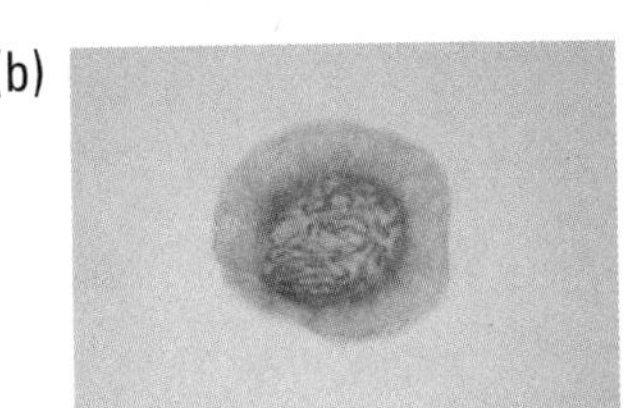

(c)

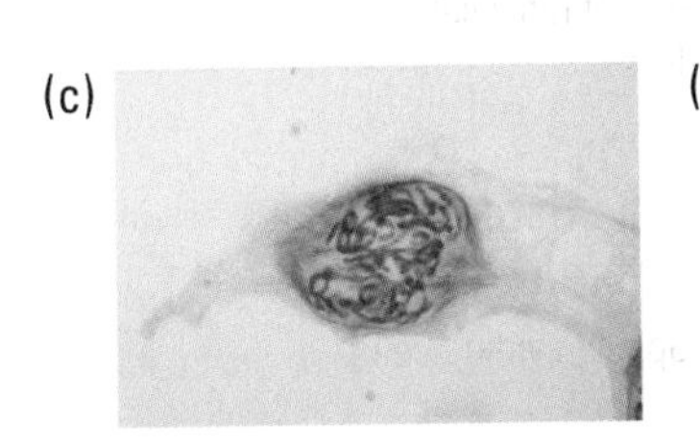

(d)

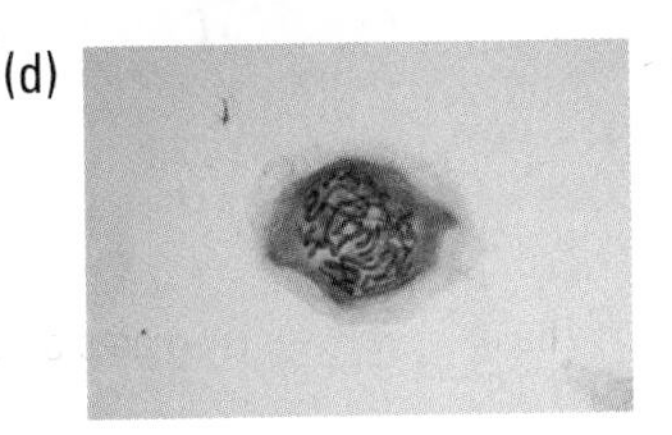

(e)

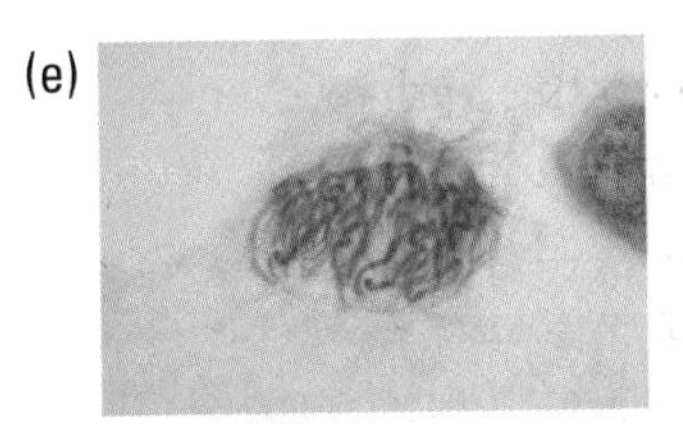

(f)

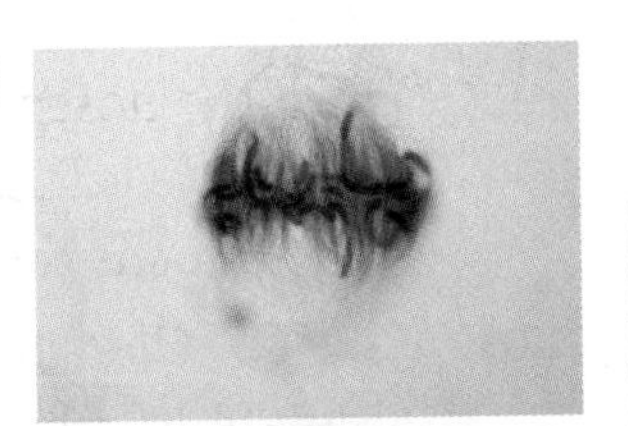

(g)

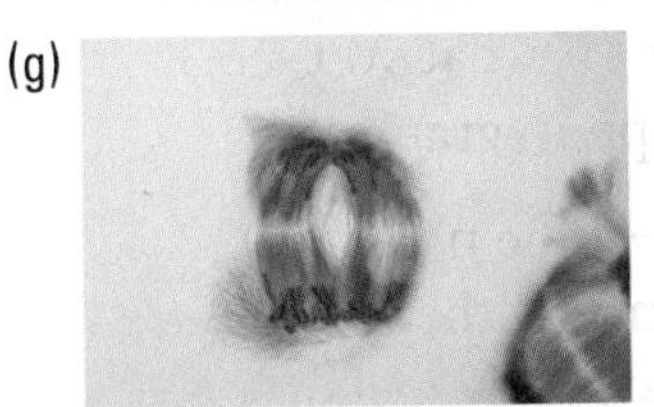

(h)

(i)

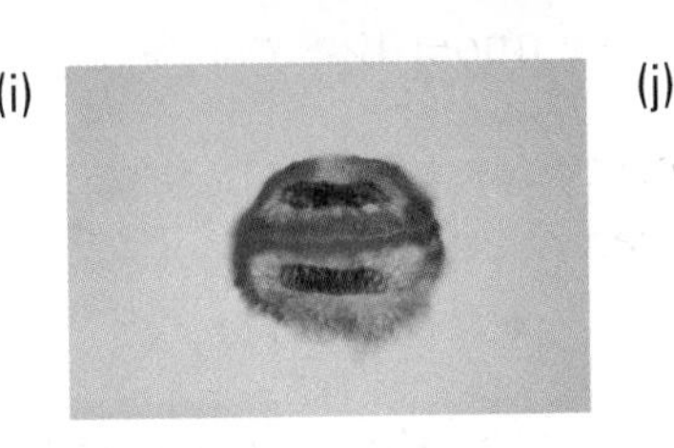

(j)

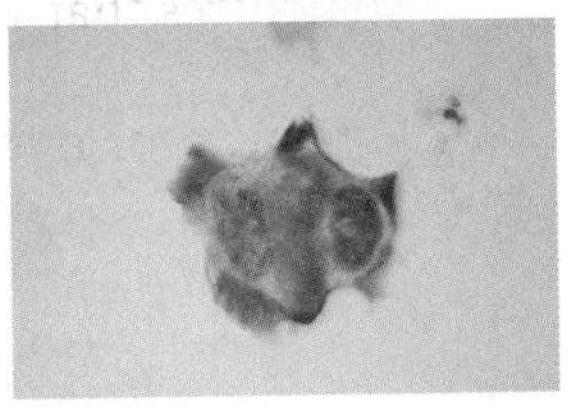

Figure 2

Cell division in onion root tip cell

Understanding Concepts

1. Why were plant root tip cells and animal embryo cells used for viewing cell division?
2. Explain why the cells that you viewed under the microscope do not continue to divide.
3. Use a table to list the differences and similarities between the appearance of the dividing animal cells and the dividing plant cells.
4. If a cell has 10 chromosomes, how many chromosomes will each cell have following cell division by mitosis? Explain.
5. (C2) Predict what might happen to each daughter cell if all the chromosomes moved to only one side of the cell during anaphase.

Exploring

6. (O) Search the Internet for pictures of plant and animal cells that are dividing. How closely did your microscope observations resemble these pictures? Explain any differences.
7. In this investigation, you observed cells in various phases of cell division. Did you observe any cells that did not appear normal? What might have gone wrong with these cells?

Work the Web

To research onion root cell division, visit www.science.nelson.com and follow the links from *Nelson Science 9: Concepts and Connections*, 2.3. What proportion of cells is undergoing mitosis at any point in time? Explain.

Challenge

2 Radishes grow very quickly. Explain this in terms of the rate of cell division.

Cell Division and Growth

Humans grow most quickly during the nine months before and the first three months after their birth. If they continued to grow at the same rate, the average 14-year-old would be over 8 m tall!

Growth of the Body

Figure 1 shows how body proportions change as a person grows from a **fetus** (developing unborn baby) to an adult. You will notice that the figures are all the same size but are not drawn to the same scale. The lines behind the figures help to show the size of the different body parts in comparison with the entire body. Some body parts grow at different rates than others. For example, the head of a newborn represents two of the eight sections ($\frac{2}{8}$ or $\frac{1}{4}$ of the total height). However, the head of an adult represents only one of the eight sections ($\frac{1}{8}$ of the total height). Imagine how funny we would look if our adult head was in the same proportion as a newborn baby's head!

(a) Which parts of the body appear to grow the most between a two-month-old fetus and an infant?

(b) Which parts of the body appear to grow the most between infancy and adulthood?

(c) Which parts of the body grow the least during each time period in (a) and (b)?

(d) Why do you think an infant's head is so large compared with the rest of its body?

Growth of Organs

Figure 2 compares the rate of growth of the brain and the heart with the overall growth of the body. (F2) At the age of 2, the masses of the brain and heart have doubled, whereas the mass of the body is almost four times the mass at birth.

(e) By how many times has the body mass increased by age 20? By how many times has the heart mass increased in the same period?

(f) At what approximate age does the brain reach its maximum mass?

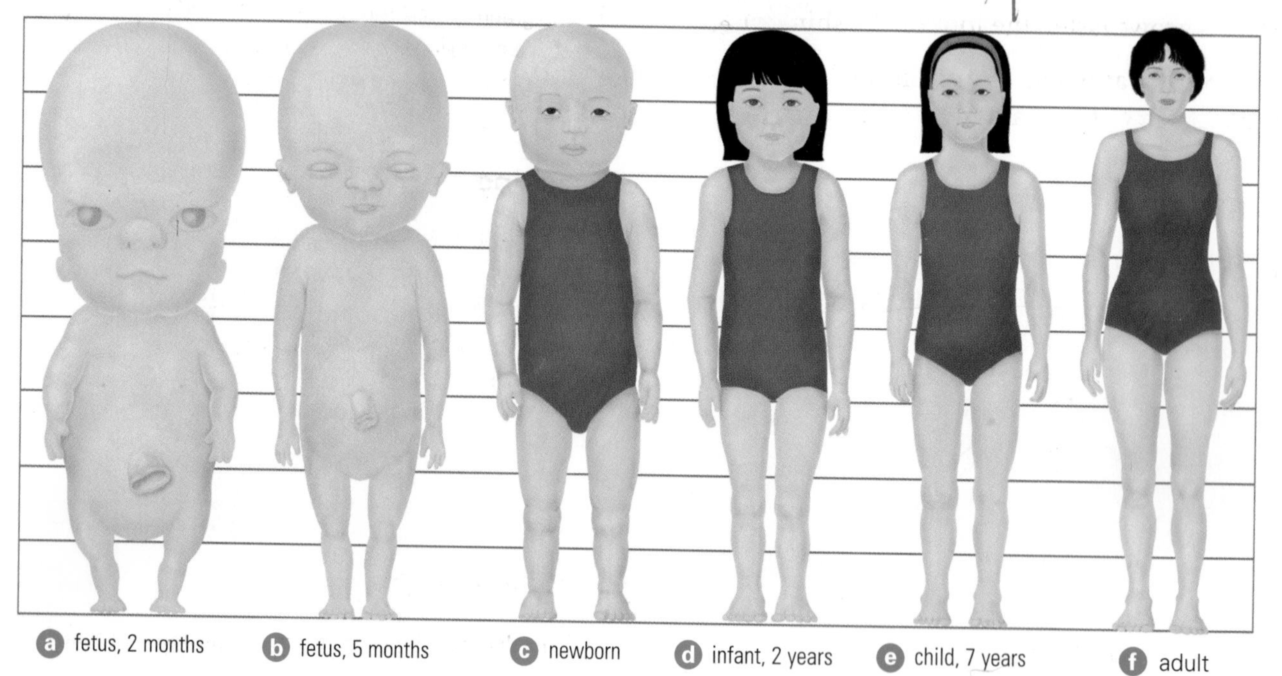

Figure 1

Not all body parts grow at the same rate.

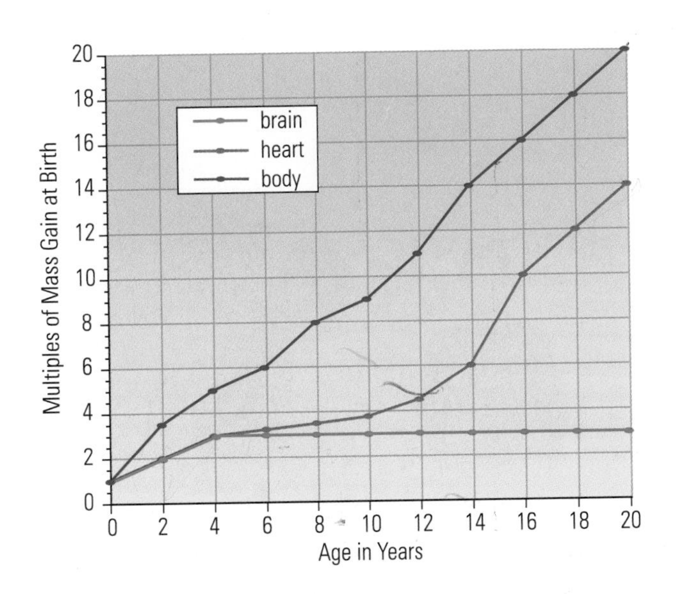

Figure 2

Growth rates of the brain, heart, and body

(g) How does the growth of the heart compare with that of the brain?

(h) Would you expect the increase in the mass of the heart and the body to continue at the same rate after 20 years of age? Explain your answer.

Growth of Bones

Figure 3 shows changes in the growth rate of the foot and the shin bone (tibia).

(i) Which grows faster, the foot or the shin bone?

(j) Describe what would happen if the bones in the feet grew at the same rate as the tibia.

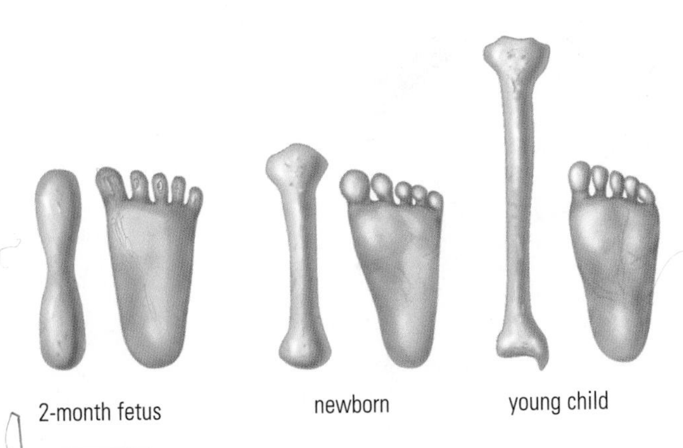

Figure 3

The bones of the foot and the tibia do not grow at the same rate.

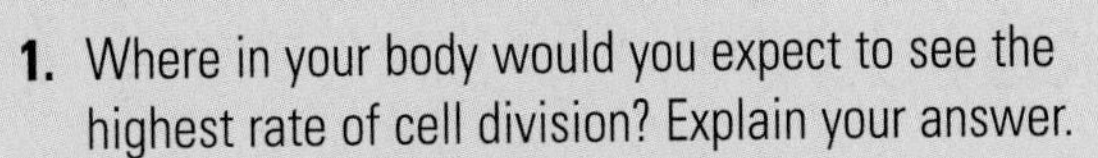

Understanding Concepts

1. Where in your body would you expect to see the highest rate of cell division? Explain your answer.

2. Based on the information in **Figure 2**, what can (F2) you conclude about the growth of your brain?

Applying Skills

3. (G) An experiment measured the rate of growth of a seedling root. Lines were marked on the root 1 mm apart. After 48 h, the root appeared as shown in **Figure 4**. Draw a conclusion based on these results.

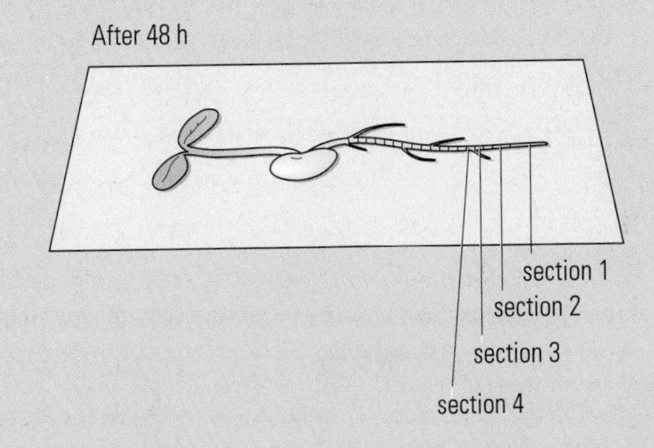

Figure 4

Making Connections

4. The graph of the growth rates in **Figure 2** is taken from data collected from a large group of people. Why do scientists obtain data from many people rather than from a single person?

Reflecting

5. What evidence do you have from your own growth patterns that suggests that all parts of your body do not grow at the same rate? If possible, use photographs of yourself at different ages as evidence.

Challenge

3 How is normal rapid cell division different from cancer?

(G) Concluding

Cancer

The DNA in the nucleus of each of the cells in your body is identical. DNA is like software that determines what you look like. It also controls all of the functions within the cells and in your body. The information encoded in this software came from both of your parents. Usually, the software runs smoothly and the program works as it should. However, sometimes "bugs" develop in the software and problems occur.

These bugs in the genetic software are called **mutations** (**Figure 1**). Mutations may be harmful to the cell. One set of harmful mutations causes cancer.

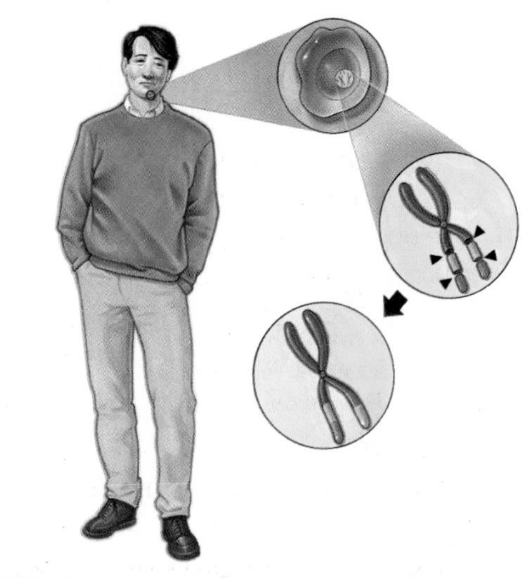

Figure 1

Mutations can occur when a chromosome is broken and repaired incorrectly.

Cancer

Genes on the chromosomes control cell division. If these genes undergo mutations, then the cell does not divide at the normal rate. When cell division goes out of control, the cell continues to divide until a lump or **tumour** develops. A tumour can be either benign or malignant. A **benign** tumour does not spread to other parts of the body and is generally harmless. However, a **malignant** tumour, or **cancer**, is more dangerous because it has the ability to spread to other parts of the body. Unlike many diseases that cause the death of cells, cancer cells divide more quickly than they should. Normal body cells grow and divide, and eventually die. Cancer cells just continue to grow and divide.

The causes of cancer are not completely understood. However, we do know that certain viruses can cause cancer, too much exposure to the Sun can cause skin cancer, and hazardous chemicals such those found in cigarette smoke can cause lung and other cancers (**Figure 2**). These cancer-causing agents create mutations that change the instructions for normal cell functions. We also know that hereditary factors are associated with some types of cancer—for example, breast cancer.

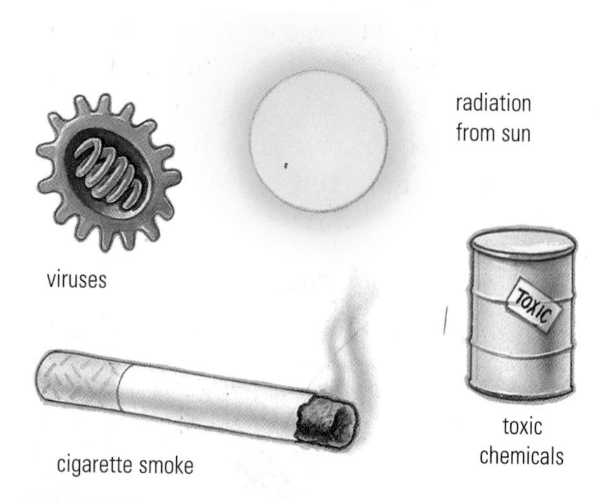

Figure 2

Known causes of cancer

Cancer is found not only in humans. Many plants and animals also develop cancers. Sunflowers and tomatoes often show a form of cancer called a gall. These plant tumours are caused by viruses, bacteria, fungi, or insects. Evidence of cancerous tumours has been found in dinosaur bones and even in the cells found in the linen wrapped around ancient mummies.

Cancer Cells

Cells are usually in contact with other cells and tend to stick together. This contact is required for cells to divide. Normal cells cannot divide when they are separated from one another, but cancer cells can. In a laboratory situation, cancer cells have been known to divide once every 24 h. At this rate, a single cancer cell would generate over 1 billion cells in a month. Fortunately, cancer cells do not divide that quickly in an organism.

There are many different types of cells in the human body. Each type carries out a specialized function—for example, muscle cells for movement, nerve cells for carrying signals, or red blood cells for carrying oxygen around the body. One important difference between cancer cells and normal cells is that cancer cells do not become specialized as they grow (**Figure 3**). They use up energy from food but do not carry out the work of normal cells. Also, if a tumour grows large enough, it can interfere with the normal function of other cells, tissues, and organs.

Another problem with cancer cells is that they do not stick together or stick to normal cells very well. Cancer cells may separate and move and begin dividing in other parts of the body. This makes the cancer hard to control.

(a) normal cells

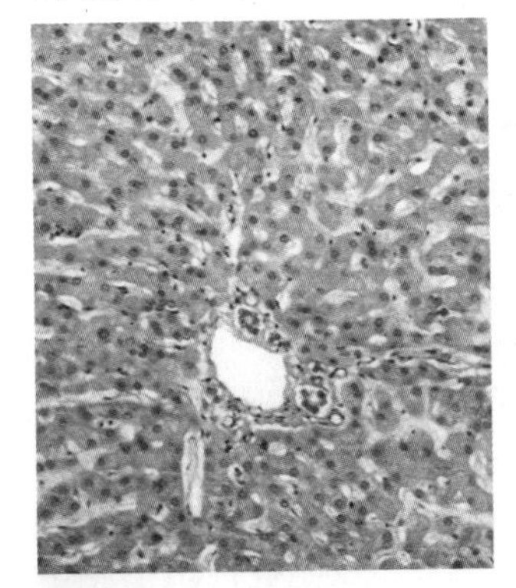

(b) cancer cells

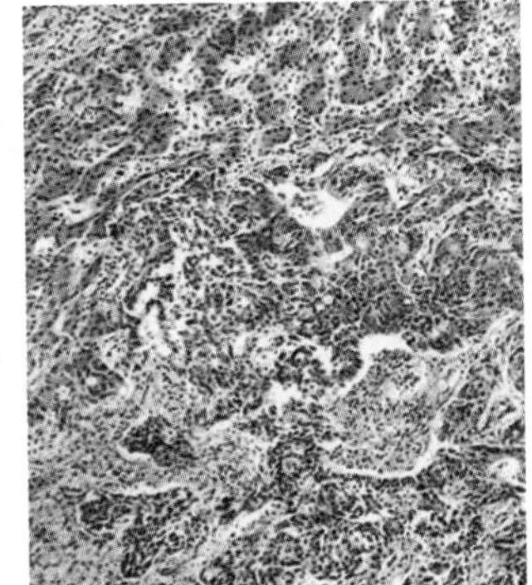

Figure 3

Cancer cells can often be identified by an enlarged nucleus and reduced cytoplasm. Why do you think cancer cells might have a large nucleus?

Challenge

3 What risk factors should be considered in your assessment of cancer risk?

Understanding Concepts

1. What is a mutation?
2. Explain the difference between a benign and a malignant tumour.
3. What is cancer?
4. In what ways are cancer cells different from normal cells?

Making Connections

5. Not all rapid cell growth is cancerous. A certain virus causes skin cells to divide quickly, producing a wart. Imagine what would happen if a cell and its descendants divided every hour. Make a table similar to the one below and fill in the blanks. Explain the pattern you observe in the number of cells.

Time (h)	Number of cells
0	1
1	2
2	4
3	?
4	?
5	?
6	?

Exploring

6. (N) (O) Choose one type of cancer and research to prepare a brief summary that considers the following points:
   - What causes this type of cancer (i.e., virus, chemicals, radiation, unknown)?
   - What treatments are available?
   - How dangerous is the cancer?

Work the Web

Cancers may be linked to certain environmental conditions. Visit www.science.nelson.com and follow the links from *Nelson Science 9: Concepts and Connections*, 2.5. What kinds of environmental conditions are linked to cancer? What types of cancer are most commonly linked to each environmental cause?

2.6 Activity

Lifestyle and Cancer

Why do some people develop cancer while others do not? Is it something people do or something that they don't do that causes cancer? Or is it something that they have no control over?

Research has established definite links between certain factors and cancer. We know, for example, that long exposure to sunlight can cause skin cancer and that smoking causes lung and other cancers. Any substance or energy that can cause cancer is called a **carcinogen**.

Most experts believe that cancer rates can be reduced dramatically with changes in lifestyle. In this activity, you will interpret data on cancer in order to describe trends.

Materials

- graph paper
- pencil

Procedure

1 Study the pie graph in **Figure 1** that shows the risk factors associated with cancer.

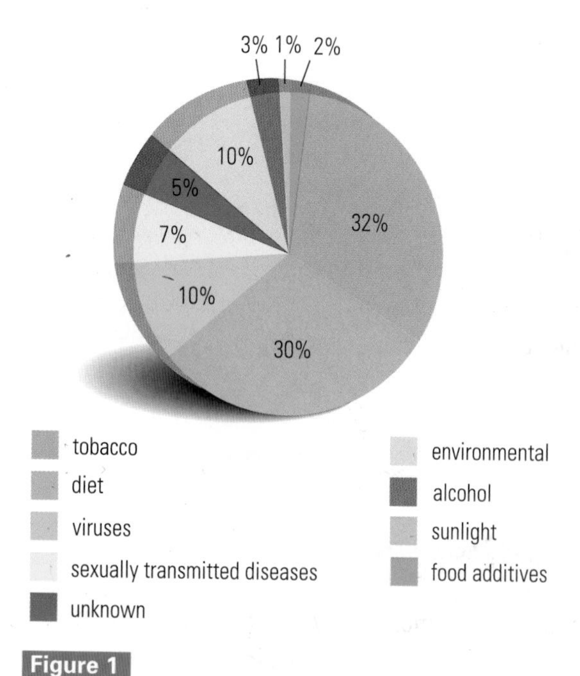

Figure 1

Estimates of cancer risk factors

(F2) (a) Which factor causes most cancer cases?

(b) Which of the cancer causes could be reduced by changes in lifestyle?

(c) List three lifestyle changes that could reduce cancer rates.

2 Copy this table in your notebook and complete the calculations for survival rates.

Type of cancer	New cases	After five years	
		Deaths	Survival rate
lung	19 600	16 600	15%
breast	17 000	5400	?
colon	16 300	6300	?
prostate	14 300	4100	?
bladder	4800	1350	?
kidney	3700	1350	?
leukemia	3200	1110	?

Source: Statistics Canada, 1994, based on latest available data from British Columbia, Saskatchewan, and Ontario.

(d) Draw a bar graph of the survival rate data (W1) from the table above.

(e) How do you think these estimates of survival rates were obtained?

(f) Based on these data, which type of cancer is most deadly? Which has the best survival rate?

3 Since the link between smoking and cancer has been established, data have been collected annually on the percentage of the population that smokes. **Figure 2** shows the data for the Canadian population for selected years from 1985 to 1999.

(g) In which age group has there been the greatest (F2) decrease in smokers since 1985?

(h) Based on your experience, suggest a possible explanation for data for the 15–19 age group.

SKILLS HANDBOOK: (F2) Interpreting Graphs (W1) Types of Graphs

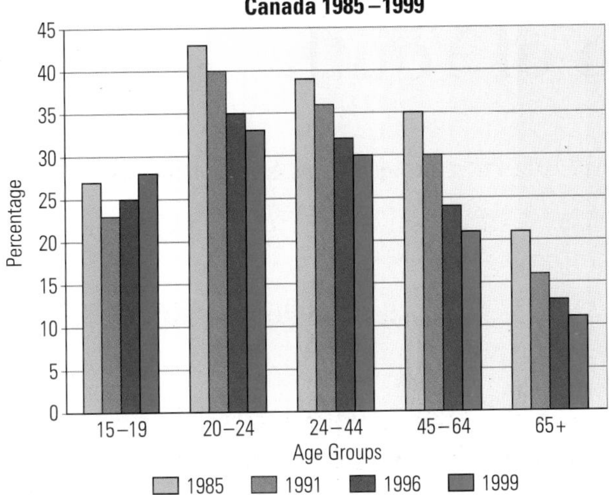

Figure 2

Percentage of population that smokes

4 There are differences in the percentage of smokers among different segments of the population (**Figure 3**). For example, a greater percentage of 20- to 22-year-olds smoke than any other age group.

(i) Describe the overall trend in smoking for the different age groups. (F2)

(F2) **(j)** Describe the differences between the sexes.

(k) Propose a possible explanation for the differences between the sexes in the 15–17 age group.

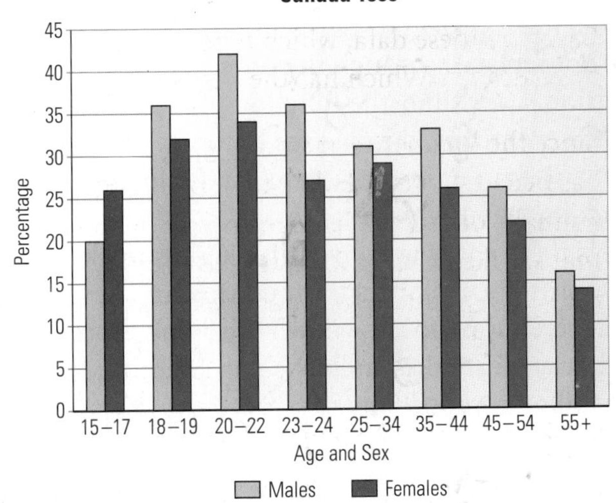

Figure 3

Percentages of male and female smokers of different ages

Understanding Concepts

1. What is a carcinogen? Give three examples.
2. The chemicals in the tar from cigarette smoke are known to cause cancer. Calculate the amount of tar absorbed by a smoker in one week. The following information will help you with your calculations:
   - Assume that a smoker smokes 10 cigarettes a day.
   - There are 20 mg of tar in most cigarettes. (1000 mg = 1 g).
   - Approximately 25% of the tar is released in the form of smoke or is exhaled. The remaining 75% is absorbed through the smoker's lungs.

Exploring

3. The money spent on cancer treatment continues to grow every year. One politician has suggested that cancers caused by smoking should be given a lower priority for treatment. Explain why you agree or disagree with this suggestion.

Reflecting

4. What changes could you make in your lifestyle that would reduce your cancer risk? Will these changes be easy? Explain.

Work the Web

Diet is the second-highest risk factor for cancer. Visit www.science.nelson.com and follow the links from *Nelson Science 9: Concepts and Connections*, 2.6. Find information to determine what features of our diet are responsible for the higher risk of cancer.

Challenge

3 What recommendations can you make regarding lifestyle choices that will reduce cancer risk? How do the statistics presented in this section affect your recommendations for lowering cancer risk?

DECISION-MAKING SKILLS
- ● Define the Issue
- ○ Identify Alternatives
- ● Research
- ● Analyze the Issue
- ● Defend a Decision
- ○ Evaluate

Search for the Fountain of Youth

In the early 1500s, a Spanish nobleman named Juan Ponce de Léon travelled with Christopher Columbus on his second voyage to the West Indies. When Columbus returned to Spain, Ponce de Leon decided to stay in what is now the Dominican Republic. There was a tale among the natives about a spring where you could regain your youth by drinking the water. Ponce de Léon decided to stay to search for gold and this magical "fountain of youth."

Of course, we now know that there was no fountain of youth. Despite all the scientific and medical advances, we have not been able to stop the aging process but we have been somewhat successful at delaying it. People are generally living longer, and our life expectancy is increasing.

In 1920, the life expectancy in Canada was 59 years for males and 61 years for females. By 1999, the life expectancy for males had increased to 76 years and for females to 83 years. This increase in life expectancy is due mainly to better living conditions and advances in medicine. Because of the increase in life expectancy, the percentage of the population that is over 60 years of age is increasing and will continue to increase for some time in the future (**Figures 1** and **2**).

Figure 1

Appropriate changes in lifestyle can increase both quality of life and life expectancy.

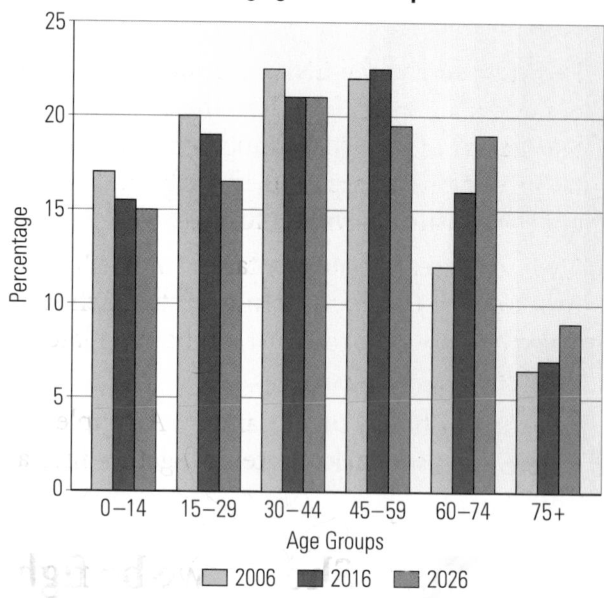

Figure 2

Population projections for the years 2006, 2016, and 2026 show that the percentage of older people will increase.

Fighting Age

Normal cells in the human body seem to have a built-in cell division counter. Different types of cells have different limits, but once they have reached their maximum number of cell divisions, the cells die.

Aging is related to cell division. Cells die and are not replaced through the process of cell division. Over time, the function of organs slows down; eventually, the damage becomes so serious that an organ or organ system can no longer function, and the person dies.

Is there a way to stop the biological clock? Scientists are exploring this question and have discovered ways to slow down the aging process and delay death. Research is being conducted into the following possibilities:

- Replacement organs. Just a few skin cells could provide the DNA needed to build a new organ (e.g., heart, lung, or liver) that might never wear out.

- Calorie reduction. Many animals live up to twice as long when they eat just 60 to 70 percent of the normal number of calories.
- New genes. Certain genes may be able to produce protection against free radicals, dangerous chemicals that can damage or kill cells.
- Resetting of the cell division counter. The cell division counter is found on the tips (called telomeres) of certain chromosomes. If these telomeres can be replaced, the cell might be able to continue dividing forever.
- Hormone treatment. Substances normally produced in the body during youth, such as growth and sex hormones, can be used later in life to slow down aging.
- Blocking of blood sugar damage. A simple sugar (glucose) sticks proteins together into a blob that damages tissues and organs. New drugs may prevent these blobs from forming.

Understanding the Issue

1. Which age groups will increase and which age groups will decrease in Canada's future population?
2. Which anti-aging approach do you think might be easiest to achieve? Explain.
3. If the average life expectancy in Canada is 79 years, does this mean that everyone will live to be 79 years old? Explain.

Challenge

3 Is there a link between cancer and aging? Do older people get cancer more than young people?

Debate Should we be fighting nature?

Statement

Hormones or drugs should not be used to stop or slow the process of aging.

Point

- The idea of reversing aging presents many difficulties. First, the cost would be immense. Hormone treatment is expensive. At current prices, injections for a 70-kg man would cost about $14 000 per year. Only the richest people would be able to pay for such treatments.
- Extending the lifespan of the average person might cause overpopulation of Earth. There might not be enough food to support the increased population. The cost of caring for elderly people would increase.

Counterpoint

- Expense has no bearing on the issue. If people spend billions of dollars on cosmetics, then they'll spend money on hormones or other treatments.
- The money spent on hormones or other treatments would produce economic benefits. For example, people would work longer and generally experience a better quality of life.
- Overpopulation can be avoided using family planning methods. Food supplies could be increased by advances in science and technology. Health costs would decrease because elderly people would be healthier.

What do you think?

- A group will be assigned to each side of the debate. You will argue for or against the statement.
- In your group, discuss the statement and the points and counterpoints. Write down additional (I) points and counterpoints that your group considers.
- Search newspapers, a library periodical index, a CD-ROM directory, and the Internet for (J)(N)(O) information on drugs or hormones used to slow aging.
- Elect your spokesperson(s) and prepare to defend your group's position in a class debate. (K)(L)

(I) Identifying Alternatives/Positions (J) Researching the Issue (K) Analyzing the Issue
(N) General Research (O) Internet Research

Cell Division and Reproduction

In sections 2.1 and 2.2, we learned about cell division that occurs for healing and repair and for growth. Cell division is also the process that allows reproduction, and therefore allows species to continue.

Organisms of all species reproduce. There are two types of reproduction: sexual and asexual. In **asexual reproduction**, a single organism produces offspring (young) with identical genetic information. Most single-celled organisms (such as bacteria) and some multicellular organisms (such as fungi and some plants) use asexual reproduction to produce offspring. In **sexual reproduction**, genetic information from two cells is combined to produce a new organism. Usually, sexual reproduction occurs when two specialized sex cells (an egg cell and a sperm cell) join to form a **zygote** (fertilized egg), which then develops into a new organism. **Figure 1** compares asexual and sexual reproduction.

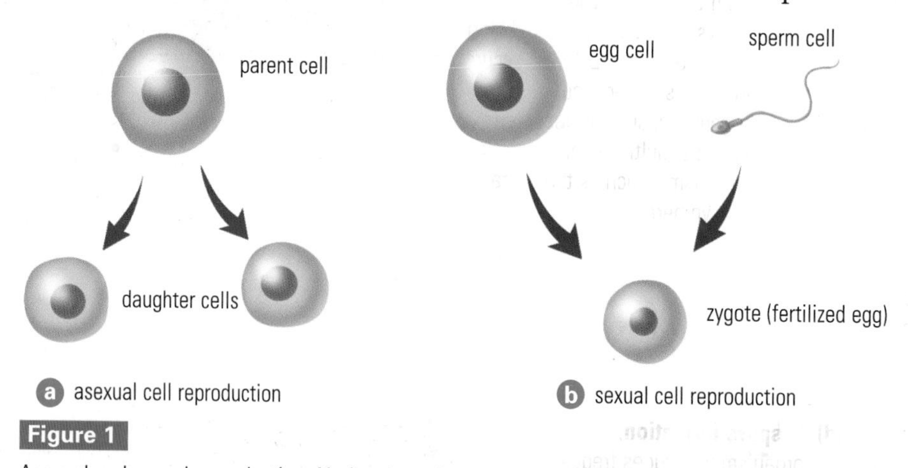

Figure 1

Asexual and sexual reproduction. Notice that when a cell reproduces asexually, the parent cell becomes two identical daughter cells. In sexual reproduction, an egg and a sperm cell join to form a zygote.

Try This Activity: How many divisions will it take?

Many single-celled organisms reproduce by simply splitting in two. How many organisms would there be after five divisions (assuming they all survived)?

Number of divisions	Number of organisms
0	1
1	2
2	4
3	?
4	?
5	?

1. What pattern relates the number of divisions to the number of resulting organisms?
2. (C2) Use the pattern to predict how many organisms are produced after 10 divisions.
3. How many divisions are required to produce over 1 million organisms?
4. The single, fertilized cell from which you began divided to produce the many cells that make up your body. Estimate how many cell divisions it took. (Assume that there are a trillion cells in your body. A trillion is 1 000 000 000 000.)

SKILLS HANDBOOK: C2 Predicting and Hypothesizing

Asexual Reproduction

There are many methods of asexual reproduction. In all cases, only one parent is required to produce offspring. **Figure 2** illustrates several different methods.

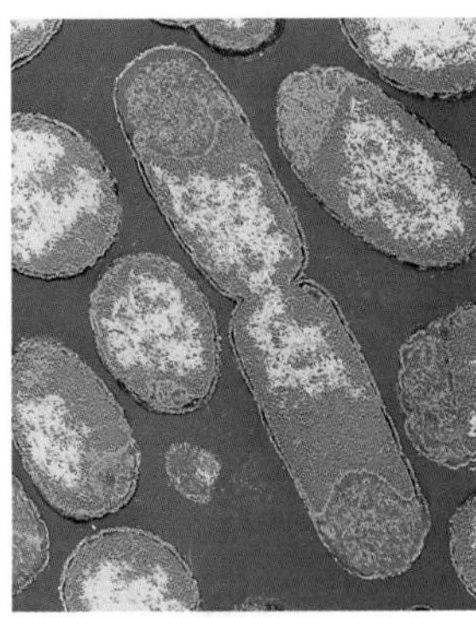

(a) Some simple organisms, such as these bacteria from the intestine, reproduce by simply splitting into two equal-sized offspring. This is known as **binary fission**. Each offspring has a copy of the parent's genetic material.

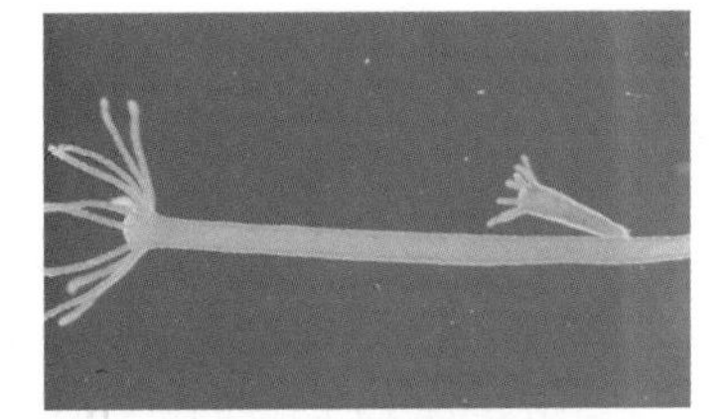

(b) In **budding**, the offspring begins as a small outgrowth from the parent. The "bud" grows and eventually breaks off from the parent. Budding occurs in some single-celled organisms, such as yeast, and in some multicellular organisms, such as the hydra shown here.

(c) Many types of algae and some plants and animals can reproduce by **fragmentation**. A new organism is formed from a part that breaks off from the parent. If a starfish is cut through its central disk, each section will develop into a new starfish.

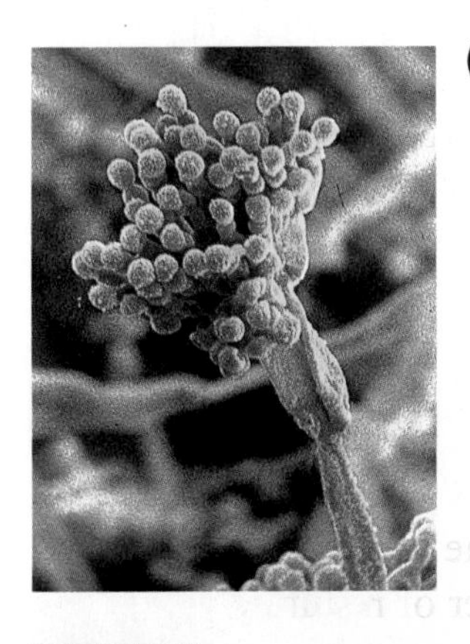

(d) In **spore formation**, the organism undergoes frequent cell division to produce many smaller, identical cells called spores. The penicillium mould shown here reproduces by forming spores. The spores are usually contained within the parent cell. Many spores have a tough coat that allows them to survive after the parent cell dies. Each spore can develop into a new organism.

(e) Many plants, such as spider plants (shown here) and strawberries, reproduce by **vegetative reproduction**. They produce runners, which are slender stems that can take root and develop into new plants.

Figure 2

Various methods of asexual reproduction. All of these methods result in offspring that have identical genetic information to the parent.

Sexual Reproduction

Cells of the human body have 46 chromosomes (or 23 pairs). When fertilization occurs, a human sperm and egg cell combine their chromosomes to make a zygote (fertilized egg) with 46 chromosomes. This is the same number that the parents each had in their body cells. Why doesn't the zygote end up with 92 chromosomes?

The cell division that allows sexual reproduction to take place is really a two-stage process (**Figure 3**) that prevents the zygote from ending up with 92 chromosomes. The first division results in daughter cells that have half the number of chromosomes (23) as normal body cells. The second division is similar to mitosis in that the new cell divides to produce two cells that are identical to the daughter cell. As a result, human sex cells (sperm and egg) have only 23 single chromosomes.

When an egg and a sperm cell join, they produce a zygote with 46 chromosomes (**Figure 4**).

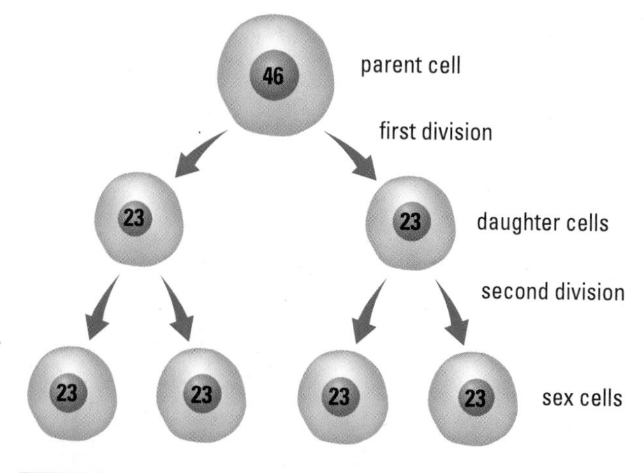

Figure 3

Cell division that produces sex cells (gametes)

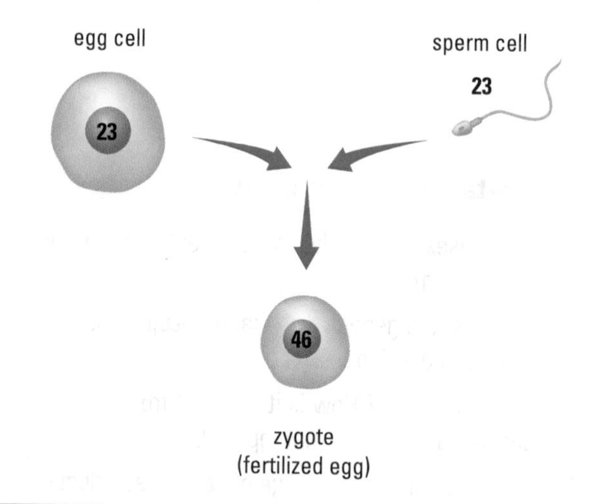

Figure 4

An egg and a sperm combine to produce a zygote with 46 chromosomes.

DNA, Genes, and Variation

The zygote from which you developed had 46 chromosomes (23 pairs). The egg from your mother contributed 23 chromosomes, and the sperm cell from your father contributed 23 chromosomes. One chromosome of each pair comes from each parent. The genetic material (DNA) in every cell in your body is therefore contained in 46 chromosomes. Each chromosome of the pair carries genes for specific characteristics. For example, on a chromosome from your mother there was a gene for hair colour. On the matching chromosome from your father there was also a gene for hair colour. These two genes together determined your hair colour (**Figure 5**).

All physical characteristics are determined by the combination of genes inherited from the parents. It is obvious that sexual reproduction, where genetic material comes from two different parents, results in greater variation among individuals in a species. This variation can include new combinations of genes that may allow organisms to adapt better to a given environment.

Figure 5

Hair colour is determined by the genes from both parents.

DNA Fingerprinting

Just like your fingerprints, the DNA contained in your chromosomes is unique to you. No one else, except an identical twin if you have one, has exactly the same DNA. Therefore, one way to identify a person is by his or her DNA. **DNA fingerprinting** is the process of using DNA material from a few cells to identify an individual.

Information from DNA fingerprinting is used for medical purposes, such as diagnosing inherited disorders. It is also used for identification and is valuable evidence in some criminal cases. DNA fingerprinting has been used to convict people of crimes and to prove that people who have been charged with crimes are innocent.

Work the Web

Visit www.science.nelson.com and follow the links from *Nelson Science 9: Concepts and Connections*, 2.8. Briefly describe the process of DNA fingerprinting. Report on a criminal case that used DNA fingerprinting as evidence.

Challenge

2 Describe the life cycle of the plant you are experimenting with. Does it reproduce asexually, sexually, or both?

Understanding Concepts

1. How is asexual reproduction different from sexual reproduction?
2. Why must the genetic material of a cell be duplicated before cell division begins?
3. What is a zygote? How is it different from daughter cells produced by asexual reproduction?
4. What is the main advantage of sexual reproduction?
5. (a) Briefly describe the five types of asexual reproduction.
   (b) Choose one type of asexual reproduction. Explain how a plant nursery could make use of it.
6. Starfish feed on mussels and other shellfish. Operators of mussel farms have been known to cut up starfish and throw them overboard. Is this a smart practice? Explain.
7. Identify the method of asexual reproduction in each of the following situations:
   (a) A multicellular algae is struck by a wave. The algae breaks up and each new piece grows into a new organism.
   (b) A new tree begins to grow from the root of a nearby tree.
   (c) A small cell begins to grow on the outside of another cell. Eventually, it breaks away from the larger cell and continues to grow.
8. Why is DNA fingerprinting a useful tool in criminal investigations?

Reflecting

9. What advantages might an organism that can reproduce asexually have? Make a list of the advantages. Add to your list or modify it as you progress through this unit.

Exploring

10. Describe what happens during each of the two cell divisions that produce sex cells.

Asexual Reproduction: Cloning

Often, a home gardener misses an onion or a potato while harvesting the crop in autumn. The following spring, a new onion or potato plant pushes up through the soil. Onions and potatoes are examples of plants that store food. Although the top of the plant dies in the fall, it leaves behind a food-containing organ. As spring approaches, the soil warms, and new shoots grow from the food-storage organ. These plants are growing without seeds by vegetative reproduction (**Figure 1**).

As you learned in section 2.8, all methods of asexual reproduction result in new organisms that have an exact copy of the genetic material of the parent. The advantage of this method of reproduction for gardeners is that they know what the new plants will be like. The new plants will have the same desirable characteristics as the parent.

Figure 1

The potato plant dies in the fall, leaving the potato tuber (not roots) in the ground. The tuber sprouts again in the spring, producing more plants. This is vegetative reproduction.

Cloning

When people think of cloning, they often think of a mad scientist working in a laboratory, creating duplicates of an original human being. Humans cannot be easily cloned, although other animals have been cloned.

Cloning is not a new process. It's always been around in nature. For example, an identical twin is an example of a clone. The potato example described above is also a form of cloning. **Cloning** is the process of forming identical offspring from a single cell or tissue. Because the clone comes from a single parent, its genetic material is identical to that parent's genetic material. Therefore, cloning is referred to as asexual reproduction.

Many plants (e.g., carrots, strawberries, potatoes) can be easily reproduced by cloning (**Figures 2** and **3**), but others (e.g., grasses) cannot. No one really knows why. The secret seems to be hidden somewhere in the genetic characteristics of the plant.

Figure 2

Strawberry plants produce runners that produce new plant clones. These plants get a head start by relying on the parent for nutrients until their own root system is established.

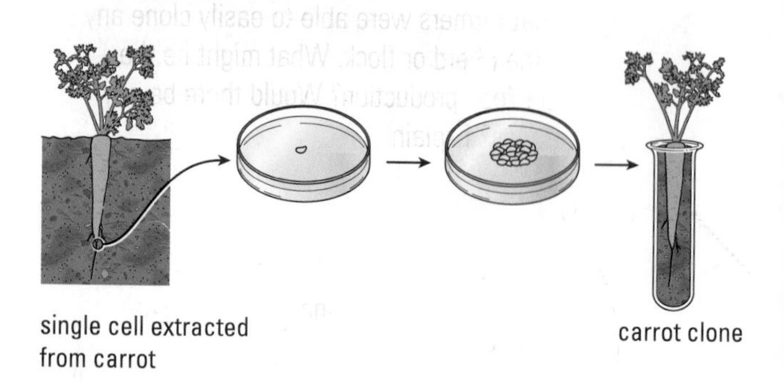

Figure 3

A single cell from a carrot root, near the tip, can grow into an entire carrot plant.

Cloning Animals

Much scientific research is being conducted into the cloning of animals. Animals have been cloned before using the nucleus of an embryo cell in the very early stages of development. The new cell that receives the nucleus can then develop into a new individual that is genetically identical to the embryo.

The biggest breakthrough happened in 1997 when Dr. Ian Wilmut, of the Roslin Institute in Scotland, announced that he had cloned a sheep using genetic material from an adult body cell rather than a cell from an embryo. Some cells were taken from the udder of an adult Finn Dorset sheep and grown in a cell culture. Then a nucleus was taken out and placed in an egg cell (from a Poll Dorset sheep) from which the nucleus had been removed. Finally, the dividing embryo was placed into the womb of a third sheep. The offspring, named Dolly, looked nothing like the birth mother. She carried the genetic information identical to the Finn Dorset adult. Dolly was a clone (**Figure 4**).

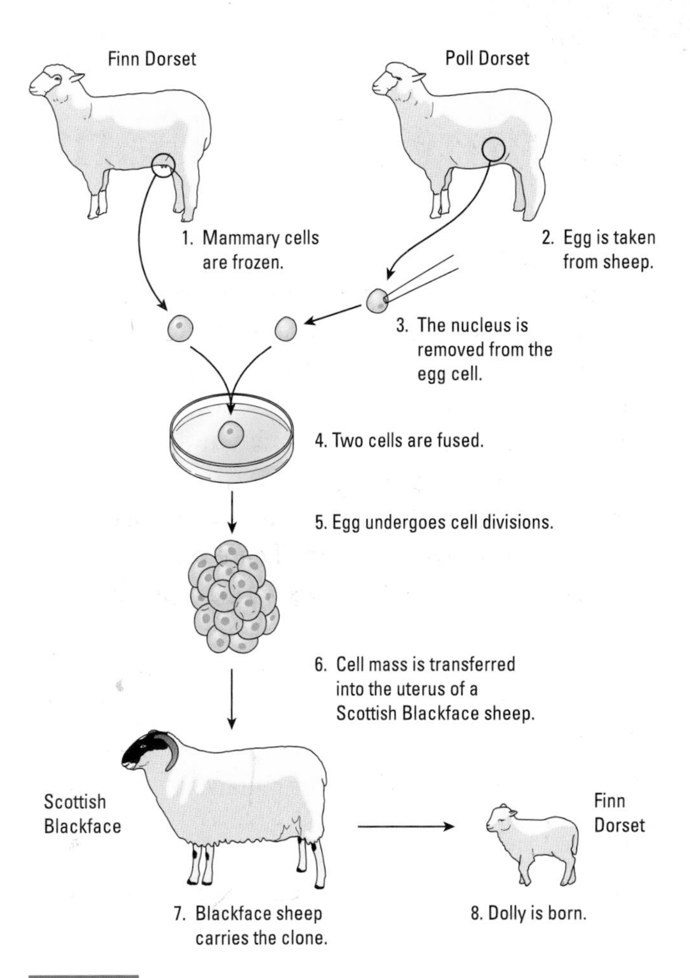

Figure 4

Dolly could claim three different sheep as mothers. In fact, her genetic mother died before Dolly was born.

Understanding Concepts

1. What is cloning?
2. In what ways are plants cloned?
3. Explain why cloning is considered asexual reproduction.
4. Dolly was not the first cloned animal or the first cloned mammal. What made her cloning so special?

Making Connections

5. Imagine that farmers were able to easily clone any animal in their herd or flock. What might be the benefits for food production? Would there be any disadvantages? Explain.

Reflecting

6. What ethical issues can you list that relate to cloning? Write a paragraph sharing your views on one of these issues.

Work the Web

There is much debate about whether humans should be cloned. Visit www.science.nelson.com and follow the links from *Nelson Science 9: Concepts and Connections,* 2.9. Identify two countries that have banned human cloning research, and describe the measures and reasons for banning this research. Identify two countries that are supporting human cloning research. What are their arguments? What is Canada's position on human cloning research?

Challenge

1 Cloning is a very common reproductive technology, especially for reproducing plants. What are the advantages and disadvantages of cloning?

2 Can your experimental plant be cloned? If so, explain.

2.10 Investigation

INQUIRY SKILLS
- ○ Questioning
- ○ Hypothesizing
- ● Predicting
- ○ Planning
- ● Conducting
- ● Recording
- ● Analyzing
- ● Concluding
- ● Communicating

Cloning from Plant Cuttings

Some plants naturally reproduce asexually when part of the plant, such as a stem or leaf, breaks off and drops to the ground. Roots develop on the broken part and penetrate the soil. The broken part grows into a new plant.

Cuttings from seedless-grape vines have been grafted to grapevines all over the world. You wouldn't be able to eat seedless grapes if it were not for cloning.

In this investigation, you will grow a plant from a cutting, thereby making a clone of the original plant.

Question

What is the process for cloning plants?

Prediction

(a) Write a prediction for this investigation.

Design

In this investigation, two methods of asexual reproduction—stem cuttings and tuber cuttings—will be used to reproduce plants.

Materials

- lab apron
- knife
- coleus plant (or alternative)
- potato
- small beaker or jar
- potting soil
- flowerpots (or alternative)

Caution: Be careful when using a knife. Wash your hands after handling soil or plants. Wear gloves if you are allergic to plants.

Procedure

Part 1: Stem Cuttings

1 Carefully cut off the tips of three coleus stems. Include two or three leaves on each stem.

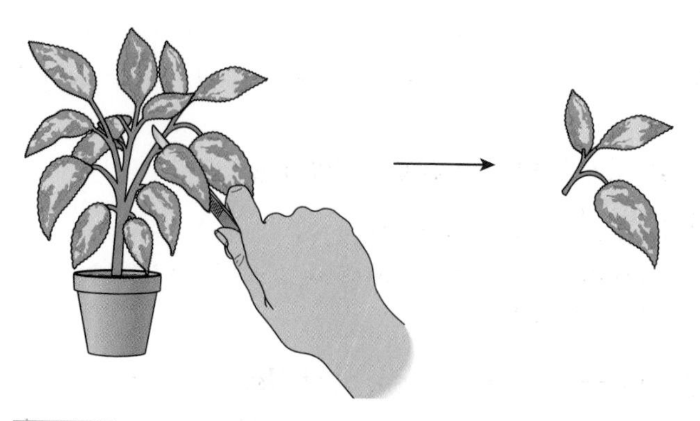

Step 1

2 Place each stem into a small beaker containing water. The water must cover as much of the stem as possible without covering any of the leaves. You may need to support the stems to keep them upright.

Step 2

3 Put the beaker in a sunny place and check it daily to maintain the water level.

(E1) **(b)** Record your observations daily.

4 After the roots appear on the stems, allow an additional week's growth and then transplant each cutting into a flowerpot filled with moist potting soil.

Step 4

(c) In a table similar to **Table 1**, record the growth (E2) of each plant for three weeks after transplanting.

 SKILLS HANDBOOK: (E1) Qualitative Observations (E2) Quantitative Observations

Table 1

Day	Height of plant	Comments
0		
2		
4		
6		

Part 2: Potato Tuber Cuttings

5 Obtain a potato from your teacher. Cut the potato into pieces, making sure that each piece has at least one eye.

6 Plant each piece in a pot filled with moist potting soil. (*If you want to continue the activity and actually grow potatoes, plant each piece in a bucket. Your teacher will provide the necessary information.*)

7 Put the pots in a sunny location and keep the soil moist.

(d) (E1) Observe the pots every other day for several weeks, and record your observations in a table similar to **Table 1**.

Analysis and Conclusion

8 (F1) Analyze your results by answering the following questions:

(e) What evidence suggests that coleus has the ability to regenerate parts of the plant lost to injury?

(f) In what ways would the new coleus resemble the parent plant?

(g) Suggest two ways to prove that the roots from the coleus cuttings are growing.

(h) (C2) If you continue growing a potato plant, would you expect to get more than one new potato? Explain.

(i) (G) Was your prediction for this investigation correct? Explain why or why not, based on your observations.

(j) (G) Based on your observations, what is present in the eye of the potato?

Understanding Concepts

1. A hailstorm can shred most plants. After a hailstorm, what advantage would a coleus plant have over plants that cannot reproduce vegetatively?
2. Explain the advantage of growing potatoes from tuber cuttings.
3. Explain why you should keep the leaves of the coleus cuttings out of the water.

Exploring

4. (C) (D) Plan and carry out an investigation to determine whether the size of the piece of cutting is a factor in cloning potatoes. Or, find out whether you can grow a coleus from a single leaf.
5. (E1) Grow a pineapple clone using the method shown in **Figure 1**. Record your observations.

Figure 1

Challenge

1 Cloning of food plants is a common reproductive technology. What are the issues with cloning food plants? Do you think most people realize that the grapes, apples, and oranges they eat come from clones? What other misconceptions might people have?

Work the Web

Visit www.science.nelson.com and follow the links from *Nelson Science 9: Concepts and Connections*, 2.10. Research to find other plants that can be easily cloned by vegetative reproduction. Grafting is another common method of cloning plants. Describe this method .

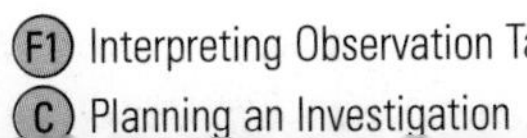
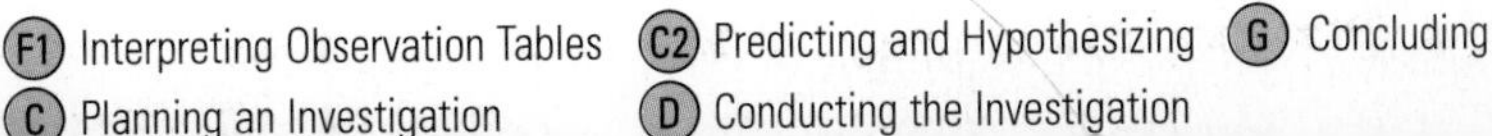
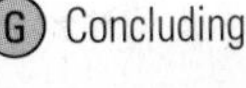

Sexual Reproduction

If animals were cloned like potatoes, we would all look the same!

Most multicellular organisms reproduce sexually. Genetic information from two specialized cells combines to form a unique composition for a new organism.

Conjugation

The simplest form of sexual reproduction is called **conjugation** (**Figure 1**). Two bacteria exchange different genes and then go on to reproduce asexually through binary fission. This creates variation within a bacteria species.

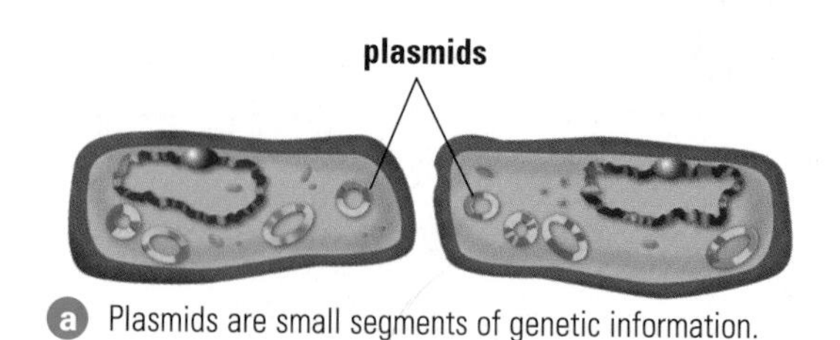

(a) Plasmids are small segments of genetic information.

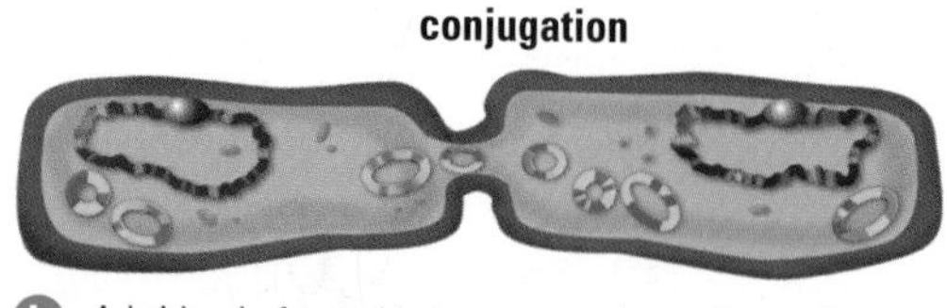

(b) A bridge is formed between two bacterium cells, and plasmids are exchanged between the cells.

Figure 1

Unlike human cells, bacteria do not have nuclei. Their DNA floats in their cytoplasm. It usually consists of one large ring and several smaller ones, called plasmids.

Separate Sexes

Most multicellular animals and some plants have separate sexes: males and females. Males produce sperm cells, and females produce egg cells. These specialized sex cells, known as **gametes**, combine to form a zygote.

In humans, sex is determined by the X- and Y-chromosomes (**Figure 2**). Females have a pair of X-chromosomes in each cell, one from each parent. Males have a single X-chromosome (from their mother), and a much smaller Y-chromosome (from their father) in each cell.

Sex cells combine in a process called fertilization. There are two different methods of fertilization. In most land animals, including humans, the male deposits sperm inside the body of the female, where the sperm fertilizes her eggs. This process is called **internal fertilization**. Most water animals, including fish, use a process of **external fertilization**. The female releases her egg cells, the male releases sperm, and the sex cells unite outside the female's body.

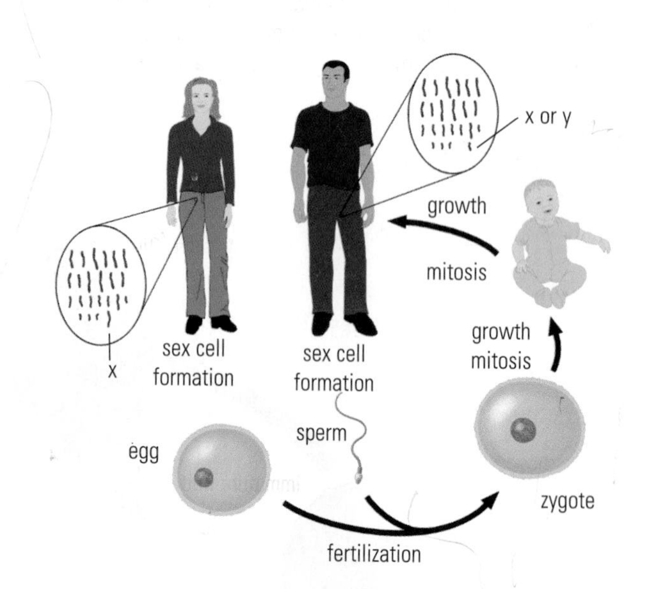

Figure 2

Humans contain organs that produce either egg cells or sperm cells.

Hermaphrodites

Sexual reproduction may be a problem for organisms that cannot move to find a mate or for those that come in contact with few members of their species.

An organism that creates both male and female sex cells within the same body is called a **hermaphrodite**. Tomato plants and earthworms contain male sex organs that produce sperm and female sex organs that produce eggs. They can reproduce with any other member of their species.

In some cases, two hermaphrodite animals can join together and each deposits sperm into the other animal, as shown in **Figure 3**. Some animals that live in water simply release their sex cells into the water.

Many plants are hermaphrodites, with both male and female sex organs. The life cycle of the tomato plant is shown in **Figure 4**.

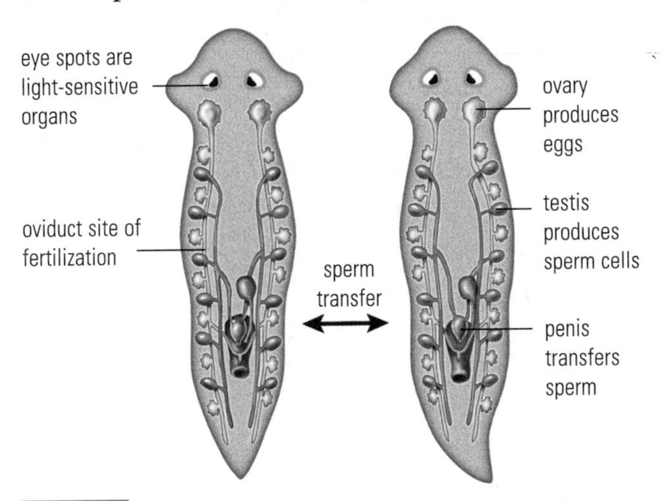

Figure 3

Sperm is transferred between two flatworms. Each flatworm acts as a male by releasing sperm and as a female by receiving the sperm to fertilize eggs cells. Once fertilized, the egg cells are released inside a capsule.

Understanding Concepts

1. Describe how conjugation is different from human reproduction.
2. How many chromosomes do the following cells in **Figure 2** have:
   (a) egg? **(b)** sperm? **(c)** zygote?
3. How are the chromosomes in a female zygote different from those in a male zygote?
4. Using keywords (sex cells, egg, sperm, fertilization, zygote), explain how external and internal fertilization are the same and how they are different.
5. What makes an earthworm a hermaphrodite, and how does this help it reproduce?
6. How is a plant a hermaphrodite? Use the information in **Figure 4** to answer this question.

Making Connections

7. List as many differences as you can between sexual and asexual reproduction, using information presented in the sections so far.

Challenge

3 From what you have learned from your research, what effect does gender (male or female) have on cancer rates?

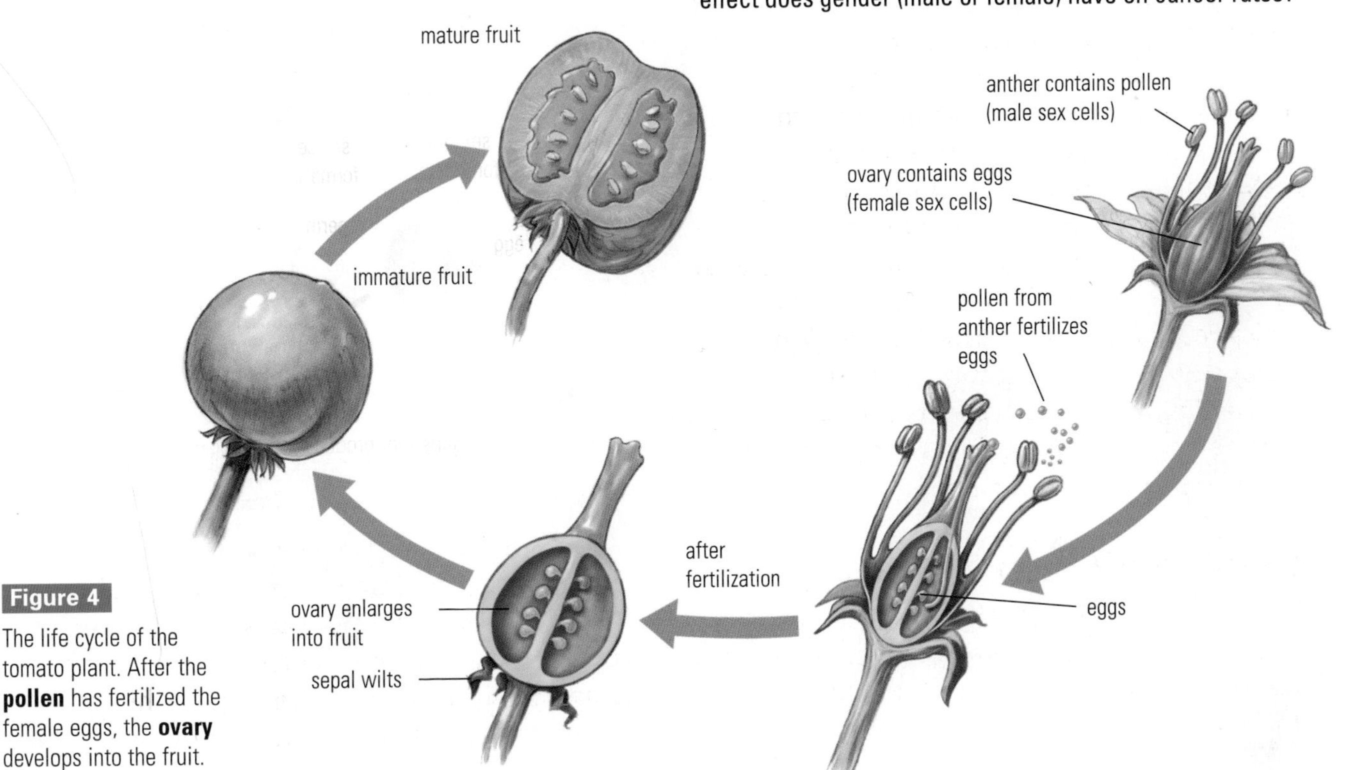

Figure 4

The life cycle of the tomato plant. After the **pollen** has fertilized the female eggs, the **ovary** develops into the fruit.

INQUIRY SKILLS
- ○ Questioning
- ○ Hypothesizing
- ● Predicting
- ○ Planning
- ● Conducting
- ● Recording
- ● Analyzing
- ● Concluding
- ● Communicating

Sexual Reproduction in Plants

Many plants have both male and female sex organs, and as you learned in the previous section, are considered hermaphrodites. How are the male and female gametes produced within a single organism, and how do they unite? Flowers are key to these processes.

Plants that have brightly coloured, sweet-smelling flowers and a sugary nectar attract animals to pollinate them. **Pollination** is the process where pollen is moved from the anther to the egg cells and fertilizes them. Most pollinators are insects, but bats and hummingbirds also pollinate flowers. As the animal crawls into the flower to collect nectar, pollen from the anther falls on its body. When the animal moves on to the next flower in search of more nectar, some of the pollen brushes off onto the pistil. Animals may deposit pollen over a wide area. Wind also carries pollen. Many people are bothered by wind-carried pollen, such as from ragweed or oak. If you have asthma, it may be caused by pollen.

In this investigation, you will examine different flowers to see which structures they have in common and which are different. You will also look at the functions of the parts of a flower.

Question

How do flowers differ from one another?

Prediction

(a) Create a prediction for this question.

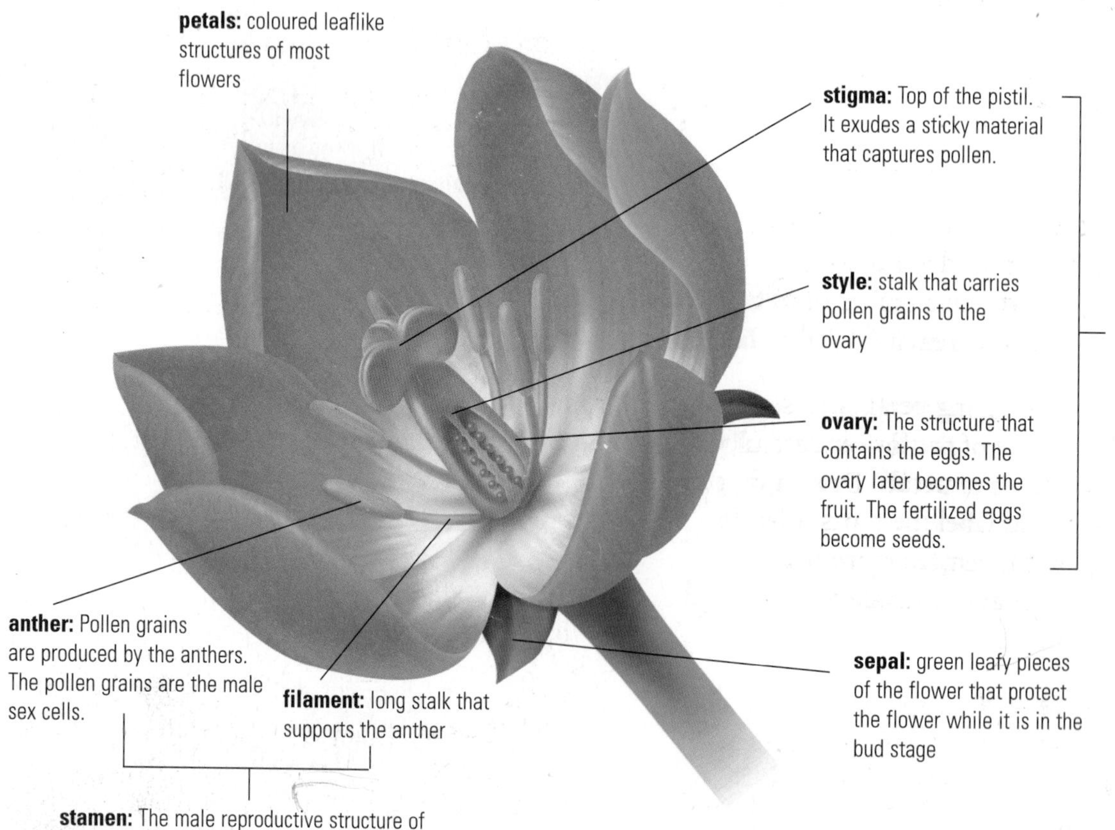

Figure 1
A typical flower

Materials

- whole flowers (various kinds)
- coloured pencils
- hand lens or dissecting microscope
- tweezers
- small knife
- gloves (if needed)
- disposable masks (if needed)

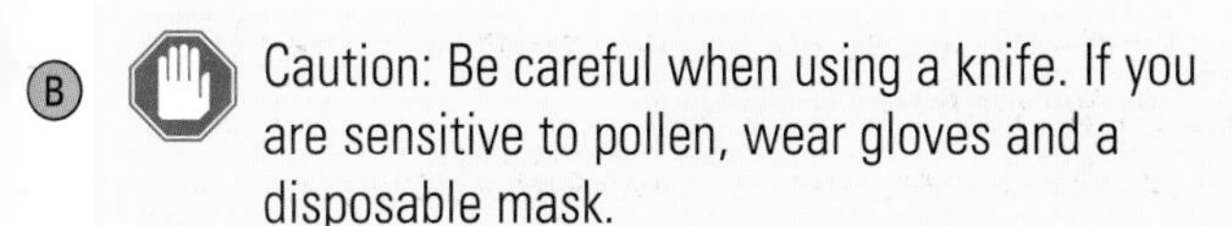

(B) Caution: Be careful when using a knife. If you are sensitive to pollen, wear gloves and a disposable mask.

Procedure

1 Examine **Figure 1** and read the descriptions of the various parts of the flower. Not all flowers are alike. Your flowers may be different from the diagram.

(C6) **2** Prepare a table similar to **Table 1**.

Table 1 **Flower Comparison**

Characteristic	Flower 1	Flower 2
number of petals		
number of sepals		

3 Obtain two flowers and examine them closely (D5) with a hand lens or under a dissecting (E2) microscope. Record the number of petals, sepals, and stamens on each flower in **Table 1**.

4 Remove a few adjoining petals and sepals. (E2) Examine the inside of the flower carefully. (E3) Record the number of petals. Draw a diagram of each flower and label the parts. Identify which parts are the female reproductive system and which are the male reproductive system.

5 Carefully cut vertically through the pistil of (D5) each flower. Using a dissecting microscope, (E3) draw a diagram of the inside of the pistil and label the structures.

Analysis and Conclusion

6 Compare the flowers you examined by answering the following questions:

(b) Was your prediction correct? Explain why or (G) why not, based on your observations.

(c) List the ways in which the two flowers were similar. Why would this be the case?

(d) List the ways in which the two flowers were different. Why would this be the case? (Hint: Consider how the flower is pollinated and how its seeds are dispersed.)

(e) Were any of the parts in **Figure 1** missing in your flowers? Describe any differences.

Understanding Concepts

1. What happens when the pollen reaches the egg? What do the fertilized eggs become?
2. In what part of the flower do the seeds form?
3. How do insects and birds help with the process of pollination?

Exploring

4. Hummingbirds are attracted to red trumpet-shaped flowers. They hover over the flowers and use their long beak to collect nectar. In this way, the flowers are pollinated. Use the Internet or other resources to find two examples of how flowers and animals have developed special structures to ensure pollination.

Challenge

1 Can hand pollination of plants be considered a reproductive technology? Explain.
2 Describe how the plant you are growing is pollinated.
3 What plants have cancer-fighting properties?

Work the Web

Hay fever is the name given to pollen allergies. One in eight Canadian students has asthma, which can be triggered by pollen allergies. Find out more about hay fever by visiting www.science.nelson.com and following the links from *Nelson Science 9: Concepts and Connections,* 2.12. Answer what? who? when? where? why? and discuss available treatments.

2.13 Case Study

Reproduction of Plants for Food

Imagine your school lunchroom with twice as many people but the same amount of food. The lunchroom would have two problems: not enough space and not enough food for each person.

The problems facing world agriculture are similar. As populations increase, the amount of land available for producing food decreases (**Figure 1**). Increased food production is essential to feed the extra people.

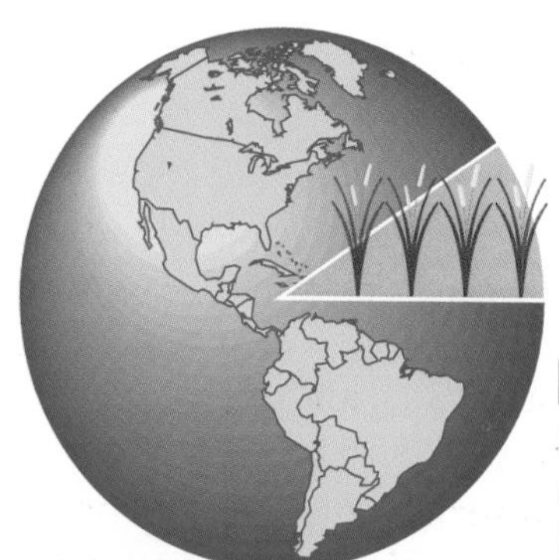

Figure 1

Of the world's 13.1 billion ha, only about 1.4 billion ha can be used for cropland.

(a) What percentage of Earth's landmass is used for cropland? Why is this is the case?

Ways of Increasing Food Production

Traditional Methods

When early farmers noticed that some corn plants were better than others in size, hardiness (durability), yield, or flavour, they wanted a way of reproducing these desired characteristics or **traits**. They did so through crossbreeding.

Crossbreeding involves taking the pollen from one plant and using it to fertilize the eggs of another plant. All the seeds are planted, and the plants showing the desired traits are selected and bred again. After several generations of **selective breeding**, all of the offspring have the desired traits. Sir Charles Saunders, a Canadian scientist, developed Marquis wheat by crossbreeding two varieties to obtain the best traits of both.

In cloning plants for food, cuttings are taken from a plant with a desired trait (**Figure 2**). These cuttings are rooted and grow into exact copies of the parent plant.

Some varieties of fruit trees, such as apple, produce good roots but poor fruit. Other varieties have good fruit but roots that are easily damaged by cold weather. Rather than planting trees from seeds, **grafting** is used. Branches from trees with the desired fruit are grafted, or attached, onto trees with the desired roots (**Figure 3**). The new trees produce good fruit and survive winter well.

Figure 2

Cloning from plant cuttings

(b) Create a table to summarize the traditional (C6) methods of increasing food production. Define and give two examples of each method.

(c) Study **Figure 3**. What are two advantages and two disadvantages of this grafting method?

Figure 3

A single apple tree can produce several varieties of apples if twigs from different varieties are grafted onto one root stock.

SKILLS HANDBOOK: (C6) Preparing Observation Tables

Newer Methods

With the discovery of DNA and research into gene mapping, crops can now be bred using **genetic engineering**. The resulting products are referred to as genetically modified (GM) foods. Genetic engineering involves transferring genes from the DNA of one organism to another. In laboratories, these transformed cells are grown into plants.

Using this technique, Canadian scientists have developed wheat and rye that can grow in very cold conditions. Corn, soybeans, and potatoes have been bred to resist pests. Foods are genetically modified to improve their quality or food value. For example, scientists have developed tomato varieties that taste good without becoming overly soft. Because the ripe tomato is still firm, it can survive handling and shipping. Canola, a Canadian success story, has benefited from both selective breeding and genetic engineering methods (**Figure 4**). Cooking oil is produced from Canola, which is grown widely across the Canadian prairies. In the 1960s, scientists from the University of Manitoba used selective breeding to improve its colour and flavour. In the 1990s, genetic engineering produced a variety of Canola that can resist disease and drought.

Figure 4

The plant pictured is Canola, originally called rapeseed.

Try This Activity: Simulated Gene Splicing

In genetic engineering, the actual transfer of a gene is carried out using a complex "cut and paste" procedure, which you will simulate. The plastic jar represents the cell nucleus; its lid is the nuclear membrane. Inside is ticker tape that represents the DNA in the chromosomes. Coloured markings on the tape represent gene segments of the DNA.

- Jar A is the donor nucleus. Through the nuclear membrane (open the lid), carefully remove a loop of DNA. The red genes have been selected for removal because they contain the desired trait. Using scissors, remove the red segment.
- Jar B is the receiving nucleus. Through the nuclear membrane, carefully remove a loop of DNA. Notice that there is no red segment in the chromosome. Search for the green segment, as this contains the trait that is going to be replaced. Using scissors, remove the green segment.
- Tape the red segment from DNA A to the open segments of DNA B. Replace the nuclear membrane of Jar B. Remove Jar B's label and replace it with the label "Jar C."

You have now successfully completed a gene-splicing procedure and can answer the following:

(a) Why was the jar relabelled in the last step?

(b) Jar A no longer represents a complete nucleus. Explain why.

(c) Summarize this activity in a labelled diagram. (E3)

(d) Make up an example to accompany this activity. For example, what plant is being modified? What does the red segment contain, and what species does it come from? How is the new plant different from the original? What will happen when the new plant reproduces?

(d) List three ways in which genetic engineering has improved crops.

(e) What disadvantages of genetic engineering can you think of?

(f) How did Canadian scientists modify rapeseed to make it more pleasing to the consumer?

Comparing Food Production Methods

Selective breeding has been practised for centuries, and its strengths and weaknesses are well known. This type of breeding is a natural process that is easy to do on the farm. The process requires more knowledge than equipment. It is often necessary to grow many generations of plants to select for desired characteristics. For example, it can take up to 12 years to breed disease-resistant crops.

Because genetic modification of foods is a relatively new practice, its long-term effects are unknown. In the short term, it is a faster, more precise way to introduce new traits into living things than traditional breeding, where all the traits of two parents are mixed. New traits can be introduced without changing other desirable traits. Desirable traits can be taken from outside the species, something that is not possible with traditional breeding methods. For example, the Flavr Savr tomato was engineered from the genes of a fish to improve its storage qualities, human and pig genes have been mixed to create leaner pork, and chicken genes have been spliced into a potato to increase the potato's resistance to disease.

Like other people in Canada and around the world, you may have questions about genetically modified foods. Is it ethical or right to mix the genes of plants and animals? Could GM foods affect the food chain and damage the environment? Will these foods have an effect on our health years from now? Who regulates genetic engineering? Should GM foods be labelled as such so that consumers have a choice?

There are no easy answers to these questions, but government officials, legislators, and consumers are now wrestling with the issues that new technologies raise.

(g) Compare traditional breeding methods and genetic engineering methods using a table. List three strengths and three weaknesses of each.

Understanding Concepts

1. Why is it so important to increase food production?
2. How do selective breeding, cloning, and grafting increase food production?
3. List three ways in which genetic engineering of plant crops could increase food production.

Reflecting

4. What is your personal opinion of GM foods? Write Ⓡ a paragraph that includes some of the information presented in this case study. Remember to write an introduction and a conclusion.

Work the Web

Find out more about a specific food that has been changed using selective breeding or genetic engineering (examples of genetically engineered foods include corn, wheat, canola, McIntosh apples, tastier tomatoes, and seedless watermelons). Visit www.science.nelson.com and follow the links from *Science 9: Concepts and Connections*, 2.13. What improvements were made, and how were they done?

Challenge

1 Design a model (see Try This Activity) to illustrate the reproductive technology you are researching.

2 You must double the yield of your plant crop. What method of increasing food production would you use, and why? At what cost?

SKILLS HANDBOOK: Ⓡ Writing for Specific Audiences

Horticulturist

Jeanine Cole

Can you identify a shrub from a leaf sample, or detect a disease from the appearance of a leaf?

Horticulturist Jeanine Cole can. Diagnosing customers' problems at Aubin Nurseries in Carman, Manitoba, is only one part of her job. Finding solutions is equally important. Jeanine explains the safe use of chemicals like insecticides, herbicides, and fungicides. She tests customers' soil and water samples, and gives advice on how to bring fertility back to their soil. She recommends propagation methods such as budding, grafting, soft- and hardwood cuttings, seed, and root division. Her vocabulary includes the common and the botanical names for trees, shrubs, and perennials.

Jeanine loves working with nature and seeing things grow. The road to becoming a horticulturist was not a straight one, though. Jeanine dropped out of high school and worked in a variety of sales and marketing jobs until enrolling at the University of Manitoba. After two years in the School of Agriculture, she earned a Horticulture Diploma.

Even with such good qualifications, Jeanine realizes that her field is constantly changing. She keeps up with new techniques through continuing education. Next she will become a certified arborist. As Jeanine says, "Agriculture is a large industry: if you tire of one aspect, it is very realistic to move into another branch."

Agricultural employment may be seasonal or full-time. Stamina is needed, as is physical fitness and the ability to work in all kinds of weather conditions. Like the plants, customers need to be well-tended. Jeanine's good people skills and her sense of humour are as necessary as her marketing ability and her knowledge of leaf beetles. Drop in and choose something for your garden!

Jeanine Cole handles bare root stock that has been over-wintered in cold storage to survive Canadian Prairie conditions.

Understanding Concepts

1. Use a dictionary to find the most appropriate definitions for new terms in this section: (a) arborist, (b) fertility, (c) fungicide, (d) herbicide, (e) horticulturist, (f) insecticide, (g) perennial.

Making Connections

2. Review section 2.10, and explain what you think is meant by "soft- and hardwood cuttings."
3. What interests and skills would you need to succeed as a horticulturist?

Work the Web

Find out more about careers in agriculture across Canada. What diploma courses in horticulture are available? List other opportunities in agriculture that do not require a university degree. Visit www.science.nelson.com and follow the links from *Science 9: Concepts and Connections*, 2.14.

Challenge

2 How would you market your plant? Write a nursery description of your plant species.

Reproductive Strategies for Survival

Asexual and sexual methods of reproduction have been studied in this unit. The concept map in **Figure 1** compares the two methods.

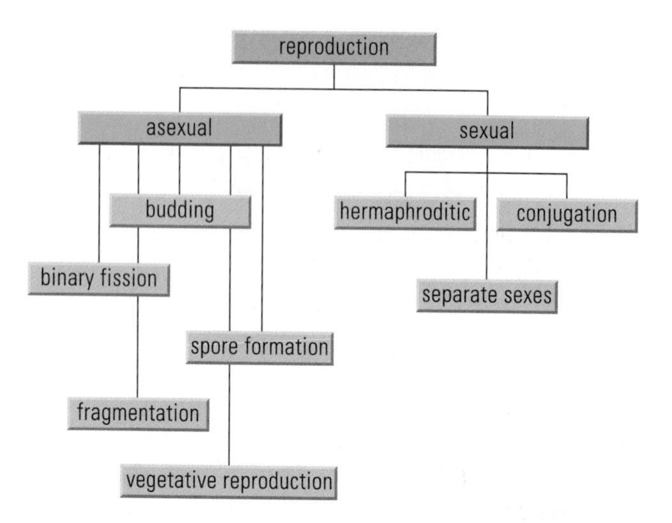

Figure 1

A concept map showing reproductive strategies

Most unicellular organisms and the body cells of multicellular organisms reproduce asexually. A single cell (parent cell) duplicates genetic information and becomes two identical daughter cells. The offspring have the same genes as the parent. Asexual reproduction does not require a partner, and can take place whenever environmental conditions such as food, warmth, and moisture are suitable.

Sexual reproduction is common among multicellular animals and plants. Genetic information from two different sex cells combines to form the genetic code for a new organism. Offspring are not identical to either parent or even to one another. Sexual reproduction produces new combinations of genes that may allow organisms to adapt better to new situations.

If a species is to survive—regardless of whether it uses sexual or asexual reproduction—its offspring must survive. Different survival strategies are described in **Figures 2** to **5**.

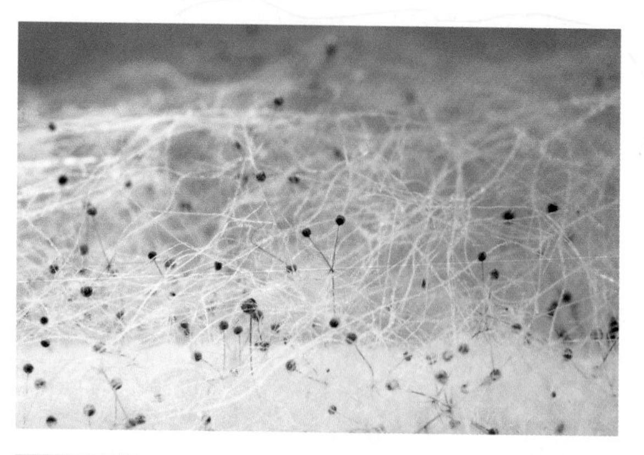

Figure 2

Spores

The black spots on bread mould are spore cases. A spore is a reproductive body inside a protective shell. Under unfavourable conditions, the spores are inactive. When conditions improve, the spores germinate, or sprout, and enter a growth phase.

Figure 3

Seeds

An apple seed contains the plant zygote wrapped in a protective food package, the apple. Unlike spores, which must wait until conditions become favourable, seeds bring nutrients to their environment and can get a head start on growth.

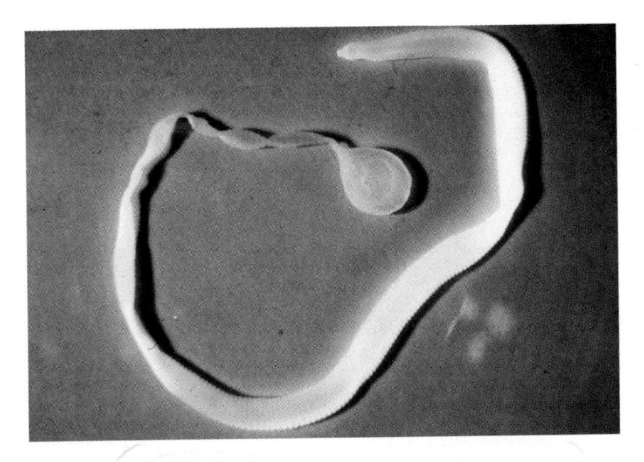

Figure 4

Eggs
The tapeworm, a parasite that lives in the intestines, produces thousands of eggs in each egg case. As an animal infected with tapeworm defecates, some of the egg cases break, releasing eggs. A resistant coating prevents the eggs from drying up, and the coating sticks to blades of grass or other plant material. If another animal eats the grass, it swallows the eggs. In the intestines, the eggs develop into adult tapeworms that can grow up to 20 m long.

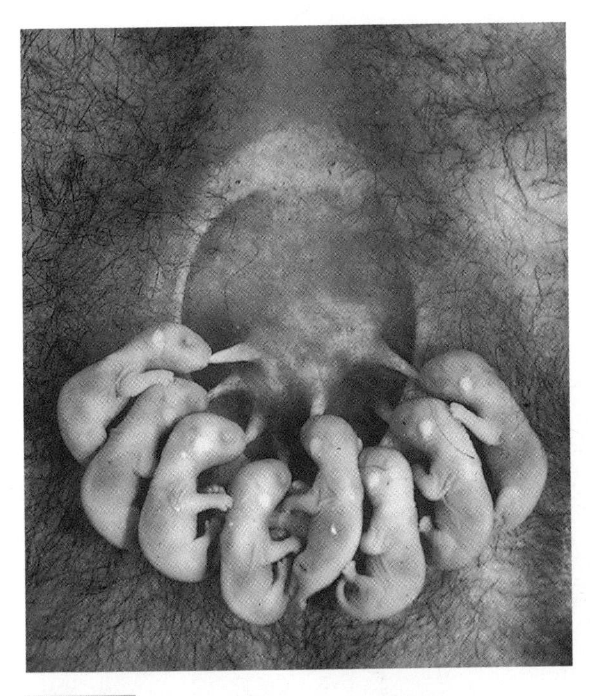

Figure 5

Young develop in the womb
Opossums are marsupial mammals. After emerging from the uterus, the young crawl into a pouch and attach to the nipple of a mammary gland. They leave the pouch when they are too big to be carried around.

Mammal young, including humans, develop in a womb and are nourished through an organ called the **placenta**.

Understanding Concepts

1. Use a table to compare and contrast asexual and sexual reproduction. Summarize information from previous sections as well.
2. Classify the following as either sexual or asexual reproduction:
   (a) A small piece of a cactus breaks off the plant, falls to the ground, and begins to grow.
   (b) Pollen from a male poplar tree fertilizes sex cells on a female poplar tree.
   (c) Two earthworms each produce sperm and eggs and fertilize each other. Eggs are laid.
   (d) A flatworm is cut in half and grows into two flatworms.

   Add three examples of your own.
3. How can two sisters with the same parents have different hair colours?
4. Identify an organism that uses each of the following strategies to make its offspring survive. Describe the strategy more fully and explain why it is used.
   (a) The zygote is wrapped in a food package.
   (b) This package can be fertilized externally or internally.
   (c) The young develop within the adult organism.
   (d) Under unfavourable conditions, the zygote remains inactive.

Making Connections

5. Consider the place of bacteria on the concept map (**Figure 1**). They can reproduce both asexually and sexually. Describe each method they use. What advantage is gained by using both methods?
6. If the entire planet became tropical through climate change, which survival strategy presented here would best ensure the survival of humans? Explain your answer.

Challenge

1 As a reproductive technologist, design a new survival strategy for the 21st century.

2.16

Human Sex Cell Development

You have learned about the stamen and the pistil—reproductive structures of plants—and how plant sex cells and zygotes are formed. Animals also have sex cells. What do they look like, and how are they formed?

Male Sex Cell Development

Structure of Sperm

The male sex cell (**sperm**) is well-designed for its job (**Figure 1**). Its streamlined design allows it to move well. Only a small amount of cytoplasm surrounds the nucleus, so the sperm carries no extra weight. However, because of the limited cytoplasm, the sperm has a limited energy reserve.

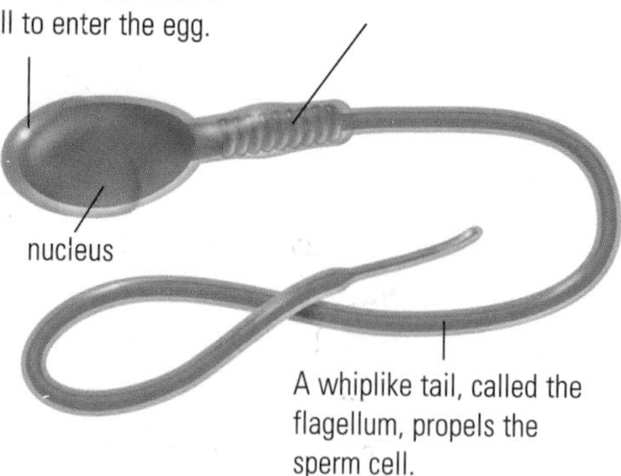

Figure 1
Human sperm cell

Sperm Production and Development

The reproductive organ of the male mammal is the **testis** (plural *testes*). The insides of the testes are filled with tiny twisting tubes called **seminiferous tubules**. It is here that the male sex cells, or sperm, are produced and fed (**Figure 2**). Remember that the important difference between sex cells and other cells is the number of chromosomes. Mature human sperm contain 23 chromosomes; all other body cells contain double that number, or 46 chromosomes.

As the sperm cells develop, they begin to grow a long tail called a **flagellum**. As well, the amount of cytoplasm inside the sperm is reduced. Their development is completed in the **epididymis**, which lies near the testes (**Figure 2**).

Sperm are not built to last. Those not released for reproduction during ejaculation die within a few days and are removed by white blood cells. New sperm continually replace old ones. Males can produce millions of sperm cells every day, throughout life. Even males over 90 years old have been known to father children.

The male sperm cell is much smaller than the larger female egg cell. In humans, the egg cell is 100 000 times larger than the sperm cell.

If you untwisted a seminiferous tubule, it would measure about 70 cm long!

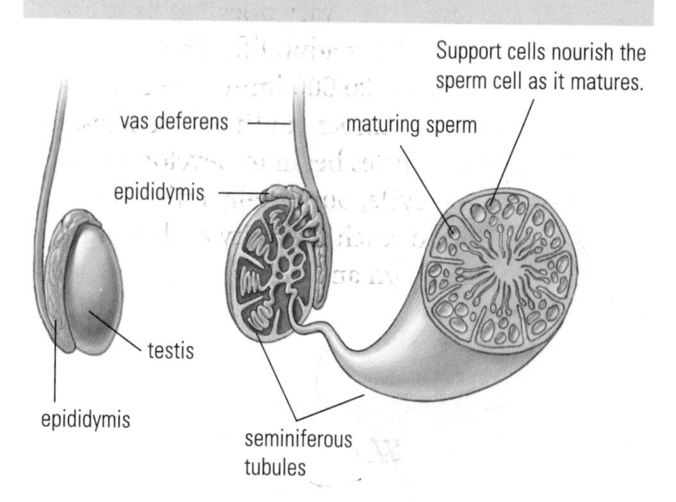

Figure 2
Male sex cells are produced in the testes.

Female Sex Cell Development

The female reproductive system is more complicated than the male system because it has an extra responsibility. Not only must the female produce sex cells (**eggs**), but she must also nurture the embryo from conception to birth.

Structure of the Egg

Like the sperm, the egg is well-designed for its job. The female sex cell is much larger than the sperm. It is packed with nutrients so that it can divide quickly when fertilized. It is also capable of building a barrier after fertilization to prevent other sperm from entering.

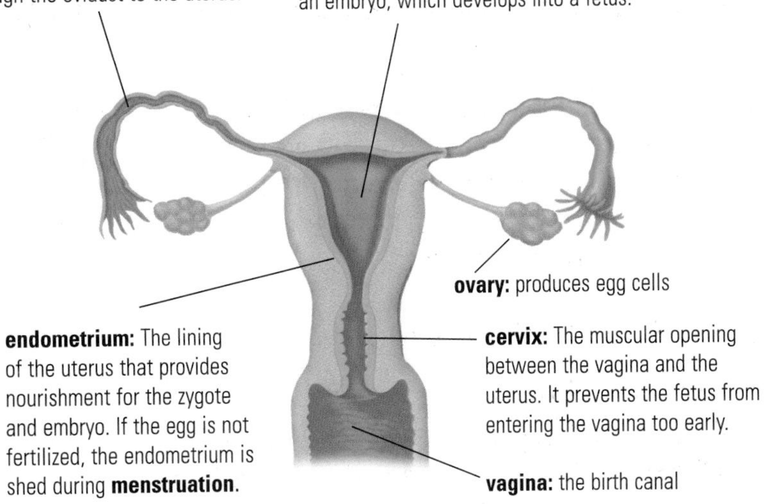

Figure 3

Frontal view of the female reproductive system

Production and Development of the Egg

The primary reproductive organ of the female is the **ovary** (**Figure 3**). There are two ovaries. Here the eggs form and mature. The ovary contains small groups of cells, called **follicles.** One type of cell in the follicle produces the egg. In humans, this reproductive cell has a full number of 46 chromosomes, yet it produces a sex cell with half that number, or 23 chromosomes. Meanwhile, another type of cell in the follicles divides to feed the future egg. The nutrient cells surrounding the egg cell develop into a fluid-filled cavity. When the egg is ready, the ovary wall bursts and the egg cell is released into the **oviduct.** The release process is called **ovulation.** Chemicals (**hormones**), which act as messengers between cells, control ovulation.

Unlike the testis, the ovary does not produce new egg cells throughout adult life. The human ovary contains about 400 000 immature follicles at puberty, and the number continually decreases. Hundreds of the follicles begin to develop during each reproductive cycle, but usually only a single follicle is allowed to reach maturity each month. The others break down and are absorbed into the ovary.

Work the Web

Hormones control both the male and the female reproductive systems. What are the names of these hormones, and how do they affect the development of sperm and egg? For your research, visit www.science.nelson.com and follow the links from *Science 9: Concepts and Connections*, 2.16.

Challenge

2 Describe the male and female reproductive structures in your plant.

3 Include cancers of the male and female reproductive systems in your research.

Understanding Concepts

1. Describe the development of sperm.
2. Why do sperm die only a few days after they are produced?
3. Compare the sperm and the egg. How are they similar? How are they different? Consider size, shape, number, formation, length of life, and release.
4. What two types of follicle cells are found in the ovary? Explain the function of each type.
5. Describe ovulation.

Making Connections

6. If the ovaries of a woman are removed, can she still give birth to a baby? Support your answer.
7. Ectopic pregnancies occur when the embryo becomes implanted in the oviduct rather than the uterus. Why is this a dangerous situation?

Human Conception and Implantation

During human intercourse, about 150 to 300 million sperm cells travel through the vagina into the uterus. However, only a few hundred actually reach the oviducts and the egg (**Figure 1**).

Sperm cells can survive as long as five days in the oviduct of the female. The egg cell can survive only 48 h after ovulation if it is not fertilized.

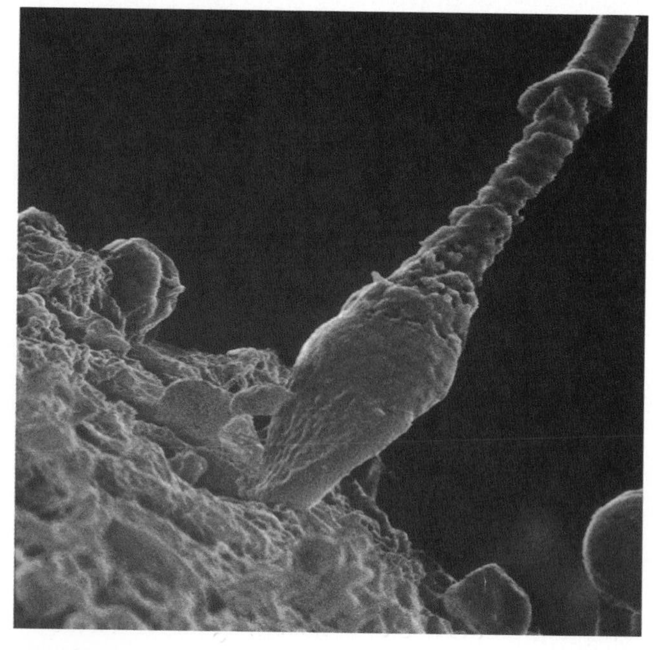

Figure 1

Human sperm cell and egg cell

During fertilization, also called **conception**, the head of the sperm breaks through the cell membrane of the egg (**Figure 2**). Even though several sperm may become attached to this membrane, only a single sperm cell enters the egg. The egg and sperm nuclei combine. The rest of the sperm cell and its flagellum are pinched off by the egg's cell membrane. The union of the egg and sperm create the zygote.

Within hours of fertilization, tiny hairlike **cilia** that line the oviduct move the zygote toward the uterus. As it travels, the zygote undergoes cell division (**Figure 3**). Around the fourth day, a 16-cell mass enters the uterus, where it floats freely for about two days (**Figure 4**). Around the sixth day after conception, the dividing cell mass has more than 100 cells and is a hollow ball. It implants (attaches) in the wall of the endometrium. It is now called an **embryo**.

Rapid growth continues. A yolk sac develops beside the embryo (**Figure 5**) and provides early nourishment. Then a membrane, called the

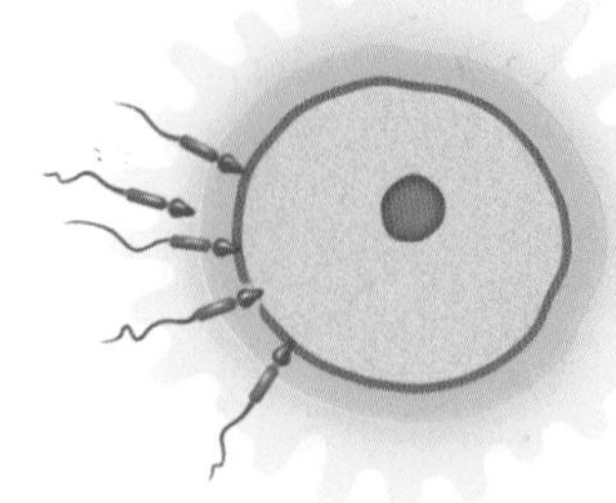

a Sperm cells attach to the egg cell. A single sperm cell penetrates the cell membrane.

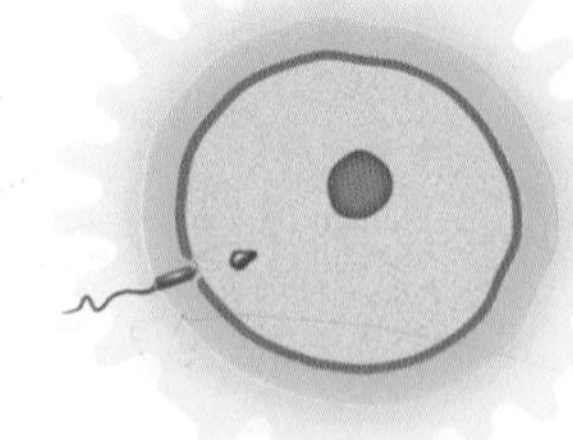

b The body and flagellum of the sperm cell are pinched off by the cell membrane of the egg cell.

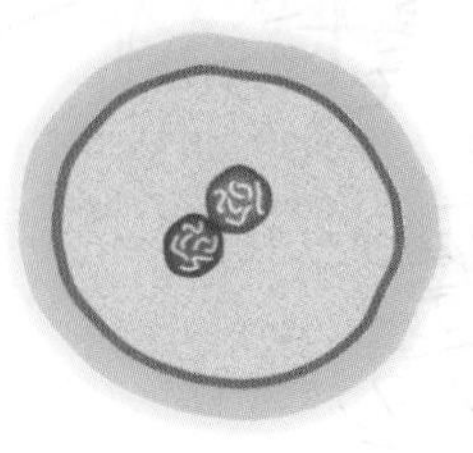

c The nucleus of the sperm cell finds the nucleus of the egg cell, and the 23 chromosomes from the sperm cell combine with the 23 chromosomes from the egg cell.

Figure 2

Fertilization

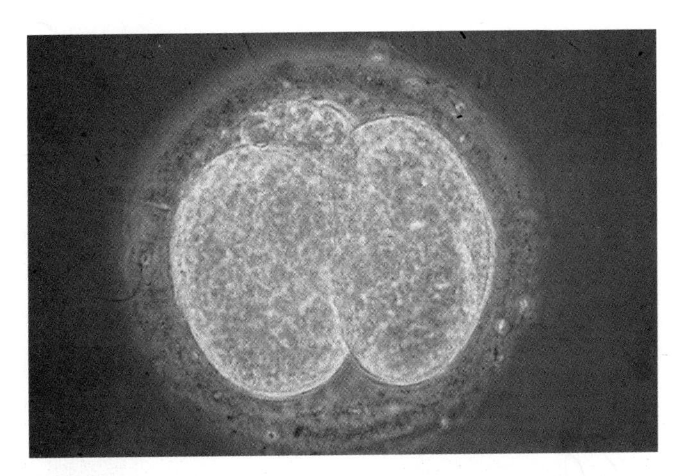

Figure 3

The zygote divides quickly. Here it has completed the first division.

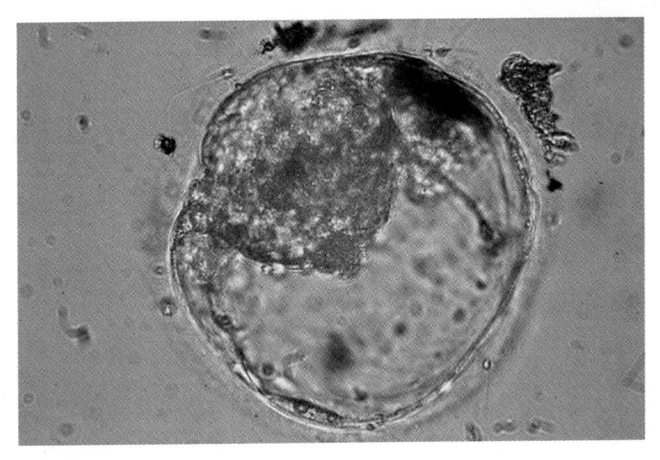

Figure 4

Cell division occurs for six days (from conception to implantation).

amnion, develops into a fluid-filled sac. The amniotic fluid surrounds the embryo, protecting it from infection, dehydration, impact, and changes in temperature.

Specialized cells in the embryo combine with cells in the endometrium to form the **placenta**. Here food and wastes are exchanged. Nutrients and oxygen move from the blood vessels of the mother into the blood vessels of the embryo. Wastes move in the opposite direction, from the embryo to the mother, who disposes of the wastes. The **umbilical cord** connects the embryo to the mother.

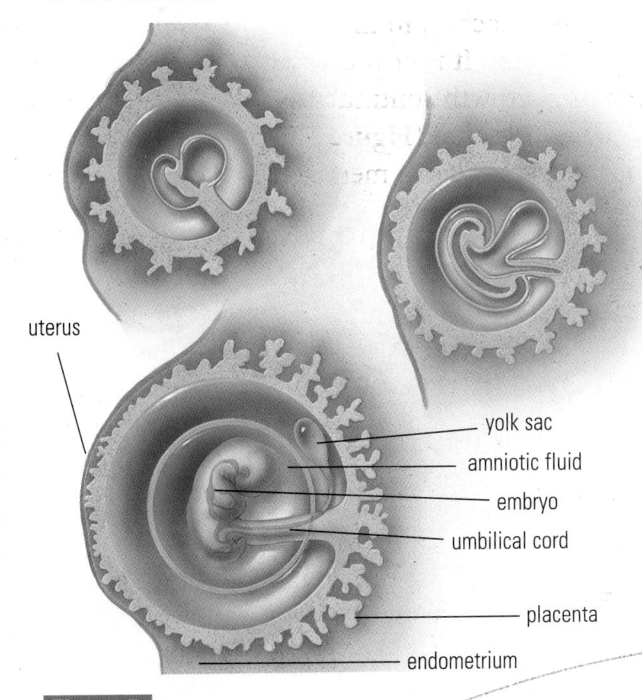

Figure 5

Formation of the membranes that protect the embryo

Understanding Concepts

1. Describe the journey of the egg from the ovary to (BB) the uterus. Use a time line to show what happens when.
2. Where does the placenta come from?
3. Use a table to compare the functions of the membranes and structures protecting the embryo: uterus, endometrium, placenta, yolk sac, amnion, amniotic fluid, and umbilical cord.

Making Connections

4. The placenta begins releasing hormones after three months of pregnancy. The hormone progesterone does the following:
   - keeps the endometrium healthy
   - prevents ovulation
   - stops the uterus from contracting

   (a) Predict what would happen during pregnancy if (C2) the placenta became damaged and could not maintain progesterone levels. Give reasons for your prediction.

   (b) Why don't women conceive again later in their pregnancy?

Challenge

2 How does fertilization occur in your plant? Are variations possible? Explain.

2.18

Human Reproductive Technology

Many couples are unable to have a baby. Much can go wrong: a man may not produce enough sperm, a woman's oviducts may be blocked, enough hormones may not be produced, a woman's follicles may not function properly. Some people go to fertility doctors for help to conceive a child.

Reproductive technology—using scientific and medical advances to alter natural processes—is changing the way some humans reproduce. It is also affecting the laws that determine parenthood and responsibility.

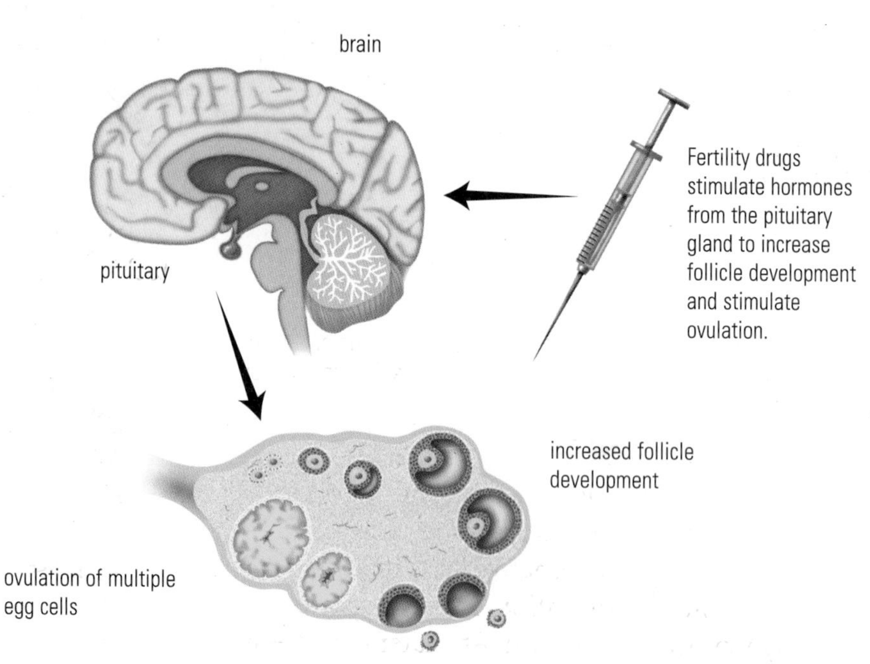

Figure 1
Fertility drugs

Reproductive Technologies

Fertility Drugs

If a woman is unable to release eggs, she may be given fertility drugs to stimulate her hormones. (**Figure 1**). Because fertility drugs increase the likelihood that more than one egg will be released, the chance of having multiple births (twins, triplets, etc.) increases.

Intrauterine Insemination

Normally, most sperm cells die as they travel through the uterus to the oviduct. If a male has a low sperm count, insemination ensures that more sperm cells reach the egg. Sperm cells are transferred directly into the oviducts of the woman following ovulation (**Figure 2**).

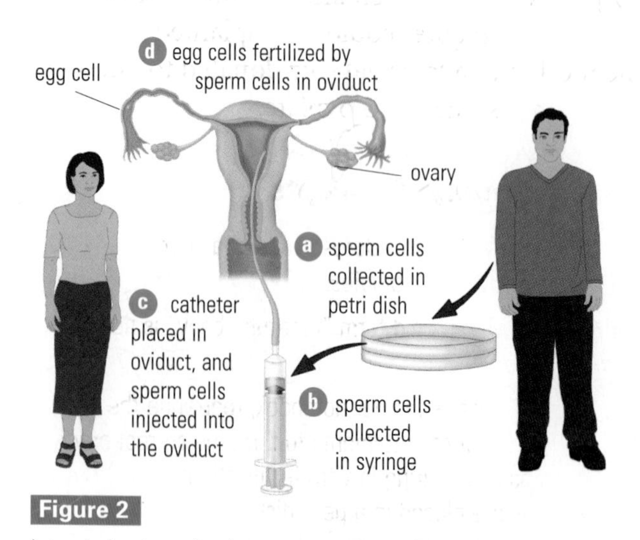

Figure 2
Intrauterine insemination

In Vitro Fertilization

In vitro, from Latin, means "in glass" and refers to the process that takes places outside the living organism. In vitro fertilization is used by some women who experience difficulty conceiving (**Figure 3**). Hormones are given to a woman to stimulate the release of eggs. During ovulation, a doctor inserts an instrument called a laparoscope into the woman's abdomen to locate the ovary. Mature eggs are removed from the ovary using suction. The eggs are placed in a petri dish and fertilized by the partner's sperm. Following a brief incubation period, one or more embryos are transferred into the uterus by a small catheter. If at least one of the embryos implants, a baby will be born nine months later.

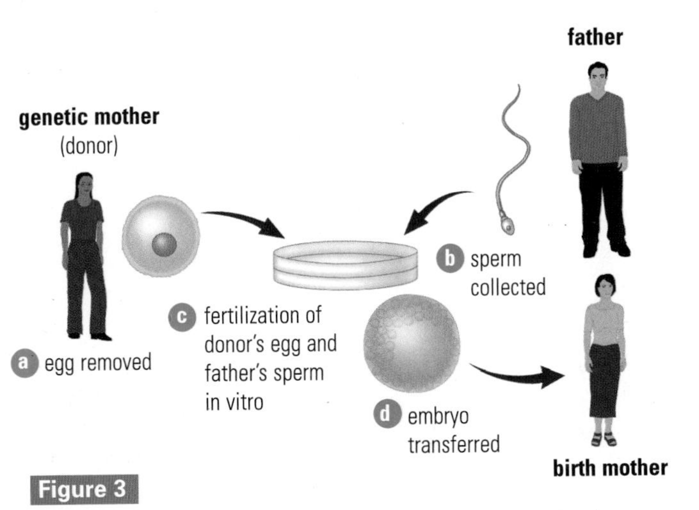

Figure 3

In vitro fertilization. In this example, the egg is taken from the donor, but the genetic mother and the birth mother can be the same person.

Egg Freezing and Egg Donations

In vitro fertilization is expensive and stressful. Fertilized eggs do not always transplant successfully. In this alternative, fertility drugs may stimulate multiple ovulation. Excess eggs are usually removed and frozen. At a later date, these eggs may be thawed and fertilized for another attempt.

Zygotes can be frozen after fertilization. Some of the eggs or zygotes could be implanted into the same mother at a later date, or donated to another woman who is unable to ovulate.

Embryo Transfer

Embryo transfer is a procedure in which a woman with a defective cervix or uterus asks another woman to give birth to her genetic child (**Figure 4**). The egg from the first woman is combined with the sperm of her partner. Fertilization occurs in vitro. The zygote is transferred to a **surrogate** mother, who carries the baby to term and returns it to the genetic parents. Dolly, the cloned sheep (see section 2.9), had a surrogate mother. Like the other procedures described in this section, this technology raises legal and ethical questions.

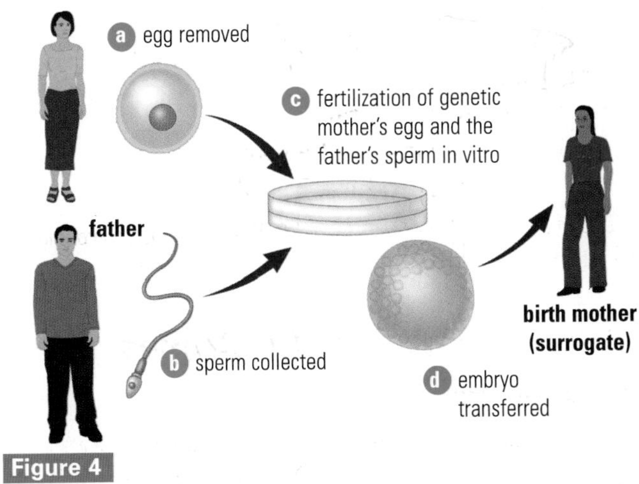

Figure 4

Embryo transfer

Understanding Concepts

1. Summarize the procedures used in each of the five reproductive technologies.
2. Explain why the old term "test-tube baby" is not accurate.
3. Study **Figure 3**. In your notebook, rearrange the following list of statements to form the correct order of steps used during in vitro fertilization:
   - Eggs are placed in a petri dish.
   - Eggs are extracted from the ovary.
   - Eggs are fertilized by sperm.
   - Hormone fertility drugs are given to the woman.
   - The embryo is transferred to the uterus.
   - The embryo is incubated.

Making Connections

4. (BB) (R) Reproductive technologies create controversy because of the ethical questions involved. Use a dictionary to define "ethical." Draw a concept map of this section listing all the ethical questions that could be raised by each reproductive technology. Prepare a summary paragraph outlining your viewpoint on using reproductive technologies.

Challenge

1 Create a real-world story, or scenario, for the technology you have chosen. This will help the people you survey understand your questions.

Work the Web

Infertility affects both males and females. Visit www.science.nelson.com and follow the links from *Science 9: Concepts and Connections*, 2.18, to find possible causes and risk factors of infertility.

SKILLS HANDBOOK: (BB) Graphic Organizers
(R) Writing for Specific Audiences

Pregnancy and Birth

Your life began as a single cell the size of the period at the end of this sentence. In only nine months, you grew into a baby about 50 cm long with a mass of about 3 kg. One cell became trillions of very different cells. How does it all happen?

First Trimester

Human pregnancy can be divided into three **trimesters** (three-month stages). The first trimester lasts from fertilization to the end of the third month (**Figure 1**).

Each cell in the embryo knows just what to do and what role it has to play during the important first month. At four weeks, the embryo is 7 mm long, 500 times larger than the fertilized egg. A basic heart has formed and begins to beat. A large brain is visible, as are the beginnings of arms and legs, called limb buds. Like all animals with a spinal column, the human embryo has a tail at this stage.

By the end of the second month, the cartilage of the embryo's skeleton begins to be replaced by bone. The embryo is now called a **fetus**, which means "young one." It is about the size of a cashew nut. Facial features, limbs, hands, feet, fingers and toes become visible. The nervous system responds to stimuli, and many of the internal organs begin to function.

During the third month, the external parts of the fetus continue to take shape. Eyelids, fingernails, and toenails form. Muscles begin to develop. The fetus can hiccup, flex its arms and legs, and move around. A sucking reflex appears in the mouth area. The ribs and backbone are very soft, and the fetus's skin is almost see-through. Sex organs form, but gender is not yet obvious. By the end of the third month, the fetus is 100 mm long and weighs just over 50 g (**Figure 2**).

While all this happens, a woman may not realize she is pregnant. Some signs of pregnancy are no menstrual period; nausea or vomiting; tender, enlarged breasts; and frequent urination. Not all women experience symptoms, so the best confirmation is a pregnancy test. Spontaneous abortions, or **miscarriages**, occur most often during this trimester. The risk of birth defects is also greatest in these first three months.

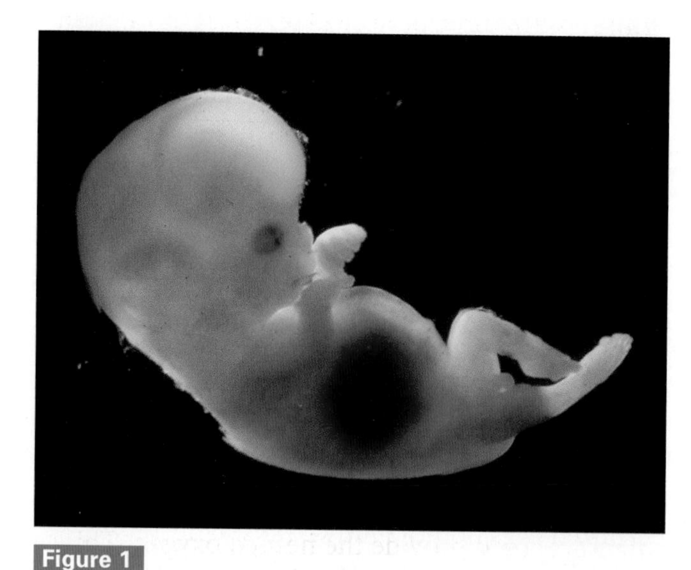

Figure 1

Early first trimester: A limb bud and the beginning of an eye can be seen.

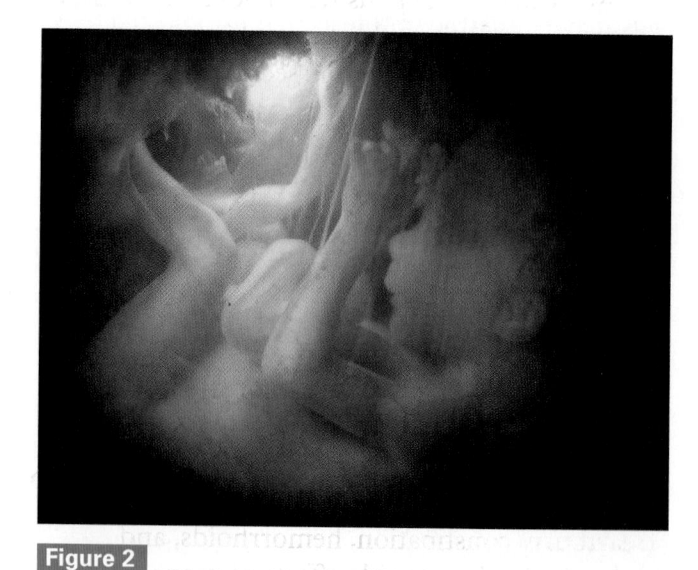

Figure 2

Late first trimester: All muscles and organs have formed and are beginning to function.

Second Trimester

The fetus is growing fast during this trimester, and the mother now appears pregnant as her abdomen begins to swell. Around the fourth or fifth month, the mother will feel fetal movement, called "quickening."

During this period, the fetus's external sex organs are clearly visible and its gender is obvious. Hair grows on the head and body. Eyelashes form, and the tongue develops taste buds. Most of the skeleton is replaced with bone. The fetus can blink, grasp with its fingers, and suck its thumb. It can also swallow, hear, and cry.

Inside the amniotic sac, the fetus drinks the amniotic fluid and urinates back into it. The fluid is completely replaced and cleaned about every three hours. The fetus also "breathes" in the fluid, but it doesn't drown because the placenta and umbilical cord provide the needed oxygen. This early breathing helps to develop small air sacs in the fetal lungs.

At 24 weeks, the fetus resembles a miniature infant except for its reddish, wrinkled skin (**Figure 3**). By the end of the second trimester, the fetus is 350 mm long and weighs about 680 g. If born at this time, there is a 50% chance the fetus will survive under intensive medical care.

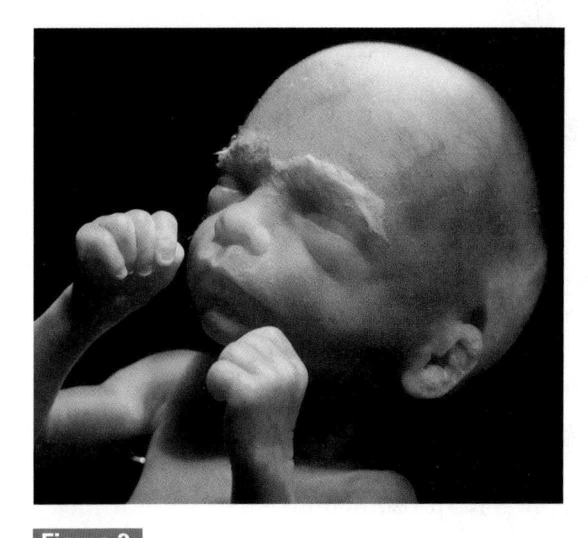

Figure 3

Second trimester: Facial features are present and internal systems are maturing.

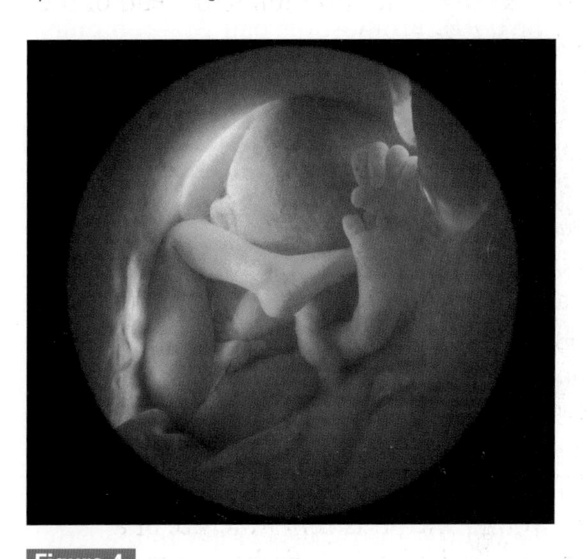

Figure 4

Third trimester: Movement is restricted by the space.

Third Trimester

During this final trimester, the fetus continues to grow and mature in preparation for life outside the womb. The mother may feel increasingly uncomfortable and short of breath as the uterus pushes up against her diaphragm. As weeks go by, the fetus will settle or drop into her pelvis. Heartburn, constipation, hemorrhoids, and abdominal pains are side effects of pregnancy during this final trimester.

The fetus puts on most of its weight in the last three months (**Figure 4**). The organ systems, established during the first two trimesters, begin to function properly. The fetus is very active, kicking and stretching as it alternates between sleep and wake cycles. As it gets bigger, it has less room to move around. The body hair begins to disappear. Bones in the body begin to harden, although the bones in the head stay soft to help the head fit through the birth canal.

A fetus is considered full term at 37 weeks of development. At this time, the baby's lungs are able to breathe on their own. The average baby is approximately 530 mm long and has a mass of 3400 g, or 3.4 kg, at birth.

Genetic Screening

The older the mother, the greater the risk of passing on genetic disorders. Many genetic disorders, such as Down syndrome, can be detected before the baby is born. By using ultrasound (**Figure 5**), physicians can locate the position of the developing fetus in the uterus.

Using a second technique called amniocentesis, the doctor removes a small amount of amniotic fluid with a syringe. By treating the cells with special stains, the chromosomes can be seen under a microscope. A microphotograph is then taken. Chromosomes are identified, arranged in pairs in a chart, and examined for irregularities. For example, if a fetus has Down syndrome, it will have an extra chromosome for chromosome pair 21—too much genetic information. Children with Down syndrome have specific physical features, a range of mental disabilities, and lower life expectancy.

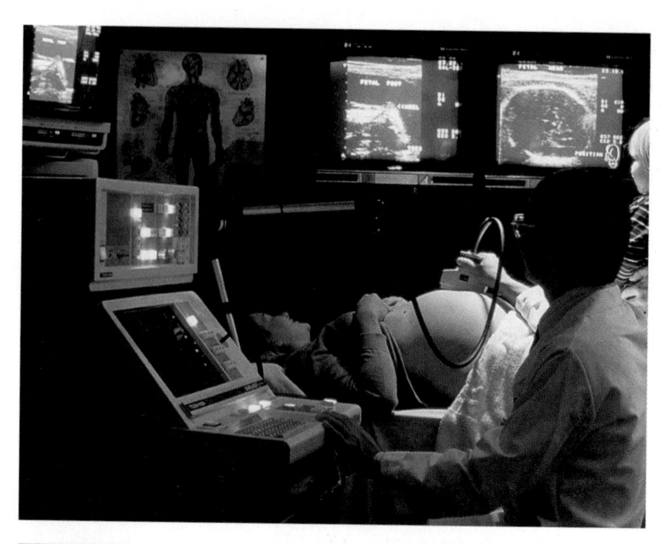

Figure 5

Sound waves are used to create an image of the fetus.

Risk Factors and Prenatal Care

"Do this, don't do that." Couples who are expecting a baby are exposed to many myths about birth. Here is sound advice for mothers—much of which applies to fathers as well.

Cautions

- Don't smoke or inhale second-hand smoke. Nicotine reduces the oxygen supply to the fetus and raises the risk of miscarriage, premature birth, and low birth weight.
- Don't drink alcohol. Alcohol poisons fetal cells and is the major cause of preventable birth defects in Canada, including mental disability.
- Don't use other drugs—either prescription or nonprescription—unless they have been proven safe for pregnant women. Some drugs increase the risk of miscarriage, premature birth, birth defects, deformities, and addiction in the newborn. Be aware that caffeine (found in tea, coffee, soft drinks, and chocolate) is a stimulant and could affect fetal development.
- Don't eat raw or undercooked red meat and fish, or clean a cat's litter box. These can be sources of toxoplasmosis, an infection that harms the baby.
- Don't sit in a sauna, hot tub, or steam room. Extreme moist heat raises the risk of miscarriage and birth defects.

Recommendations

- Do contact a physician early, and see a health care provider regularly throughout pregnancy.
- Do have blood samples checked for blood type, anemia, and Rh factors (which determine whether the baby's blood is compatible with the mother's blood). If the blood samples are incompatible, a miscarriage could occur.
- Do eat a well-balanced diet high in iron and calcium. Take folic acid supplements if advised by the doctor.
- Avoid exposure to radiation and toxic substances such as chemicals, solvents, and gases. These may lead to cancer or other types of genetic defects.
- Avoid exposure to infections.
- Do get enough sleep and continue regular physical activity.
- Do take childbirth classes.

Everything the mother does during pregnancy can have an effect on the baby.

Birth

About nine months after the embryo implants, the muscles of the uterus contract and the process of birth begins. Hormones control this process, which is described in **Figure 6**. The umbilical cord no longer functions after birth. It dries up and falls off.

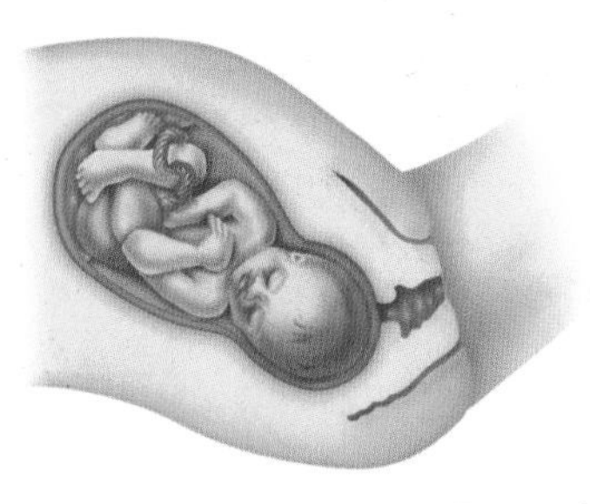

a The cervix begins to dilate, or open up. The membrane surrounding the baby is forced into the vagina (now called the birth canal).

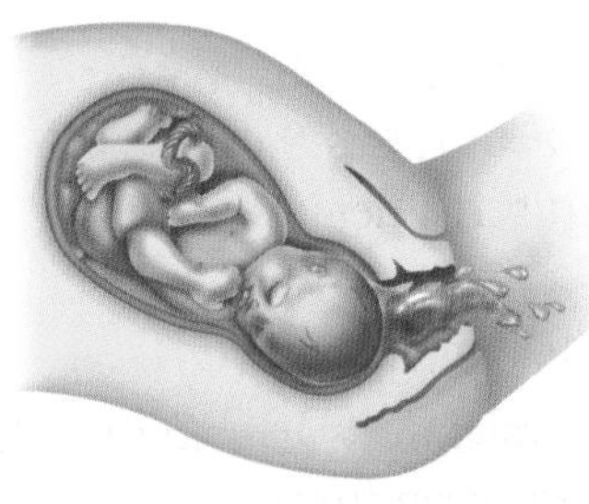

b The amniotic membrane breaks and amniotic fluid lubricates the canal. This event is called the "breaking of the water."

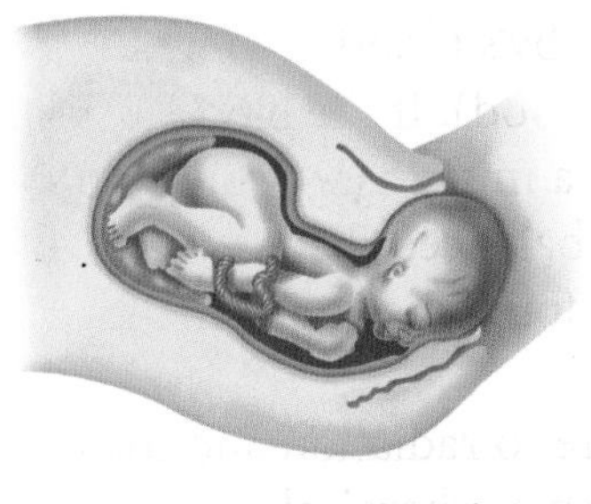

c Once the cervix has dilated enough, uterine contractions push the baby's head into the birth canal, followed by the rest of the body.

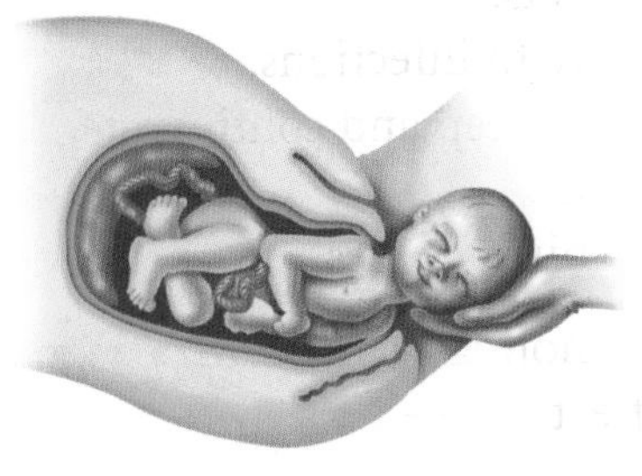

d When the head and shoulders are free of the birth canal, the baby slips out easily. The baby is born. A short while later, the placenta is pushed out of the uterus.

Figure 6

Understanding Concepts

1. Prepare a time line of the events from conception (BB) to birth. Organize this line vertically down a full sheet of paper. Summarize all the information from sections 2.17 and 2.19.
2. Choose one of the pieces of prenatal advice. (E3) Illustrate this advice on a sheet of blank paper, using as many drawings and as few words as possible.
3. What are the differences among a zygote, an embryo, a fetus, and a baby?

Making Connections

4. The circulatory system is the first system to function in the embryo. Use a dictionary to define "circulatory system," and then explain why you think this is the case.
5. Review the female reproductive system in section 2.16. Why does the risk of genetic disorders increase with the age of the mother?

Exploring

6. Physical signs of pregnancy are described in this (N) section. Summarize what happens in the mother (O) during each trimester, and use an encyclopedia or medical reference for additional details. There are also chemical signs of pregnancy. Research pregnancy tests and how they work.

Challenge

1 How does the reproductive technology you are studying affect the mother or her pregnancy?

2 Describe the embryonic development of your plant. What risk factors affect germination and growth?

Work the Web

Visit www.science.nelson.com and follow the links from *Science 9: Concepts and Connections*, 2.19. Choose a genetic disorder from the websites. Describe the causes, likelihood, risk factors, prevention, and symptoms of the disorder.

SKILLS HANDBOOK: (BB) Graphic Organizers (E3) Scientific Drawings (N) General Research (O) Internet Research

2.20 Explore an Issue

DECISION-MAKING SKILLS

- ● Define the Issue
- ○ Identify Alternatives
- ● Research
- ● Analyze the Issue
- ● Defend a Decision
- ○ Evaluate

Fetal Alcohol Syndrome

When a mother drinks alcohol, it crosses the placenta and enters the blood of the embryo (**Figure 1**). The effects on the embryo are the same as those on the mother: alcohol depresses the functioning of the nervous system. Alcohol is also a poison. Like other poisons, it is broken down by the liver. Unfortunately, the liver of an embryo is not fully developed until the last months of pregnancy, so it cannot break down the alcohol quickly. This means that alcohol is harmful much longer in the embryo than it is in the mother. Not only can alcohol kill many embryonic cells, but it may also change the genetic information in some cells, producing mutations.

Alcohol is the most common cause of fetal damage in the country, and the leading cause of preventable mental disability. Every day in Canada, one child is born with fetal alcohol syndrome (FAS) or its milder form, fetal alcohol effects (FAE). About 1 to 3 per 1000 live births in Canada are affected.

Children with FAS/FAE may have low birth weight, growth deficiencies, and facial deformities such as a small head; small, widely placed eyes; a thin upper lip; and a small jawbone. These children can suffer from:

- learning disabilities
- hyperactivity
- attention or memory deficits
- inability to manage anger
- poor judgement
- difficulties with problem solving

These problems may lead to others in later life. Fewer than 10% of people with FAS are successful in living and working independently.

It has been estimated that 60% to 70% of alcoholic women who become pregnant give birth to babies with FAS. A woman need not be an alcoholic to have an FAS child. A pregnant woman who drinks any amount of alcohol is at risk, since a "safe" level of alcohol consumption during pregnancy has not been determined. Despite these well-known facts, pregnant women continue to drink and the problem is getting worse.

Symptoms of FAS cannot be reversed, and there is no cure.

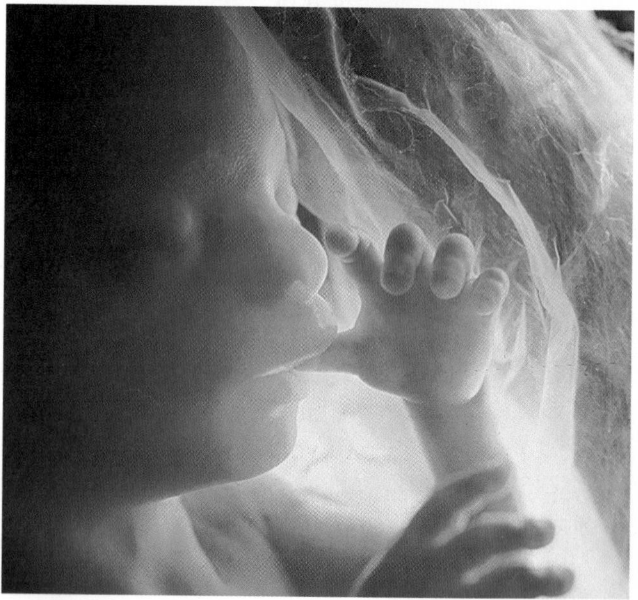

Figure 1
Alcohol can pass from the mother's blood across the placenta into the embryo.

Debate FAS: A Preventable Problem?

Statement

Pregnant women should be required to have blood tests on a regular basis to monitor drinking problems.

Point

- FAS is the third most common reason for babies being born with mental disability in North America. (The most common reasons for mental disability are genetic defects, which are not preventable.) In addition, physical problems with the heart and nervous system are common in FAS children. If testing can reduce or prevent this, it should be done.
- Most people know that alcohol is bad for the fetus, yet about one-fifth of pregnant women continue to drink even after they learn they are pregnant. Education is obviously not enough.

Counterpoint

- Most birth defects occur during the first three months of pregnancy, when the fetus's organs are forming. Alcohol is not the only factor linked to birth problems. Should women also be monitored to ensure that they eat well, do not gain too much weight, or do not become depressed?
- Removing the right to choose for any person is a serious matter. Everyone hopes that mothers will recognize their responsibility, but laws and penalties are not the answer. Changes in attitudes are best accomplished through education.

What do you think?

- In a group, carefully read the statement and the points presented for and against it. Discuss and then write down any additional points for and against you can think of from the information given on these pages.
- Search for additional information using the library, CD-ROMs, or the Internet.
- From your research, prepare a summary of additional information on the topic, and use this to form your own opinion on the statement.
- Share the research information with your group members, and then decide whether your group agrees or disagrees with the statement.
- Write a summary of your group's final position.
- Prepare to defend your group's position in a class discussion.

Understanding Concepts

1. What is the incidence of FAS/FAE in Canada?
2. Use a table to list the possible physical and mental characteristics of children with FAS.

Reflecting

3. How do you think the FAS problem could be solved?

Challenge

1 Could reproductive technologies be used to prevent FAS or cure this problem? Explain.

2 What effect does alcohol have on plant germination and growth?

3 Is there a link between alcohol consumption and cancer? Explain.

Unit 2 Summary

Key Expectations

Throughout this unit, you have had opportunities to do the following:

- Describe the basic process of cell division, including what happens to the cell membrane and contents of the nucleus (2.2, 2.3)
- Demonstrate an understanding of the importance of cell division to the growth and reproduction of an organism (2.1, 2.2, 2.3, 2.4, 2.5, 2.7)
- Demonstrate an understanding that the nucleus of a cell contains genetic information and determines cell processes (2.1, 2.2, 2.3, 2.5, 2.8, 2.16, 2.17)
- Describe various types of asexual reproduction that occur in plant species or in animal species and various methods for the asexual propagation of plants (2.8, 2.9, 2.10)
- Describe the various types of sexual reproduction that occur in plants and in animals, and identify some plants and animals, including hermaphrodites, that exhibit this type of reproduction (2.11, 2.12, 2.16)
- Compare sexual and asexual reproduction (2.8, 2.9, 2.15)
- Explain signs of pregnancy in humans and describe the major stages of human development from conception to early infancy (2.17, 2.19)
- Identify a current problem or concern relating to plant or animal reproduction (2.6, 2.10, 2.20); formulate scientific questions and develop a plan to answer these questions (2.6, 2.10, 2.20); demonstrate the skills required to plan and conduct an inquiry into reproduction, using instruments and tools safely, accurately, and effectively (2.3, 2.10, 2.12); select and integrate information from various sources (2.7, 2.20); organize, record and analyze the information gathered (2.1, 2.3, 2.4, 2.6, 2.7, 2.10, 2.12, 2.13, 2.15, 2.20); predict the value of a variable by interpolating or extrapolating from graphical data (2.4, 2.6, 2.15); communicate scientific ideas, procedures, results, and conclusions using appropriate language and formats (2.3, 2.4, 2.6, 2.7, 2.10, 2.12, 2.15, 2.20); defend orally a position on the concern or problem investigated (2.7, 2.20)
- Use a microscope to observe and identify animal and vegetable cells in different stages of mitosis, as well as cells undergoing asexual reproduction (2.3)
- Describe the use of reproductive technologies in a workplace environment and explain the costs and benefits of using such technologies (2.13, 2.18)
- Examine some Canadian contributions to research and technological development in the field of genetics and reproductive biology (2.13)
- Identify local environmental factors and individual choices that may lead to a change in a cell's genetic information or an organism's development, and investigate the consequences such factors and choices have on human development (2.5, 2.6, 2.7, 2.19, 2.20)
- Provide examples of the impact of developments in reproductive biology on global and local food production, populations, the spread of disease, and the environment (2.13, 2.18)
- Describe careers that involve some aspect of reproductive biology (2.14)

Key Terms

amnion
anaphase
asexual reproduction
benign
binary fission
budding
cancer
carcinogen
cell cycle
cell membrane
cervix
chromosome
cilia
cloning
conception
conjugation
crossbreeding
cytokinesis
cytoplasm
deoxyribonucleic acid (DNA)
DNA fingerprinting
egg
embryo
endometrium
epididymis

external fertilization
fetus
flagellum
follicle
fragmentation
gamete
gene
genetic engineering
grafting
hermaphrodite
hormone
internal fertilization
interphase
malignant
menstruation
metaphase
miscarriage
mitosis
mutation
nucleus
organelle
ovary
oviduct
ovulation
placenta
pollen
pollination
prophase
selective breeding
seminiferous tubules
sexual reproduction
sperm
spore formation
surrogate
telophase
testis
traits
trimester
tumour
umbilical cord
uterus
vagina
vegetative reproduction
zygote

What HAVE YOU *Learned?*

Revisit your answers to the What Do You Already Know? questions on page 69 in Getting Started.

- Have any of your answers changed?
- What new questions do you have?

Unit Concept Map

Use the concept map to review the major concepts in Unit 2. This map can help you begin to organize the information that you have studied. You may copy the map, and then add more links to your map. Also, you may add more information in each box.

A concept map can be used to review a large topic on a general level, or it can be used to examine a very specific topic in detail. Select one concept from this unit that you need to study more, and make a detailed concept map for it.

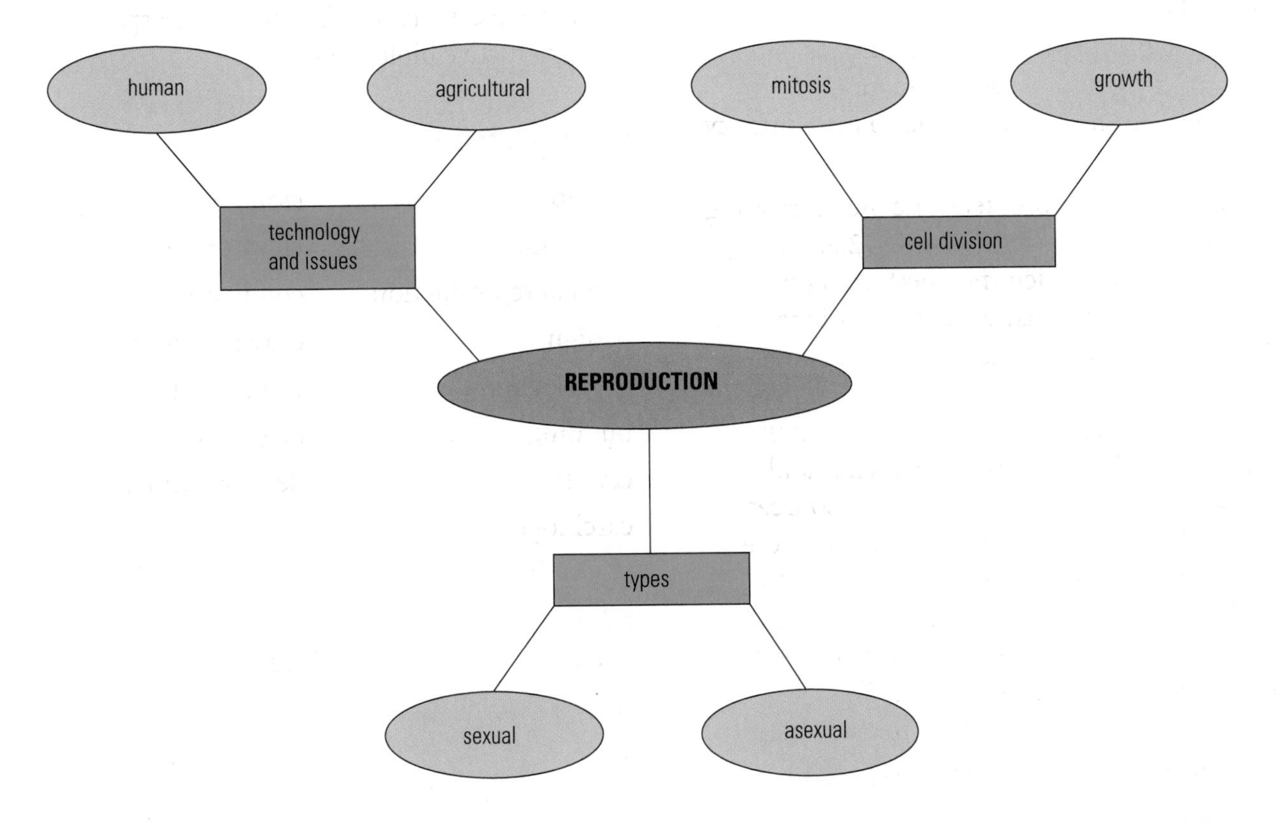

Unit 2 Review

Understanding Concepts

1. In your notebook, write the word(s) needed to complete the following sentences:
   (a) In this phase of mitosis, called ________, chromosomes line up in the middle of the cell.
   (b) The stage between cell divisions, called ____, is marked by rapid growth and the duplication of genetic material, followed by another period of growth.
   (c) The division process in which a single cell divides into two identical daughter cells is called ______.
   (d) __________ is the process by which identical offspring are formed from a single cell or tissue.
   (e) A change in a cell's genetic information is a ______________.
   (f) Animals that contain both male and female sex organs are called ________.
   (g) Bacteria exchange genetic information by way of plasmids in a process called _________.
   (h) The organ responsible for nutrient exchange between a human mother and a fetus is called the __________.
2. Indicate whether each of the following statements is true or false. If you think the statement is false, rewrite it to make it true.
   (a) The larger an organism is, the larger is the size of its cells.
   (b) If a fertilized egg from a mouse has 22 chromosomes, you should expect 22 chromosomes in the muscle cell of the same mouse.
   (c) When plants such as strawberries reproduce by sending out runners, they reproduce without sex cells.
   (d) If a sheep were cloned, you would expect the offspring to be the same sex as the birth parent.
   (e) All of the cells in the human body divide at the same rate.
   (f) Human sperm cells have half as many chromosomes as unfertilized human egg cells.
   (g) Multiple human sperm cells fertilize a single human egg cell at the same time.
   (h) Cancer cells divide at a faster rate than normal cells.
3. Use the diagram in **Figure 1** of plant and animal cells during cell division.
   (a) Identify each of the cells as either a plant or an animal cell.
   (b) Identify the phases of cell division.

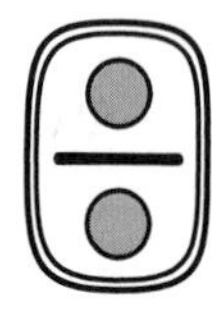
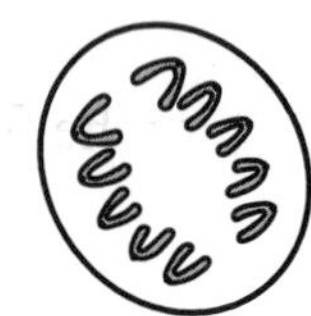
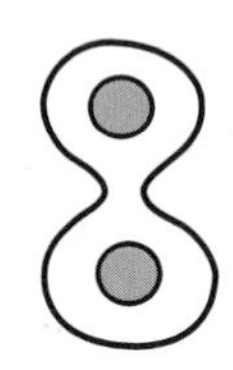

Figure 1

4. What is the cell cycle?
5. What is interphase, and why is it important for the process of cell division?
6. Sexual reproduction contributes more to variation within a species than asexual reproduction does. Use the following terms to explain this statement: egg, sperm, zygote, DNA, chromosomes, genes, traits.
7. Briefly describe two uses of DNA fingerprinting.
8. In what ways does a cancer cell differ from a normal cell?
9. List three factors that cause or contribute to the development of cancer.
10. What changes in lifestyle could reduce the occurrence of cancer?

Use **Figure 2** to answer questions 11 and 12.

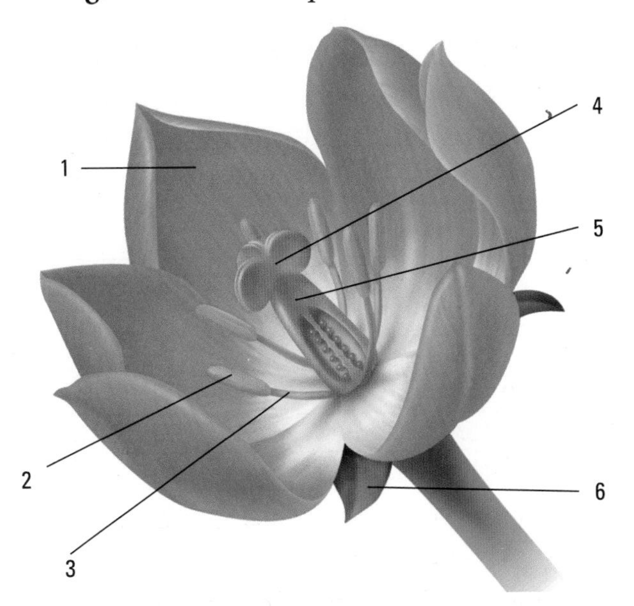

Figure 2

11. The male parts of the flower can be identified as structures
- **(a)** 1 and 6
- **(b)** 1 and 2
- **(c)** 2 and 4
- **(d)** 5 and 6
- **(e)** 2 and 3

12. The female structure that receives pollen from an insect during cross-pollination is labelled
- **(a)** 1
- **(b)** 2
- **(c)** 4
- **(d)** 5
- **(e)** 6

13. Describe how pollen cells fertilize egg cells in a flower.

14. Why might a gardener want to graft branches from different trees to a single stem?

15. Give examples of organisms that reproduce by way of
- **(a)** spores
- **(b)** seeds
- **(c)** eggs
- **(d)** zygotes developing within the parent

16. A human sperm cell contains
- **(a)** 23 chromosomes, of which two are X-chromosomes
- **(b)** 46 chromosomes, of which two are Y-chromosomes
- **(c)** 23 chromosomes, of which one is X or Y
- **(d)** 46 chromosomes, of which one is X or Y

17. Compare in vitro fertilization with the usual course of events in human reproduction.

18. In the human female reproductive system, identify the
- **(a)** organ that is the site of implantation of the embryo
- **(b)** tissue that provides nourishment for the embryo
- **(c)** site where fertilization takes place
- **(d)** organ that produces the female sex cell
- **(e)** structure that becomes the birth canal
- **(f)** fluid-filled sac that insulates the embryo
- **(g)** structure that connects the embryo with the placenta

19. Using any of the following drugs during pregnancy can harm the developing embryo. Explain how each drug affects the embryo and the damage the drug can cause.
- **(a)** alcohol
- **(b)** tobacco
- **(c)** prescription medication
- **(d)** illegal drugs

20. What causes fetal alcohol syndrome, and what are some of its symptoms?

Applying Skills

21. Use **Figure 3** to answer the following questions. In which phase(s) of cell division
    (**a**) does growth occur?
    (**b**) is the cell undergoing cell division?
    (**c**) is the cell copying genetic information?
    (**d**) are chromosomes lined up in the middle of the cell?
    (**e**) do the cytoplasm and contents split into two equal parts?
    (**f**) does each chromosome split into two halves that move to opposite poles?
    (**g**) do chromosomes shorten, thicken, and become visible?

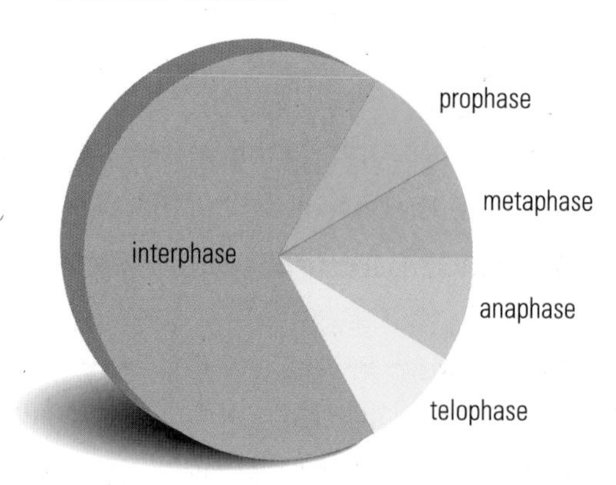

Figure 3

22. A research team studied the growth rate of a type of cancer cell in mice. Every 2 days for 60 days, the team counted the number of cells in an area of 1 mm^2. Which of the graphs in **Figure 4** represents their data? Explain your answer.

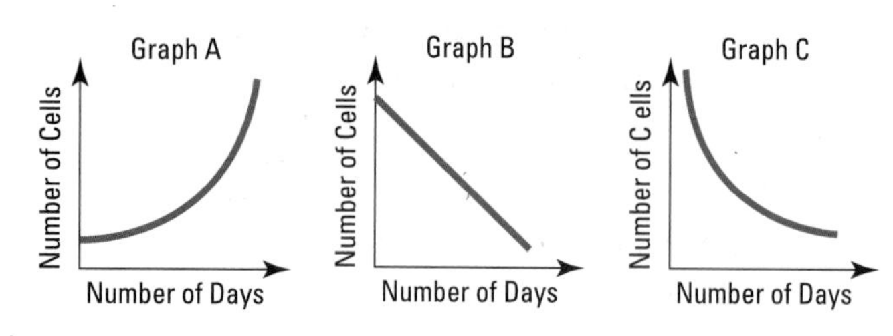

Figure 4

23. A scientist wanted to determine whether age affects body mass. The scientist predicted that body mass increases with age. To test this hypothesis, five people from each age group were selected at random and their body mass was recorded. The results are given in **Table 1**.

Table 1

Age group	Average body mass (kg)
20–29	60
30–39	65
40–49	72
50–59	75
60–69	68

The scientist concluded that the older a person becomes, the greater his or her body mass. Assess the experimental design used by the scientist. What additional information would you want to collect before accepting this conclusion?

24. Three groups of seedlings were placed in containers. Each container received 10 mL of a different nutrient solution. The root lengths of the seedlings were measured over a five-day period. The data in **Table 2** were obtained.

Table 2

Time (days)	Root Length (mm)		
	Solution X	Solution Y	Solution Z
0	2	2	2
1	2	4	4
2	3	6	9
3	4	10	14
4	4	12	18
5	5	15	28

(**a**) Graph the results obtained by plotting days on the *x*-axis and root length on the *y*-axis.
(**b**) Provide a conclusion based on the data given.

25. Use **Figure 5** to answer these questions.
  (**a**) How many chromosomes were in the sperm cell?
  (**b**) Explain how this sperm cell could be produced from a cell that had 46 chromosomes.
  (**c**) How many chromosomes would be found in each cell following mitosis?

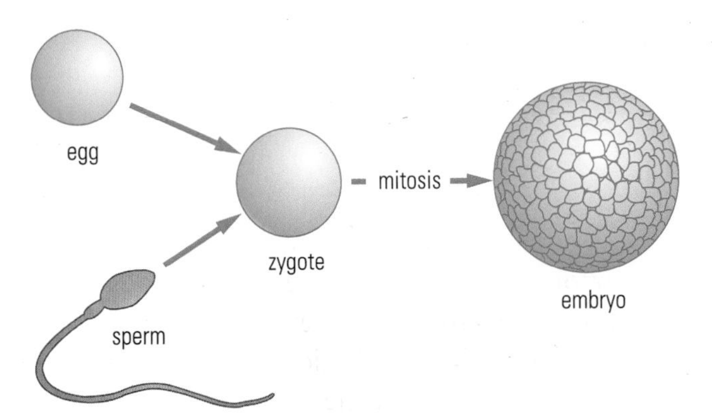

Figure 5
Fertilization in humans

Making Connections

26. Irradiation (exposing cells to X-rays) can break chromosomes apart. **Figure 6** shows the effects of irradiation on cells undergoing metaphase. Food companies sometimes irradiate fruit and vegetables to improve their shelf life. How does irradiation help to preserve food?

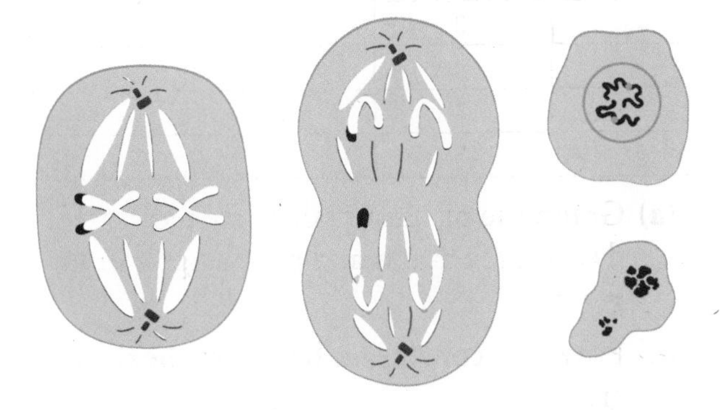

Figure 6

27. Imagine that in the year 2105 politicians decide that all food plants will be reproduced by cell cloning. Predict some of the potential problems.

28. In the movie *Multiplicity*, genetic engineers make many copies of the movie's hero, played by Michael Keaton. A number of funny situations are created because his wife is unable to distinguish him from the clones. Explain the scientific flaw in the plot.

29. The Human Genome Project is one of the largest research projects ever carried out. The object of the project is to identify the position of every human gene along each of the 46 chromosomes. Make a list of some of the possible benefits of this project. How might the research be used in negative ways?

30. A man has cancer in one testis and has to have it removed. Speculate about how this will affect his ability to father children.

31. What advantage does cross-pollination provide to flowering plants?

32. Describe the advantages of each of the following reproductive strategies:
  (**a**) A bacterium forms a spore with a resistant coating.
  (**b**) An opossum produces eight embryos, but only six find their way into the mother's pouch and attach to a nipple.
  (**c**) The wildebeest gives birth to young that begin running next to the mother a few minutes after birth.
  (**d**) The whooping crane lays two eggs. The first chick to hatch breaks the other egg. Only the first chick will survive.
  (**e**) A parasitic worm produces hundreds of thousands of eggs. The eggs are released with the solid wastes of the host animal.

33. In intrauterine insemination, sperm cells from the donor or partner are transferred by catheter into the oviducts of the woman following ovulation.
  (**a**) Suggest one reason this technique might be used.
  (**b**) Give one reason someone might object to this technology.

34. Describe the effects of both parents using drugs and alcohol on the conception and development of their fetus.

Electrical Applications

Electricity is an important part of our daily lives. Think about it: When we enter a room, we turn on the light; when we are hungry, we look in the refrigerator. We use computers, watch television, and we listen to music on a stereo system or a CD player. What do all these things have in common? They all use electricity to work. It is only when the power goes out that we realize how many of our activities at home, school, and work depend on electricity.

Unit 3 Overview

Overall Expectations

In this unit, you will be able to

- understand how static and current electricity work
- design and build electrical circuits that perform a specific function
- analyze the practical uses of electricity at home and in the workplace

Static and Current Electricity

Clothes sticking together when you take them out of the dryer and pressing a button to turn on a flashlight are both explained by electricity.

Specific Expectations

In this unit, you will be able to

- explain common situations such as a balloon sticking to the wall after it has been rubbed on your hair
- compare static and current electricity
- describe electric current, voltage, and resistance, and compare each to the flow of water
- explain how electric current, voltage, and resistance are measured using an ammeter and a voltmeter
- describe the effects of varying electrical resistance on electric current in an electric circuit
- apply the following relationship to simple series circuits: voltage drop = resistance × current

Designing and Building Electric Circuits

Putting together an electric circuit will help you to understand how electricity works, what safety precautions you should take, and the importance of this energy source.

Specific Expectations

In this unit, you will be able to

- identify the electrical safety procedures to follow when designing and building electric circuits and choosing materials
- show the skills required to plan and conduct an electrical investigation
- design, draw, and construct series and parallel circuits that perform a specific function
- organize, record, and analyze information gathered while designing and building electric circuits
- collect and graph data
- communicate the results of your investigations
- describe household wiring and its typical components

Practical Uses of Electricity in Everyday Life

Have you ever wondered how turning on a switch turns on the light in a room? Electricity enables us to perform many of our everyday tasks, as well as many jobs. Without electricity, our lives would be very different!

Specific Expectations

In this unit, you will be able to

- explain how some common household appliances operate
- determine the percent efficiency of an electric device
- describe and research careers that involve electricity
- compare electrical energy production technologies
- research and report on electricity using a variety of sources
- ask questions about a problem or issue involving electricity
- identify a problem related to the use of electricity
- develop a solution to a problem related to the use of electricity

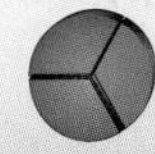

Challenge

Learning about Electricity

As you learn about how electricity works, you will analyze your electricity use and its impact on your life. You will also show your learning by completing a Challenge. For more information on the Challenges, see the following page.

1 Design and Test an Electric Circuit

Electric circuits can be used for a variety of purposes. Design and test an electric circuit that performs a job of your choice. Present the completed electric circuit to the class and explain its purpose.

2 Electrical Safety Pamphlet

Electricity can be dangerous if not used properly in the home and workplace. Create an electrical safety pamphlet outlining specific steps to follow to ensure electrical safety. Use your own home and a chosen workplace to develop this pamphlet. Include an electrical safety checklist for readers to complete.

3 Electric Game Show

Electricity is an important form of energy that we rely on. Produce a quiz board that requires information about static and current electricity, designing and building electric circuits, electricity's impact on society and the environment, and using electricity safely in our everyday lives.

Record your ideas for your Challenge as you progress through this unit, and when you see

Challenge

Electrical Applications

Electricity is a form of usable energy. Natural gas, oil, and radioactive nuclei are sources of this energy. It has many practical uses, but it can be dangerous if safety precautions are ignored. Electricity also has an impact on the environment. Learning about electricity will help you to understand how it works and its function in society. Each Challenge allows you to examine a different electrical application.

1 Design and Test an Electric Circuit

Once you understand electricity, you can design and build an electric circuit to carry out a specific task. The type of electrical device that you build will depend on your imagination and the available materials. Your electric circuit can perform any task you choose.

In building your own electric circuit, you should

- draw a diagram of the electric circuit
- write a report on how you created the circuit and tested it to ensure it worked; include the steps you followed so that others can build your circuit
- present the completed electric circuit to the class

2 Electrical Safety Pamphlet

Many people are injured, or even killed, because of their lack of knowledge about electrical safety. As you work through this unit, consider safety rules that you could include in a pamphlet to help prevent electrical accidents in the home and workplace.

Your electrical safety pamphlet should

- identify electrical dangers in the home and workplace and suggest how to prevent or avoid them
- include a safety checklist to help people determine how safe their home or their workplace is
- use pictures, words, and diagrams or charts
- suggest ways to prevent accidents with both static and current electricity

3 Electric Game Show

Electricity affects society and the environment. A game is a great way for people to learn about electricity. Your task is to design and build an electric quiz board that asks questions about what you consider the most important information about electricity.

Your electric quiz board should

- be organized around a game show that players will be comfortable with
- address information about static and current electricity, electric circuits, and uses and impacts of electricity
- include a feedback form for the players so that you can improve the quiz board

When preparing to build a model or carry out an activity, be sure to have your plan approved by your teacher before you begin.

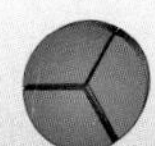

Assessment

Your completed Challenge may be assessed according to the following:

Process

- Understand the specific challenge.
- Develop a plan.
- Choose and safely use appropriate tools, equipment, materials, and computer software.
- Analyze the results.

Communication

- Prepare an appropriate presentation of the task.
- Use correct terms, symbols, and SI units.
- Incorporate information technology.

Product

- Meet the set criteria.
- Show an understanding of science concepts, principles, laws, and theories.
- Show effective use of materials.
- Address the identified situation or problem.

Unit 3

Getting Started

How Does Electricity Work?

Have you ever put on a sweater and had it stick to you? Have you ever rubbed a balloon on your hair (**Figure 1**) and stuck it to a wall? In each case, you experienced the effects of static electricity. Lightning is another example of static electricity. Do you know what safety procedures to follow during lightning storms?

Figure 1

Why does the balloon stick to the wall after it has been rubbed on your hair? The power of static electricity is at work!

Many devices need current electricity (**Figure 2**). These devices either plug into an outlet or use batteries to work. Although electricity is used for many tasks, it can harm us in many situations. Plugging in a radio near a source of water such as a tub or a pool would be very dangerous if the radio were to fall into the water while still plugged in.

Figure 2

The lights in your home are part of an electric circuit.

Electricity has many practical uses, and each one has an impact on our daily lives and on the environment. The methods we use to generate electrical energy have advantages and disadvantages (**Figure 3**). We must learn to conserve electricity to help keep the cost down.

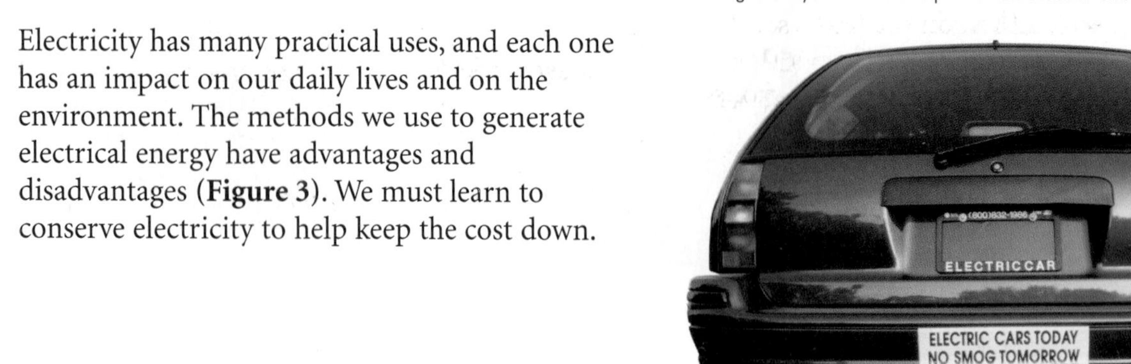

Figure 3

Electric cars would decrease the amount of air pollution caused by gas-fuelled cars.

What DO YOU ALREADY *Know?*

1. At what time of year is static electricity most active? Why do you think this is the case?
2. Describe how the following products work, in regard to static electricity. You may use pictures in your answers.
   (a) fabric softener dryer sheets
   (b) plastic cling wrap
3. What is an electric circuit?
4. What safety precautions should you follow with current electricity? List as many as you can.
5. Name two methods of generating electrical energy.

Throughout this unit, note any changes in your ideas as you learn new concepts and develop your skills.

Try This Activity Fabric Softener Dryer Sheets

How do you get rid of static electricity in clothing? In this activity you will observe and record the effects of fabric softener dryer sheets on a static electric charge.

(a) Obtain a fabric softener dryer sheet. Record the instructions on the package in your notebook.

(b) Rubbing a balloon on your head was probably your first experience with static electricity. Rub a balloon on top of your head or on your clothing to charge it. How long do you have to rub the balloon before it will stick to the wall?

(c) (E1) Try three ways to remove the static electric charge from the balloon. Record the results in **Table 1**.

(i) Place the balloon on the fabric softener dryer sheet. Does the balloon still stick to the wall? Record your observations.

(ii) Rub the charged balloon with the fabric softener dryer sheet. Does the balloon still stick to the wall? Record your observations.

Table 1

Fabric softener test	Observations
Charged balloon placed on the fabric softener dryer sheet	
Charged balloon rubbed with the fabric softener dryer sheet	
Charged balloon touched with hands rubbed with fabric softener dryer sheet	

(iii) Rub your hands on the fabric softener dryer sheet. Place your hands on the charged balloon. Does the balloon still stick to the wall? Record your observations.

(d) (F1) Which test(s) resulted in the charged balloon not sticking to the wall?

(e) How do you think a fabric softener dryer sheet works to remove static electricity from your clothes?

Do not blow the balloon too big or it will burst.

3.1 Investigation

INQUIRY SKILLS
- ○ Questioning
- ○ Hypothesizing
- ● Predicting
- ○ Planning
- ● Conducting
- ● Recording
- ● Analyzing
- ● Concluding
- ● Communicating

Investigating Electric Charges

One of the most common examples of static electricity is when you comb your hair and the air is very dry or when you pull a sweater over your head. Your hair stands on end, and you may hear a crackling noise. Static electricity also causes clothes to cling together in the dryer. You can also sense and hear its effects if you move your hand lightly over the surface of a television or computer screen after it has been on for a while.

In this investigation, you will electrically charge a variety of substances and identify some properties of electric charges.

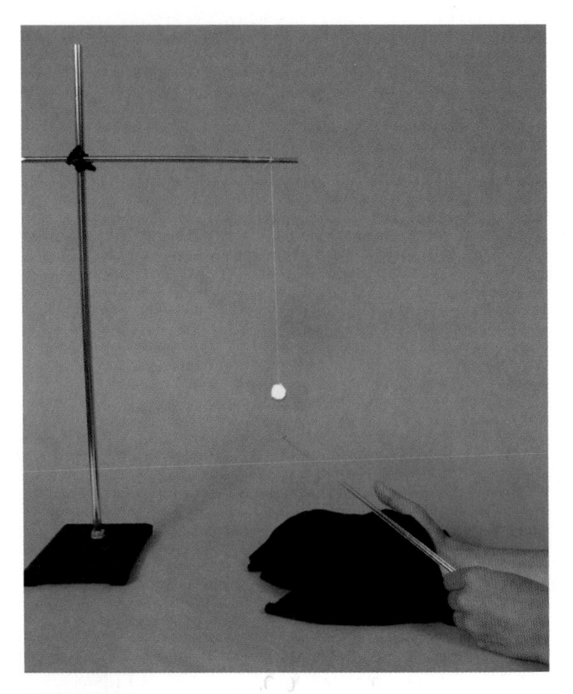

Figure 1
Materials used in this lab

Question

How are uncharged and charged substances affected when they are near one another (**Figure 1**)?

If you are allergic to fur, do not perform this investigation.

Prediction

(a) Predict what will happen when you bring two (C2) charged objects near one another.

Materials

- ebonite rod (black)
- fur sample
- Lucite rod (clear plastic or any glass rod)
- silk or polyester scrap material
- small scraps of paper (confetti size)
- thin water stream
- pith ball apparatus

Procedure

1 Before you begin this experiment, create a (C6) table like **Table 1** to record your observations.

Table 1

Electric charge test	Effect on scraps of paper	Effect on thin water stream	Effect on pith ball
ebonite rod rubbed with fur			
Lucite rod rubbed with silk or polyester			

2 Charge the ebonite rod by rubbing it with the fur. Bring the charged rod close to, but do not touch, the small scraps of paper, a thin water stream, and a pith ball. In your table, record the effect of the charged ebonite rod on each of these items.

3 Charge the Lucite rod by rubbing it with the silk or polyester. Bring the charged rod close to, but not touching, the small scraps of paper, a thin water stream, and a pith ball. In your table, record the effect of the charged Lucite rod on each of these items.

SKILLS HANDBOOK: (C2) Predicting and Hypothesizing (C6) Preparing Observation Tables

Analysis and Conclusion

Analyze your observations by answering the following questions:

(b) Was your prediction correct? Explain why, or why not, based on your observations.

(c) (F1) Compare the results of the investigation for the ebonite and Lucite rod on the scraps of paper, water stream, and pith ball (**Figure 2**).

(d) (F1) What happens when charged objects are placed near uncharged objects?

(e) (F1) What happens to the force between charged and uncharged objects as the distance between them decreases? Support your answer with observations from the investigation.

(f) List the properties of electric charges you have identified in this investigation.

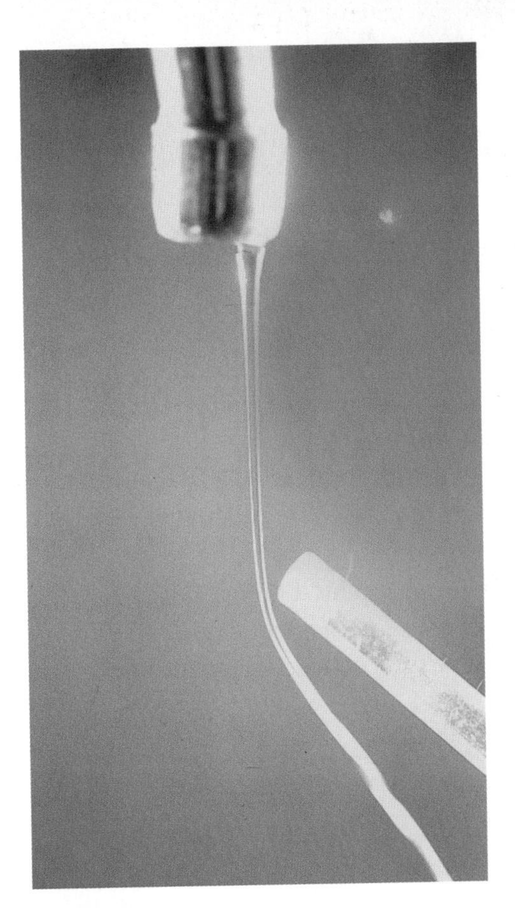

Figure 2
Charged ebonite rod and neutral water

Making Connections

1. From your observations, what can you say about what happens when charged objects are brought close to uncharged objects?
2. Describe a test that you could perform to determine whether an object is charged or uncharged.
3. How does an anti-static dryer sheet work (**Figure 3**)? Explain it in terms of electric charges.

Figure 3

4. Why do people get shocks after they drag their feet on a rug? Explain it in terms of electric charges.

Reflecting

5. Make a chart of three different situations in which you have experienced the effects of static electricity. Beside each, write down what pairs of materials you think might be responsible for producing static electricity.

Challenge

2 What information from this section could you use in your electrical safety pamphlet?

3 Use this investigation to develop questions and answers for your electric game show.

Work the Web

Visit www.nelson.science.com and follow the links from *Science 9: Concepts and Connections*, 3.1, to carry out online static electricity investigations.

Electricity and Matter

Everything around you, including the storm clouds in the sky (**Figure 1**), contains electric charges. When you rub a balloon against your hair, electric charges are not created, they simply move. Objects become charged when electrons move from one object to another.

Rubbing two objects together, like your feet on a nylon rug, can cause a buildup of charge at the point of contact. The electric charge stays where the rubbing happened on the charged objects (the charge remains "static"). The study of static electric charge is called **electrostatics.**

Figure 1

The rubbing of dust and water particles in a storm cloud can produce awesome examples of static electricity.

Types of Charges

Most objects are neutral; that is, they are electrically uncharged because they have an equal number of positive and negative charges. However, when two different neutral objects are rubbed together, one object becomes negatively charged while the other becomes positively charged. For example, when a comb is rubbed with wool, the comb gets a **negative charge** while the wool gets a **positive charge** (**Figure 2**). Positively and negatively charged objects both attract most neutral objects.

The Law of Electric Charges

Remembering the law of electric charges is easy if you remember playing with magnets. What happens when the opposite poles of two magnets are brought close to each other? when the same poles are brought together?

The **law of electric charges** states that like charges repel each other, and unlike charges attract each other.

To determine whether an object is charged and whether that charge is positive or negative, you must see the object being repelled by an object with a known charge. Charged objects will attract both neutral objects and objects with unlike charges. However, charged objects will repel only objects with like charges (**Figure 3**).

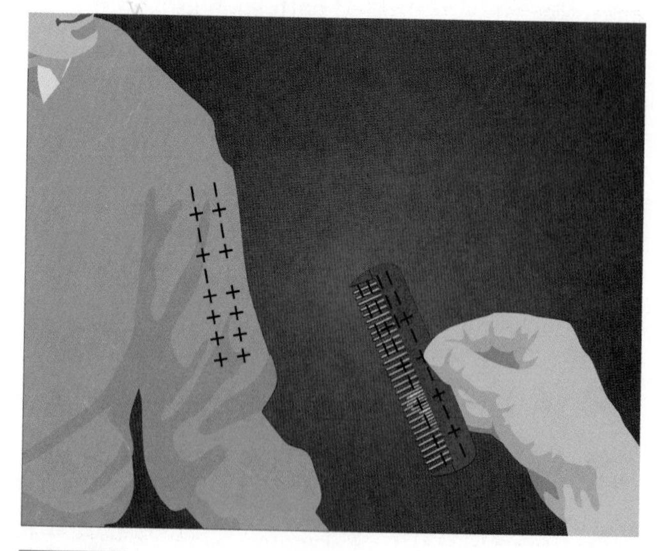

Figure 2

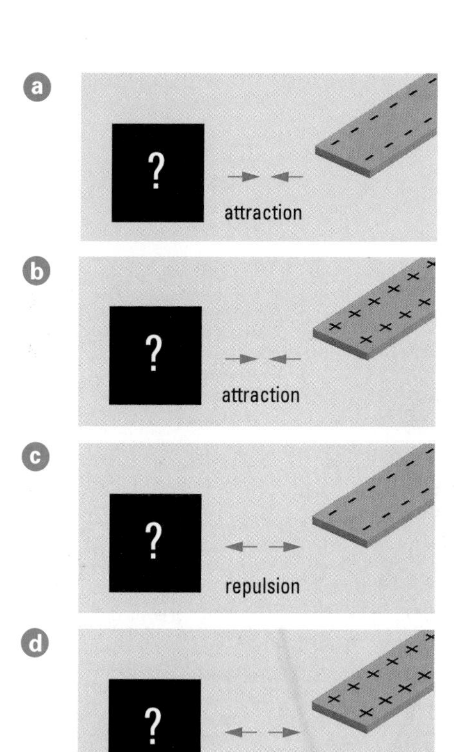

Figure 3

(a) The object being tested could be neutral or positive.
(b) The object being tested could be neutral or negative.
(c) The object being tested must be negative.
(d) The object being tested must be positive.

Static Electricity and Current Electricity

As you may have experienced, **static electricity** is electricity that does not move. However, there is another form of electricity: current electricity. **Current electricity** (**Figure 4**) is the movement of an electric charge from a source of electrical energy along a controlled path such as a wire (**Table 1**).

Figure 4

Table 1

Electricity type	Definition	Characteristic
static	stationary buildup of electric charge on a substance	electrons do not move along a path
current	electric charge that moves from a source of electrical energy along a controlled path in an electric circuit	electrons move along a path

Understanding Concepts

1. Are all objects electrically charged? Explain.
2. What is static electricity? Describe a situation involving static electricity to explain your answer.
3. Explain in detail how you could demonstrate the law of electric charges.
4. What is the difference between static electricity and current electricity? Include examples of each in your answer.

Making Connections

5. You are normally uncharged or neutral. You drag your feet on a rug and shock an unsuspecting friend by touching his or her ear.
   - **(a)** Why does your friend feel a shock?
   - **(b)** What can you do to stop the buildup of static electric charge as you drag your feet on the rug?

Reflecting

6. What happens when ebonite is rubbed with fur and Lucite is rubbed with silk or polyester? Explain your answer in terms of the law of electric charges.

Challenge

2 Make recommendations to prevent accidents with both static and current electricity for your electrical safety pamphlet.

3 Using the information in this section, develop questions and answers for your electric game show.

3.3 Investigation

INQUIRY SKILLS
○ Questioning ○ Planning ● Analyzing
○ Hypothesizing ● Conducting ● Concluding
● Predicting ● Recording ● Communicating

Charging by Contact

Why do wool sweaters crackle with static electricity when we put them on or take them off? Electrons are moving from one place to another. The process is called **charging by contact.** Charging by contact is used for many purposes, including painting cars.

An object can be either positively or negatively charged, depending on whether it loses or gains electrons. In this investigation, you will use a pith ball apparatus (**Figure 1**) to determine the kind of charge transferred from one object to another.

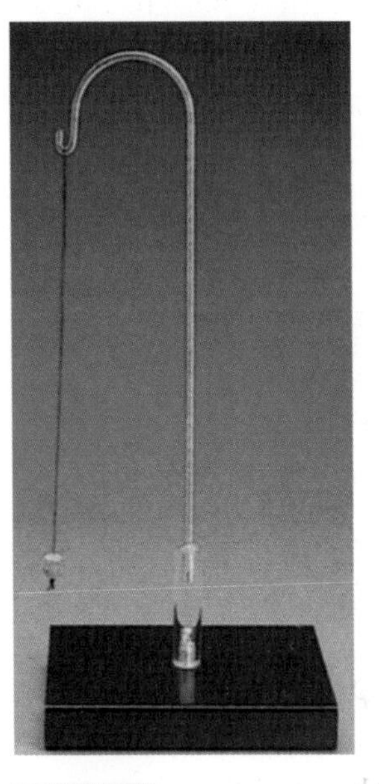

Figure 1

Question

How can we determine the kind of charge transferred to a neutral object when a charged object touches it?

If you are allergic to fur, do not perform this investigation.

Prediction

(a) Use the law of electric charges to predict what kind of charge is transferred from each object to a pith ball apparatus.

Materials

- pith ball apparatus
- ebonite rod
- fur
- Lucite rod
- silk or polyester material scraps

Procedure

1 Prepare a table (**Table 1**) to record your (C6) observations in this investigation. You can record them as drawings instead.

Table 1

Action	Observations
charged ebonite rod brought close to pith ball but does not touch	
charged ebonite rod touches the pith ball and is then brought close	
charged Lucite rod brought close to the pith ball but does not touch	
charged Lucite rod touches the pith ball and is then brought close	

2 Charge the ebonite rod by rubbing it with the fur sample. (Ebonite gains electrons and becomes negatively charged when rubbed with fur.) Bring it close to the pith ball, but do not touch it.

(E1) **(b)** Record your observations in your table.

3 Repeat step 2, except this time touch the rod to the pith ball. What happens when the charged ebonite rod is brought close to the pith ball after touching it?

E1 (c) Record your observations (**Figure 2**).

4 Discharge the pith ball by touching it with your hand.

5 Repeat steps 2 and 3 using the Lucite rod and charging it using the silk or polyester scraps. (Lucite loses electrons and becomes positively charged when rubbed with silk or polyester.)

E1 (d) Record your observations.

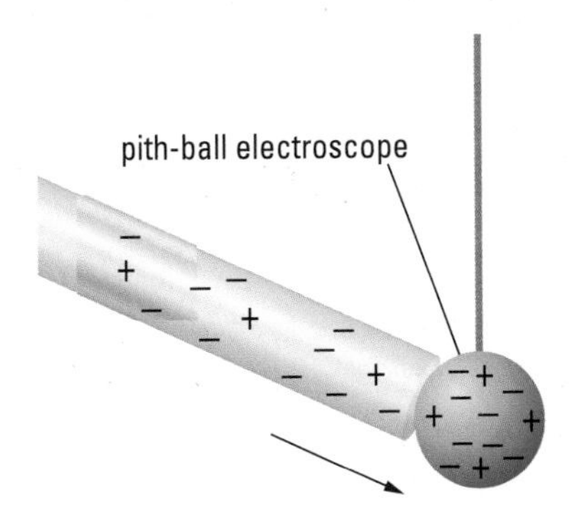

Figure 2

Analysis and Conclusion

(e) Write a statement about the transfer of electric F1 charge when the pith ball is touched by

  (i) a negatively charged object

  (ii) a positively charged object

  Explain your answers.

(f) Why was the pith ball repelled by the charged ebonite rod after being touched? Your explanation should include a drawing.

(g) Why was the pith ball repelled by the charged Lucite rod after being touched? Your explanation should include a drawing.

(h) Write a statement to compare the movement F1 of electric charges in the ebonite rod and the Lucite rod.

Work the Web

Visit www.nelson.science.com and follow the links from *Science 9: Concepts and Connections*, 3.3, to find out what safety precautions you should take if you are caught outside in a lightning storm.

Understanding Concepts

1. If your hand were negatively charged and you touched a neutral doorknob, in which direction would negative charges move? Explain your answer.

Exploring

C2 **2.** Use two pith ball apparatuses and predict, observe, E1 and explain what happens when they are brought F1 close together if

  (a) they are charged alike

  (b) they have opposite charges

  (c) one is charged and the other is not

Use simple diagrams to explain what you observe.

Reflecting

3. How could charging by contact be used to paint a car?

Challenge

3 Using the information obtained in this investigation, create two questions and answers for your electric game show.

What Is Electric Current?

When electric charges move from one place to another, we say they make an **electric current**. Two terms used when working with electric current are **voltage** and **resistance** (**Table 1**).

Table 1

Term	Definition	Units
electric current	electric charges that are moving from one place to another	amperes (A)
voltage	the force that moves electric charges in a circuit	volts (V)
resistance	the ability that tries to stop or slow the electric charges in a circuit (light bulb in **Figure 1**)	ohms (Ω)

Comparing electric current to the flow of water in a pipe (**Figure 1**) can help you to understand these electrical terms.

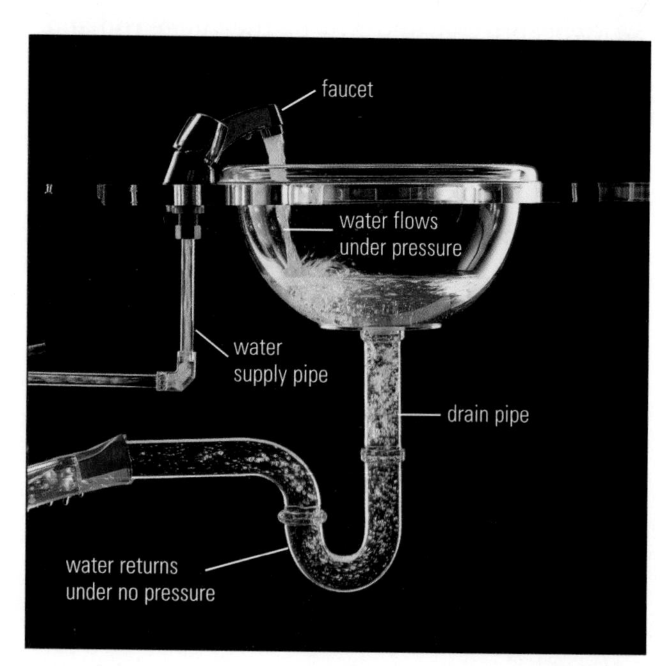

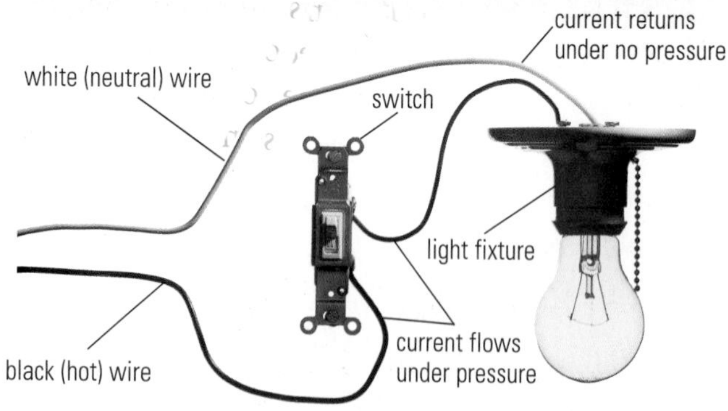

Figure 1

Comparing Electric Current to the Flow of Water

Water flows through your household plumbing system much like electrical current flows through the wires in your house. You can compare the pressure of the water in your plumbing system to the voltage in the electrical wires. You can compare the amount of water flowing in the pipes of the plumbing system to the electric current of the circuit.

When comparing water flowing in a house to electric current flowing, note the following points:

- The black wire in an electrical circuit is the **hot** or **live wire**, and it carries the voltage into the house. Compare this to the pipes carrying water.
- The white wire in an electrical circuit is the **neutral wire** and allows the current to leave the house after it has been used. Compare this to the drain that carries used water to the sewer.

Another way to compare electric current with the flow of water is to look at a water wheel.

Electric Current and the Water Wheel

Think of a water wheel (**Figure 2**) and compare the water to an electric current. When an electric current enters a house from a power line, it is like the water at the top of the water wheel. The electric current, just like the water at the top of the wheel, has the ability to do work. The water turns the water wheel, while electric current operates electrical devices in your home.

The water in **Figure 2** is like an electric current in that it can be used repeatedly. Electric current returns to a generating station to be recharged and used again. Water is pumped back up to the top of the water wheel and turns the wheel once again.

Compare **Figure 3** to the water wheel. The electric current turns an electric motor as it moves from the negative terminal to the positive terminal.

The energy provided by the electric current in a dry cell is used up as it turns the electric motor. The dry cell will eventually become completely discharged and will not be able to turn the motor. As you know, you can recharge some types of cells and batteries (**Figure 4**).

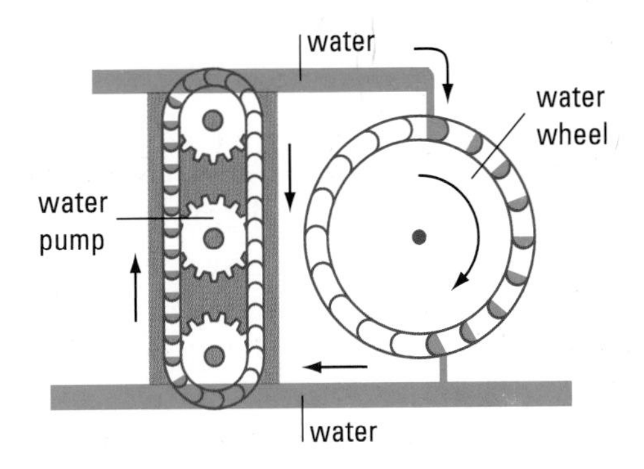

Figure 2

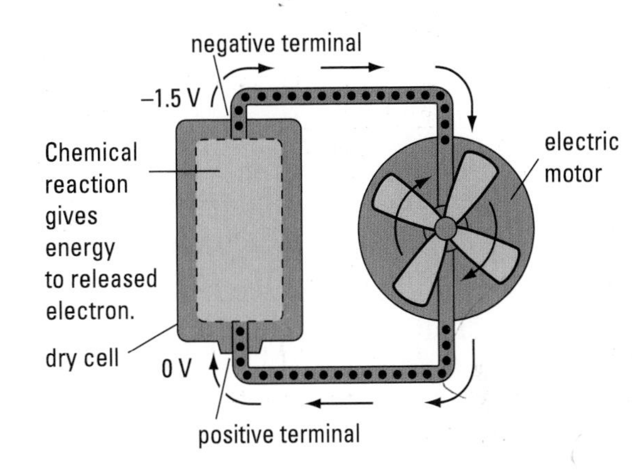

Figure 3
The energy released by the chemical reaction in the dry cell turns the electric motor.

Figure 4

Electric Current Ratings

How much current do electrical devices need to work? Slightly less than one ampere (1 A) of current flows through a 100-W light bulb in a lamp connected to a 120-V circuit. The metric SI unit used to measure current is the ampere. The symbol for the ampere, or amp, is A. **Table 2** lists the electric current required to operate some common electrical loads. The **electrical load** is the device that converts electrical energy into the needed form of energy, such as heat or light. An example of an electrical load is a light bulb.

Table 2 **The Electric Current Ratings of Some Common Electrical Loads**

Electrical Device	Electric Current (A)
electronic wristwatch	0.000 13
electronic calculator	0.002
electric clock	0.16
light bulb (100 W)	0.833
television (colour)	4.1
electric drill	4.5
vacuum cleaner	6.5
stove element	6.8
oven element	11.4
toaster	13.6
water heater element	27.3
car starter motor (V-8)	500.0

Human Response to Electric Shock

Does it take a large amount of electric current to kill a person? No! A very small amount of current is deadly—that is one reason why it is important to read the safety warnings in the operating manuals of any kind of electrical device or equipment.

Our bodies use electricity to make muscles contract. Nerve cells produce about 0.08 V of electricity to do this. When higher voltages stimulate those muscles, the contractions are even stronger.

If someone's body touches a source of electricity, an electric current may flow through his or her body. If the current is large enough, the muscles in that part of the body contract and remain contracted until the electric current stops. The chart in **Figure 5** shows the effects produced by different amounts of electric current.

Most people do not feel anything if the current is below 0.001 A, but they feel a tingling sensation if the current is about 0.002 A. When the electric current is about 0.016 A, muscles contract. This level of electric current is sometimes called the "let-go threshold," because if the current is higher than that value, the person cannot let go of the object giving the electric shock. If the electric current is flowing from one hand to the other through the chest, the breathing muscles may become paralyzed, and the victim will suffocate unless the current stops.

If a current of 0.050 A or more passes through the chest, the heart muscles stop their regular pumping action and merely flutter. This fluttering of the heart muscles is called ventricular fibrillation. The only way to restart the heart is to use a controlled electric shock. You have probably seen on television or in the movies when a doctor uses defibrillator paddles on someone whose heart has stopped.

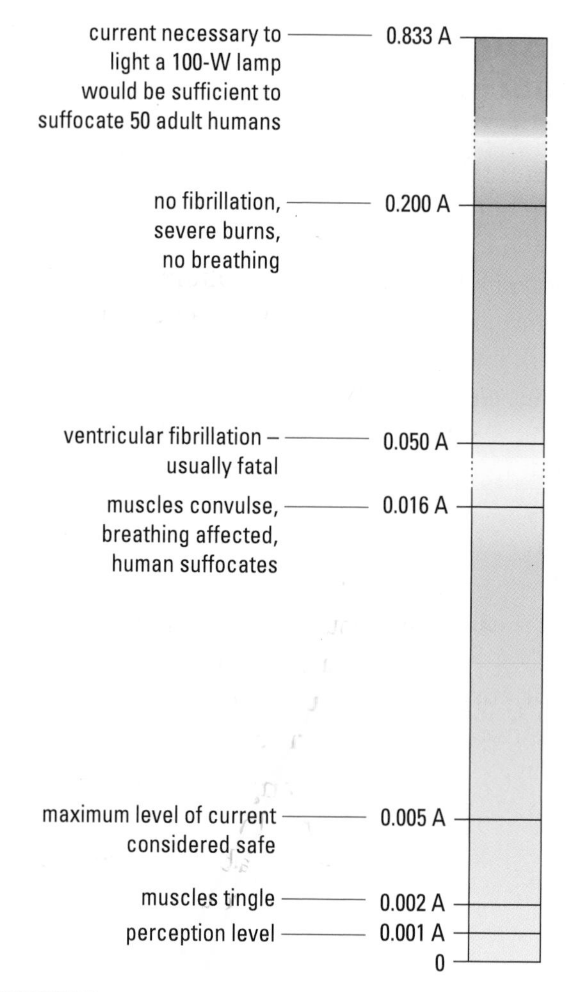

Figure 5

Since the body's nervous system operates on electric impulses, electricity can be used to ease pain as well as help healing. The ancient Egyptians applied this principle when they used the electric charges produced by torpedo fish to ease pain. The Romans evidently used electric eels to treat headaches and arthritis. Today, doctors use electrical nerve stimulation to treat certain types of pain. A 9-V battery supplies a weak electric current across the patient's skin into nerve cells under the skin's surface. The current stimulates the body's natural ability to fight pain.

Do not attempt to shock others using 9V batteries. Electric shocks should be treated by a doctor.

Try This Activity Comparing Water Flow to the Flow of Electric Current

Comparing water flow in a plumbing system to the flow of electric current in an electric circuit is best explained through a demonstration. In this activity, you will need a partner, a retort stand, two funnels of different sizes, cotton, water, two plastic containers, and a sketchpad. You must observe, record, and communicate to your partner how a plumbing system is like electric current flowing in an electric circuit.

(a) Explain how you would use the water flow to demonstrate a low-voltage electric current. How would you demonstrate a high-voltage electric current?

(b) How would you change the funnel to represent an electrical wire that has a high-current capacity—the ability to carry more electric current? How would you represent a low-current capacity?

(c) How could you use the funnel to represent an electrical line that has decreased resistance to the flow of electric current? How would you represent a line that has increased resistance to the flow of current?

Work the Web

How would you help a victim of electric shock? Start your search by visiting www.nelson.science.com and following the links from *Science 9: Concepts and Connections*, 3.4

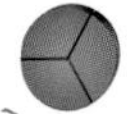

Challenge

2 What did you learn about electric shock? Add the information to your electricity safety pamphlet.

3 Develop questions that compare the flow of water in pipes to the flow of electric current for your electric game show.

Understanding Concepts

1. What are the two main parts of an electric current?
2. Explain, in your own words, how electric current, voltage, and resistance are like water flowing in a plumbed water system.
3. How much electric current is needed to kill a human being? Explain.
4. List three devices from **Table 2** that you use in your home. Order the devices from largest to smallest in terms of their electric current ratings.

Exploring

5. The use of electricity in a house can be made (N) "safer" in many ways. Research the topic of home (O) electrical safety, using the library or Internet resources, and prepare an oral report for the class.

The Electric Circuit

When you turn on a lamp, you are using a simple electric circuit. Devices such as flashlights, table saws, and computers all rely on the basic parts that make up an **electric circuit**: an energy source, connecting wires, an electrical device (load), and a switch. An electric current must always have a load to use the energy. Without a load, a connecting wire creates a "short" circuit for the electric current. Short circuits can cause sparks or fires.

In this activity, you will construct and test various simple electric circuits. The basic parts of the circuit are outlined in **Figure 1**.

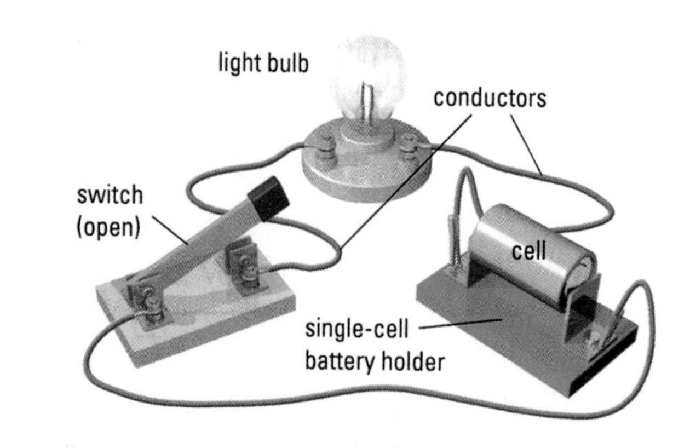

Figure 1

Question

How are the parts connected to create a working electric circuit?

Did You Know?

We often use the term "battery" instead of "cell." A battery is actually a combination of two or more of cells.

Materials

- dry cell or battery (with holder)
- switch
- connecting wires with alligator clips
- light bulb (with holder)
- small electric motor (**Figure 3**)
- LEDs (light-emitting diodes) (**Figure 2**)

Never connect both battery terminals together without another device between them. This connection would cause a short circuit. Have your teacher check your circuit before you close the switch.

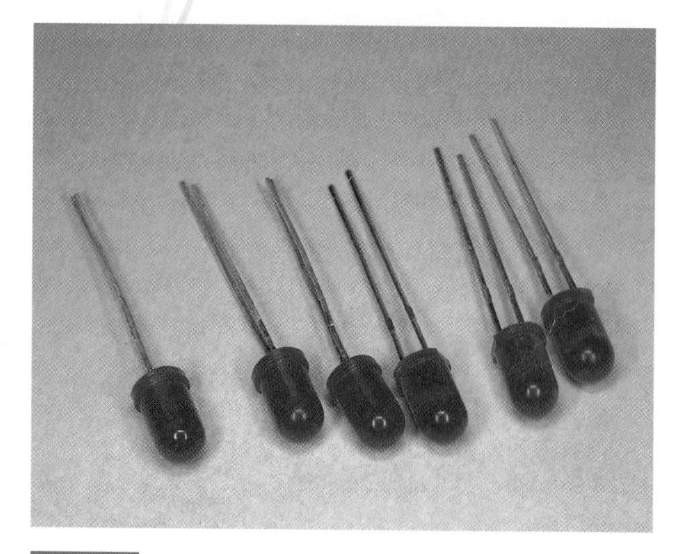

Figure 2

Procedure

1 Create a table (**Table 1**) to a record your observations and diagrams. (C6)

(a) Sketch the battery and label the positive and negative terminals. What is the voltage rating of the battery? (D3)

 SKILLS HANDBOOK: (C6) Preparing Observation Tables (D3) Drawing and Constructing Circuits

Table 1

Load used	Diagram of circuit
light bulb	
electric motor	
LED (light-emitting diode)	

2 Begin with the light bulb. Using the cell (and (D3) holder), the switch, the connecting wires, and the light bulb, build a circuit. Always build the circuit with the switch open.

(b) Sketch the circuit. To test the circuit, close the (D3) switch. If it does not work, rearrange it. Draw your circuit changes until the light bulb works.

3 The next electric circuit will use an electric motor (**Figure 3**). Use the materials from step 2 and experiment with the circuit until the motor works.

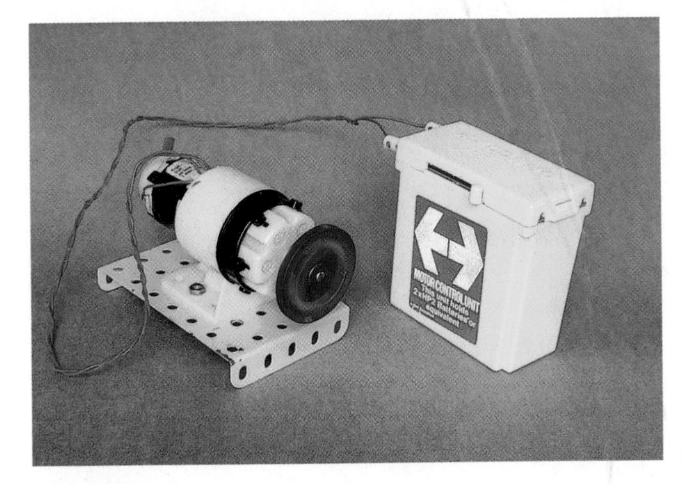

Figure 3

(D3) **(c)** Sketch each attempt in your table.

4 The last electric circuit will use a light-emitting diode (LED). Note that one of the wires on the LED is longer than the other (**Figure 2**). The LED will only work if it is connected properly.

(d) Note in your table which pole of the battery is (E1) positive and which is negative. Sketch the steps carried out to produce a working electric circuit.

5 Analyze your observations by answering the following questions:

(e) What is the function of the (F1)
- dry cell
- switch
- light bulb
- electric motor
- LED
- wires

Record your answers in a table like **Table 2**.

Table 2

Electric parts	Function
(i) dry cell	
(ii) switch	
(iii) light bulb	
(iv) electric motor	
(v) LED	
(vi) wires	

(f) Which of the four parts of the electric circuit can be removed while allowing the circuit to continue working? Why is it usually included in a circuit?

(g) List three different ways of turning the electrical devices (light bulb, electric motor, LED) on or off.

(h) Would the circuit operate differently if

- the connections on the switch were reversed?
- the switch were connected on the other side of the electrical device?

If you are not sure, try making the changes. Explain your answers. You can include diagrams in your answer.

(i) What effect would reversing the connecting wires have on the

- light bulb
- electric motor
- LED

Explain your answer. Test your answers, if possible.

6 An important part of any electric circuit is the energy source. Different energy sources are used for different purposes. Copy **Table 3** into your notebook and identify a common use for each type of battery in **Figure 4**.

Table 3

Battery type	Use
AA	
AAA	
D	
9V	
lantern battery	
small circular battery	

7 **Figure 5** is an example of how an electric circuit can be used on a bicycle. The small, specially shaped batteries on these bicycles can be recharged each night.

(j) What are the four main parts of the electric circuit in this device?

Figure 4

Figure 5

Understanding Concepts

1. How do you know whether you have a complete electric circuit?
2. Explain, in your own words, the steps to follow to build electric circuits.
3. What was the voltage rating of the battery used in this activity? Do all batteries have the same voltage? Explain.
4. Draw a working electric circuit with the four main parts labelled: (D3)
   (i) source of electrical energy
   (ii) electrical load (device)
   (iii) electric circuit control device (switch)
   (iv) connectors

Making Connections

5. Identify and describe three kinds of switches
   (a) in your home
   (b) on electrical devices you use every day
   (c) in a car
   Suggest reasons why different switches are used in different situations.
6. Think about toys that need batteries to work properly.
   (a) What problems could you have with the electric parts of these toys?
   (b) How do you know when the batteries need to be replaced? Be specific.
   (c) Have batteries become more reliable? Explain.

Exploring

7. Predict what could happen if more than one electrical device were hooked up in a row in your electric circuit. Try it, and comment on your prediction. (C2)

Work the Web

Visit www.nelson.science.com and follow the links to *Science 9: Concepts and Connections*, 3.5, to carry out online investigations of electric circuits.

Challenge

1 Record the four main components needed for an electric circuit to do something. Brainstorm ideas for an original working electric circuit.

The Intel Pentium microprocessor contains thousands of electric circuits on a silicon chip the size of a fingernail and can process hundred of millions of instructions per second.

Series and Parallel Circuits

An electric circuit can be represented in several ways. The four main parts of the electric circuit (i.e., energy source, load, control device or switch, connecting wires) are shown as pictures in **Figures 1** and **2**. Another way to represent a circuit is using a circuit diagram (**Figure 3**). **Circuit diagrams** are drawings of circuits using symbols. See section D3 of the Skills Handbook for the circuit symbols.

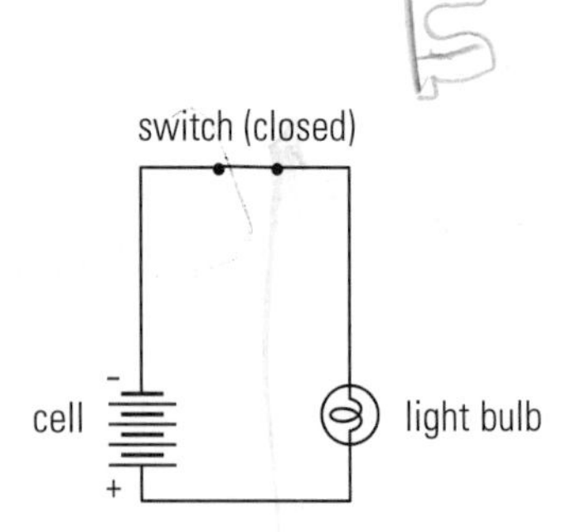

Figure 3

Circuit diagrams use symbols to represent the parts of an electric circuit.

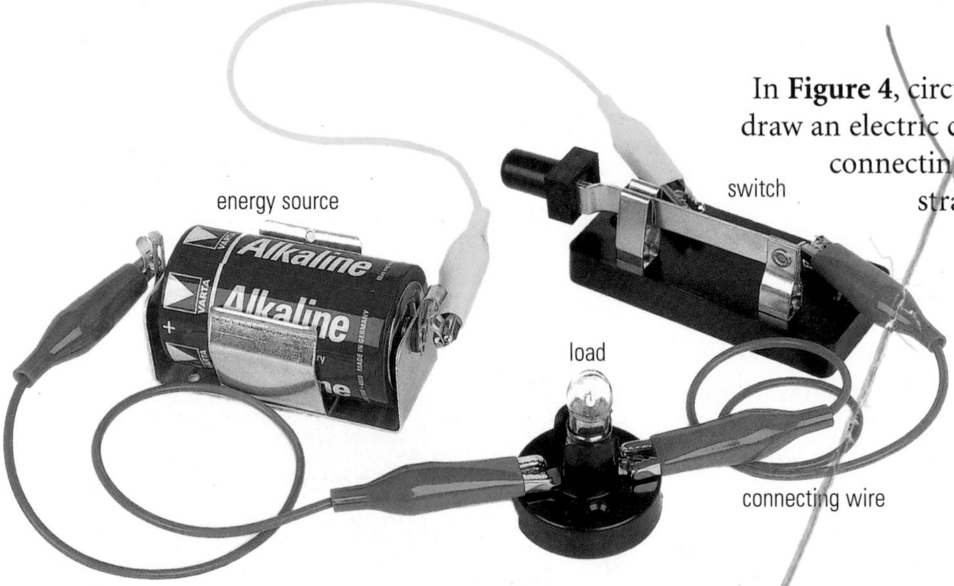

connecting wire
energy source
switch
load
connecting wire
connecting wire

Figure 1

A series circuit has only one path for electricity to follow.

In **Figure 4**, circuit symbols are used to draw an electric circuit. Notice that the connecting wires are drawn as straight lines, with right-angled corners, to make it easier to understand. Since only one path is shown, this circuit is wired in series.

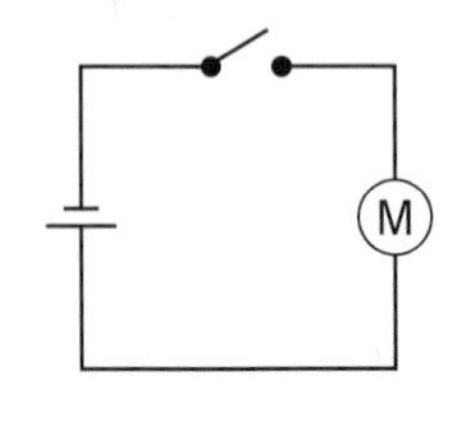

Figure 4

An electric circuit that has one dry cell, a switch, and two motors wired in parallel would look like **Figure 5**.

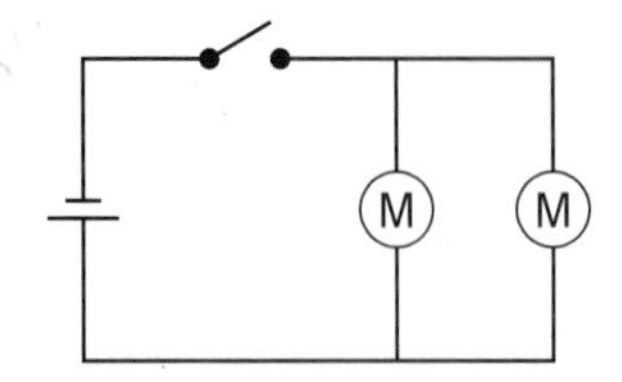

Figure 5

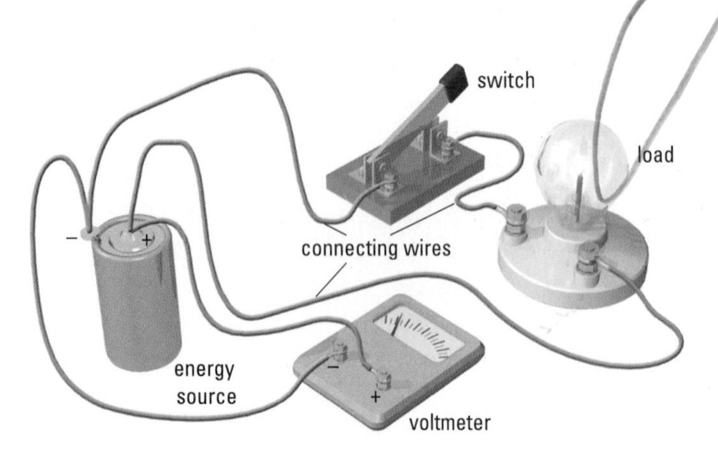

Figure 2

A parallel circuit has more than one path for electricity to follow.

Electric circuits can be wired in **series** or in **parallel**. It is easy to tell which is which when you look at three things (**Table 1**):

1. The number of paths the electricity has to follow.
2. Whether the electrical energy is shared.
3. How the devices are controlled (on or off).

Series Circuits

Figure 1 shows a simple series circuit. The electricity has only one path to follow. Many simple electrical devices, such as flashlights, are arranged in series. So are most battery-operated devices, such as toys and cordless tools. Since there is only one electrical device (load) in the circuit it has to be wired in series. A circuit can be wired in parallel only if there is more than one electrical device. However, a circuit with more than one electrical device can also be wired in series.

Parallel Circuits

Figure 2 shows an example of a parallel circuit. One path leads to the voltmeter, while the other path supplies electricity to the switch and light bulb, making two paths. Houses are wired in parallel because of the advantages this type of circuit has over series circuits. The most important advantage is that devices can be switched on or off individually. For example, you do not have to turn on the stereo to switch on the kitchen light.

Challenge

1 Draw circuit diagrams of your brainstormed ideas to make an electric circuit do something. Keep these diagrams until you make a final choice.

Table 1

Circuit type	Number of paths for electric current to follow	Electrical energy shared/not shared	Devices on/off
series	one path	electrical energy shared	all electrical devices must be either on or off at the same time
parallel	more than one path	electrical energy is not shared	each electrical device can be on or off within the circuit

Understanding Concepts

1. Explain, in your own words, the differences between a series and a parallel circuit.
2. What is a circuit diagram?
3. (D3) Draw circuit diagrams for the following circuits:
   (a) two cells, one open switch, and a light wired in series
   (b) one cell, two lights, and a clock wired in parallel
   (c) a series circuit of your own design
   (d) a parallel circuit of your own design

Making Connections

4. Why are homes wired in parallel? Give two reasons.
5. Why are battery-operated toys wired in series?

Exploring

6. (D3) Produce a circuit diagram for one room of your home. Remember that your home is wired in parallel.

Reflecting

7. Why are circuit diagrams used rather than drawings of the actual parts of a circuit?

INQUIRY SKILLS

- ○ Questioning
- ○ Hypothesizing
- ● Predicting
- ○ Planning
- ● Conducting
- ● Recording
- ● Analyzing
- ● Concluding
- ● Communicating

Building Parallel and Series Circuits

How are series circuits different from parallel circuits (**Figures 1** and **2**)? During this activity, you will build and compare series and parallel circuits.

Question

How can you determine the characteristics of parallel and series circuits?

Prediction

(a) Write one or more predictions about what you think will happen in series and parallel circuits. (C2)

Materials

- 4 light bulbs (with holders) or LEDs
- 2 D dry cells (with holders)
- 2 switches
- connecting wires with alligator clips

Procedure

1 Prepare a table to record your observations in this activity (**Table 1**). (C6)

2 Construct the circuits in **Figure 3**. (D3)

(b) Observe and record the brightness of the bulbs in the series and the parallel circuits. (E1)

3 Repeat step 2, but use two dry cells (connected in series) instead of one.

(c) Observe and record the brightness of the bulbs in the two circuits. (E1)

4 Remove one bulb from each circuit.

(d) Record your observations. (E1)

5 Build a simple series circuit using one switch, one dry cell, and one bulb. Add one bulb at a time in series to the circuit until all four light bulbs have been added. (D3)

(e) Observe and record what happened to the brightness of the bulbs. (E1)

6 Repeat step 5 but add light bulbs until four bulbs are connected in parallel. (D3)

(f) Record the effect on the brightness of the bulbs. (E1)

Figure 1
A series circuit

Figure 2
A parallel circuit

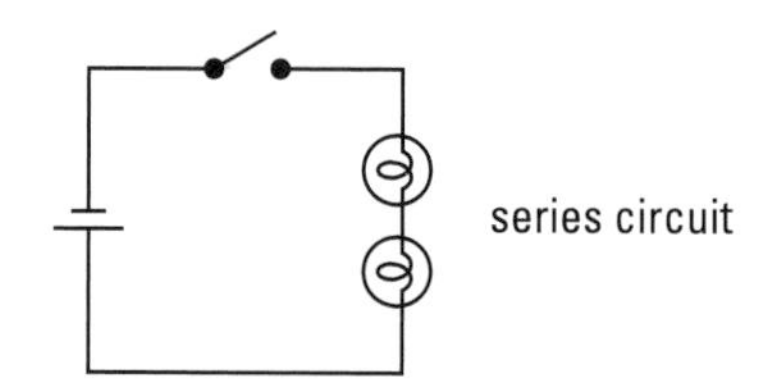

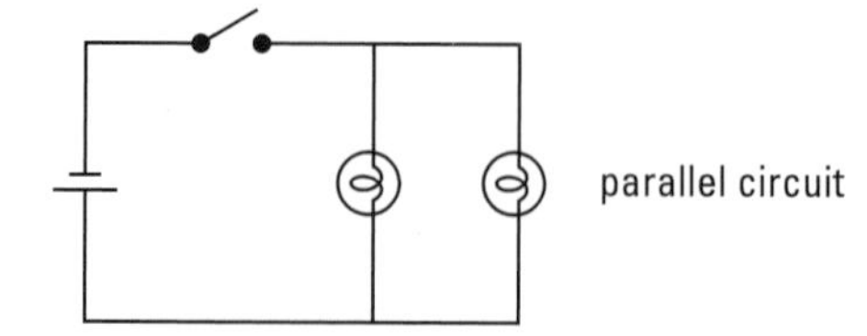

Figure 3

Table 1

Action	Observations
comparison of bulb brightness in series/ parallel circuits (two bulbs, one dry cell)	
comparison of bulb brightness in series/ parallel circuits (two bulbs, two dry cells)	
removal of one bulb in the series circuit	
removal of one bulb in the parallel circuit	

Analysis and Conclusion

(g) Compare the brightness of the two bulbs in the series and parallel circuits (one dry cell). In which circuit were the lights brighter? (F1)

(h) A light bulb was removed from the series circuit in step 4. Why did all the lights go out? (F1)

(i) A light bulb was removed from the parallel circuit in step 4. Why did the other lights stay on? (F1)

(j) In step 5, light bulbs were added to the series circuit. Explain what happened to the brightness of the light bulbs as light bulbs were added. (F1)

(k) In step 6, light bulbs were added to the parallel circuit. Explain what happened to the brightness of the light bulbs as light bulbs were added. (F1)

(l) In a series circuit, is it possible to switch a single device on or off? Why?

(m) In a parallel circuit, is it possible to switch a single device on or off? Why?

Work the Web

Want to test your knowledge about parallel and series circuits? Take an online quiz! Go to www.nelson.science.com and follow the links from *Science 9: Concepts and Connections*, 3.7.

Understanding Concepts

1. Why would a string of lights, such as the ones used to decorate trees, be wired in series?
2. What is an advantage of wiring electric circuits in parallel? What is a disadvantage?
3. What is an advantage of wiring electric circuits in series? What is a disadvantage?
4. Draw a circuit diagram of a cell connected in parallel with two lights and a motor, with each electrical device controlled by a switch. (D3)

Making Connections

5. Light bulbs represent electrical resistance in a circuit.
   - **(a)** What happens when electrical resistance is added in a series circuit?
   - **(b)** What happens when electrical resistance is added in a parallel circuit?
6. **(a)** Suppose 15 light bulbs were connected in series, and one bulb burned out. How could you find the defective bulb?
   - **(b)** How could you identify one defective bulb if the 15 bulbs are connected in parallel? Explain.

Challenge

1 Decide whether you will wire your electric circuit in parallel or series. What other wiring considerations must you make before building your device?

2 In your electrical safety pamphlet, outline the advantages and disadvantages of using series and parallel circuits.

Measuring Voltage Drop and Current

To check whether a circuit is working, electricians must measure voltage and current. Because electricity is not visible, they use special instruments. Measurements of electric current and voltage can only be made using voltmeters and ammeters.

A **voltmeter** (**Figure 1**) measures the voltage drop in a circuit. The SI unit for voltage is the volt (V).

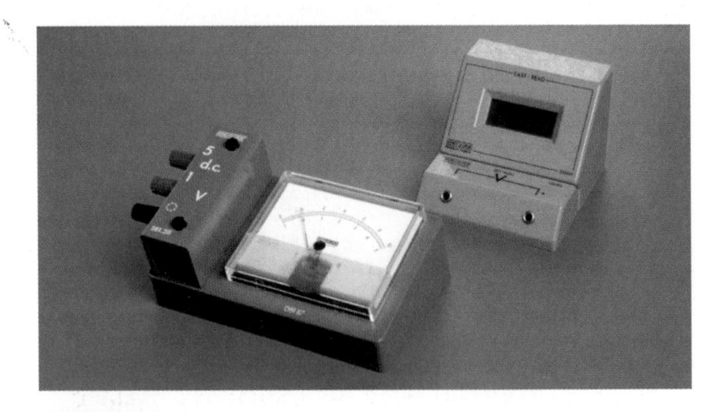

Figure 1

Analog and digital voltmeters

Figure 2 shows a typical ammeter. An **ammeter** measures the amount of electric current flowing past a point in a circuit. Depending on the electrical appliance, the current varies. The SI unit for current is the ampere (A).

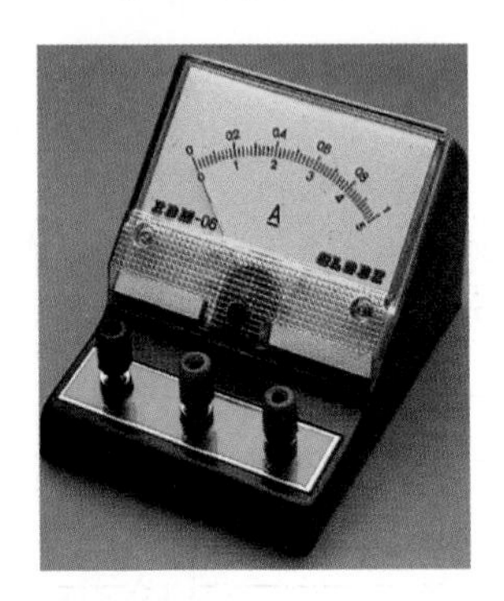

Figure 2

An ammeter indicates the amount of electric current at a point in the circuit.

How Does a Voltmeter Work?

A voltmeter measures the voltage drop between two points of an electric circuit. The voltage drop is measured by connecting the voltmeter in parallel between two points in a circuit. The positive terminal of the voltmeter is connected to the positive side of the circuit and the negative terminal to the negative side. The voltmeter indicates, in volts, the amount of energy lost or gained between the two points of the circuit. The voltmeter could be digital (giving a digital readout) or analog (using a needle moving across a scale).

Figure 3 shows two voltmeters being used to measure the voltage at two locations in an electric circuit—the voltage rise across a cell which is the source of energy, and the voltage drop across a light bulb which is using the energy. Note that the readings are both 1.5 V. The voltage within the circuit is the same. This shows that the circuit has a strong energy source and is working properly.

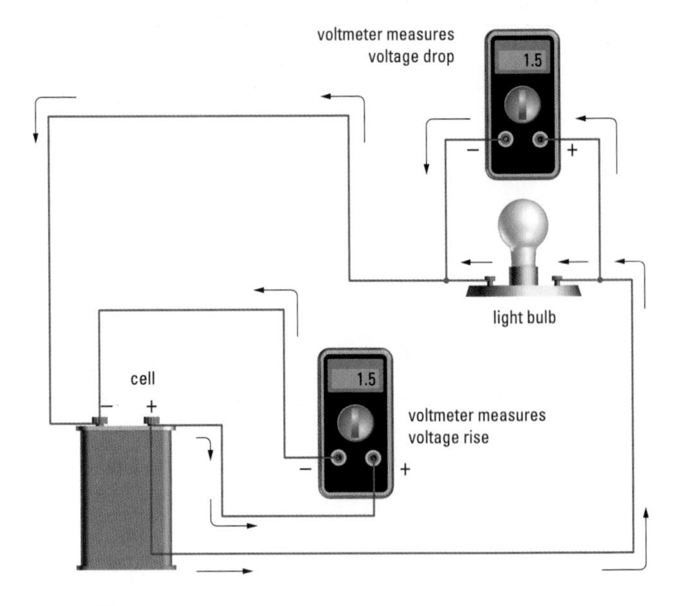

Figure 3

A voltmeter is connected in parallel in an electric circuit and measures the voltage drop (or rise) between two points in the circuit.

How Does an Ammeter Work?

An ammeter measures the amount of current flowing past a point in a circuit. An ammeter is not used in the same way as a voltmeter. An ammeter is connected to an electric circuit in series. In **Figure 4** an ammeter is connected to a circuit that includes a dry cell and a light bulb. The positive terminal of the ammeter is connected to the positive terminal of the battery. The reading in this circuit is 1.5 A. Reading digital or analogue ammeters is very similar to reading digital or analog voltmeters.

Some typical current ratings for some electrical appliances are shown in **Table 1.** Notice that devices that produce heat, such as a toaster, use a great deal of electric current to work.

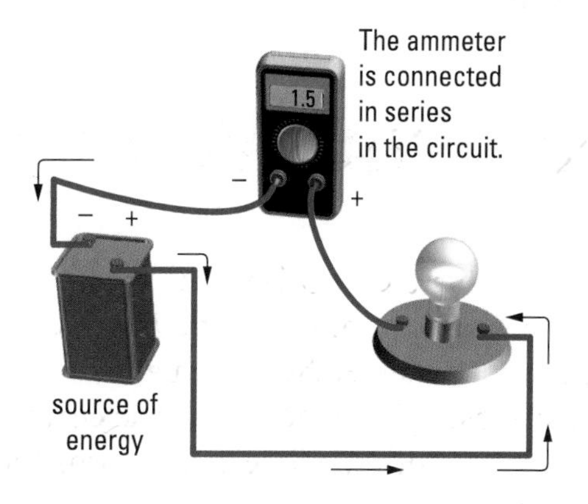

Figure 4

An ammeter, connected in series, measures the amount of electric current flowing in an electric circuit.

Table 1

Appliance	Current rating (A) (with 120-V supply)
toaster	8.3
fluorescent lighting	0.5
power drill	2.5
television	3.0
vacuum	4.3
microwave oven	5.0
iron	10.0
clothes dryer	40.0
stove	*

*Stoves operate at 240 V and draw currents in the range of 25 A to 40 A.

Challenge

3 Develop questions and answers for your electric game show that focus on the proper use of voltmeters and ammeters.

Understanding Concepts

1. How are a voltmeter and an ammeter arranged differently in a circuit?
2. (D3) Draw a circuit diagram of a dry cell, an open switch, a light bulb, and a voltmeter connected across the light bulb.
3. (D3) Draw a circuit diagram of the same electrical circuit in question 1, but insert an ammeter after the light bulb.

Making Connections

4. When would an electrician use a voltmeter? When would an electrician use an ammeter?
5. Homes in Canada are supplied with electricity at 120 V. What might a reading lower than 120 V mean?

Exploring

6. Electrical devices are labelled with their current ratings. Find the current ratings of four different devices in your home. Make sure that the devices are unplugged first. Which device requires the most current?

3.9 Investigation

INQUIRY SKILLS

- ○ Questioning
- ○ Hypothesizing
- ● Predicting
- ○ Planning
- ● Conducting
- ● Recording
- ● Analyzing
- ● Conducting
- ● Communicating

Comparing Current and Voltage Drop in an Electric Circuit

We can measure electric current and the voltage drop of a circuit using an ammeter and a voltmeter. In this investigation you will design, build, and measure electric current and voltage drop for series and parallel circuits.

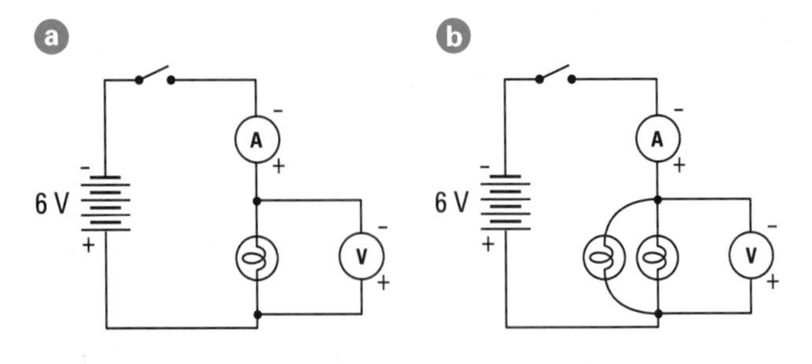

Figure 1

Question

How can you accurately measure voltage drop and electric current in series and parallel circuits?

Prediction

(C2) **(a)** Write a prediction for this investigation.

Materials

- 3 light bulbs with holders
- 4 dry cells with holders or one 6-V lantern battery
- 1 voltmeter
- 1 ammeter
- 12 connecting wires with alligator clips
- 1 switch

Make sure you have correctly connected the terminals to avoid a short circuit. Have your teacher check your circuit before closing the switch.

Procedure

Part 1: Electrical Loads in a Parallel Circuit

1 Construct the circuit shown in **Figure 1a.** The ammeter will remain in the same position for the entire investigation. Ask your teacher to inspect your circuit before continuing. (D3) (D4)

(b) Draw a table for your observations like **Table 1**. (C6)

Table 1 Light Readings

Number of bulbs	Voltage (V) across battery	Voltage (V) across bulb	Current (A)
1			
2			
3			

2 Connect the voltmeter across the bulb as in **Figure 1a**. Close the switch after connecting the voltmeter. (D4)

(c) Record the voltmeter and ammeter readings. Open the switch. (E2)

3 Connect the voltmeter across the battery. Close the switch. (D4)

(d) Record the voltmeter reading. Open the switch. (E2)

4 Connect the second bulb to the circuit, as shown in **Figure 1b**. Repeat steps 2 and 3.

5 Connect the third bulb in parallel to the circuit. Repeat steps 2 and 3.

6 Remove one bulb from its socket, then close the switch.

(E1) (E2) **(e)** Record your observations in your notebook.

7 Open the switch, and replace the bulb in the socket.

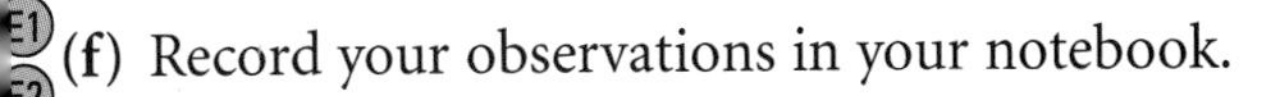

E1 E2 **(f)** Record your observations in your notebook.

8 Repeat steps 6 and 7 for each of the other two bulbs.

Part 2: Electrical Loads in a Series Circuit

9 D3 D4 Construct the circuit shown in **Figure 2a**. Ask your teacher to inspect your circuit before continuing.

C6 **(g)** Draw another table for your observations.

10 D4 Connect the voltmeter across the bulb. Close the switch.

(h) E2 Record the voltmeter and ammeter readings. Open the switch.

11 D4 Connect the voltmeter across the battery. Close the switch.

(i) E2 Record the voltmeter reading. Open the switch.

12 D4 Connect the voltmeter across the first bulb, and connect a second bulb in series with the first bulb, as shown in **Figure 2b**. Close the switch.

(j) E2 Record your voltmeter and ammeter readings in your table. Open the switch.

13 Connect a third light bulb in series with the other two bulbs. Close the switch.

E1 E2 **(k)** Record your observations.

14 With the switch closed, remove the first light bulb from its socket. Open the switch.

E1 E2 **(l)** Record your observations.

15 Close the switch and repeat step 14 for each of the other two bulbs. Open the switch.

Analysis and Conclusion

(m) Analyze your observations by answering the following questions for Parts 1 and 2:

- **(i)** F1 How does the voltage drop across the dry cell compare with the voltage drop across each of the three bulbs?
- **(ii)** F1 What happens to the brightness of the light from the bulbs as a new bulb is added?
- **(iii)** F1 What happens when you unscrew one of the bulbs?
- **(iv)** How many paths for current flow does each circuit have?

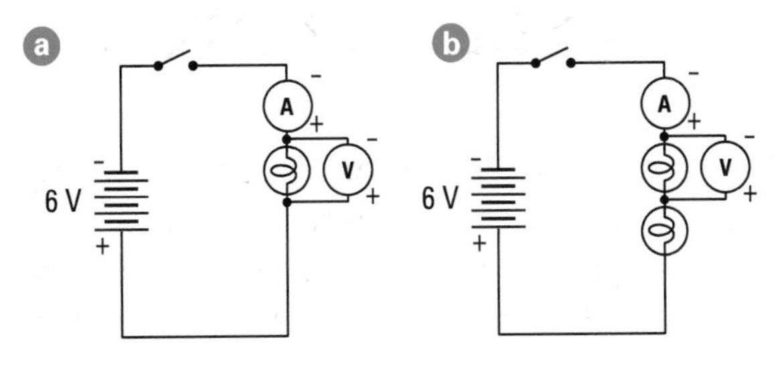

Figure 2

Understanding Concepts

1. Is electric current shared in a series circuit? Explain your answer by referring to your observations.
2. Which part of the sign in **Figure 3** is wired in series? Which part is wired in parallel?

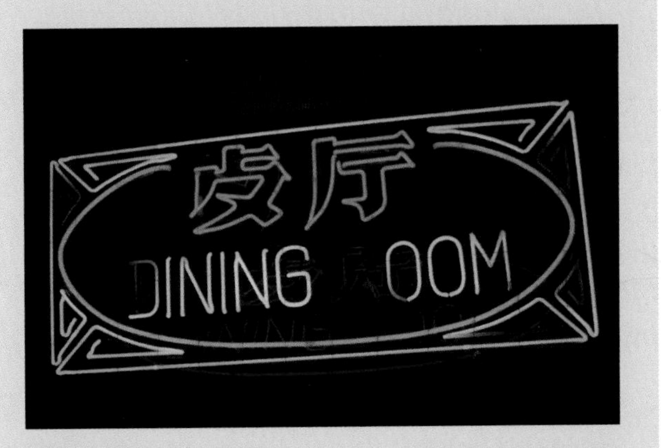

Figure 3

Making Connections

3. Explain, in your own words, how to measure current in a circuit.
4. Explain, in your own words, how to measure the voltage drop across an electrical load in a circuit.

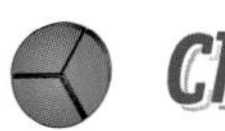

Challenge

1 Will your circuit be wired in series, parallel, or both? Will you need a voltmeter or ammeter?

Electrical Resistance

Why do we use electric circuits? Think for a moment about the ways that you use electricity in a typical day. Each time you use electrical energy, it faces electrical resistance as it moves through a circuit and meets various loads. As you learned in section 3.4, resistance is the force that tries to stop or slow the electric charge in the circuit.

When electricity meets resistance, the electrical energy is changed into one of four types of energy:

1. heat energy
2. light energy
3. sound energy
4. mechanical energy (energy of motion)

Thousands of different kinds of loads exist, and each has been designed for a specific source of electrical energy. A digital watch or flashlight is a load that uses a particular size and type of dry cell. An electric hair dryer (**Figure 1**) has been designed so that the heating element (coil) inside the hair dryer is the correct size to dry hair quickly and safely when plugged into a 120-V outlet.

We use electrical energy to power many different electrical devices at home, school, and work. **Figure 2** shows four examples of how we use electricity at home and at work.

(**a**) Resistance changes electrical energy into one of four forms of energy. What are they?

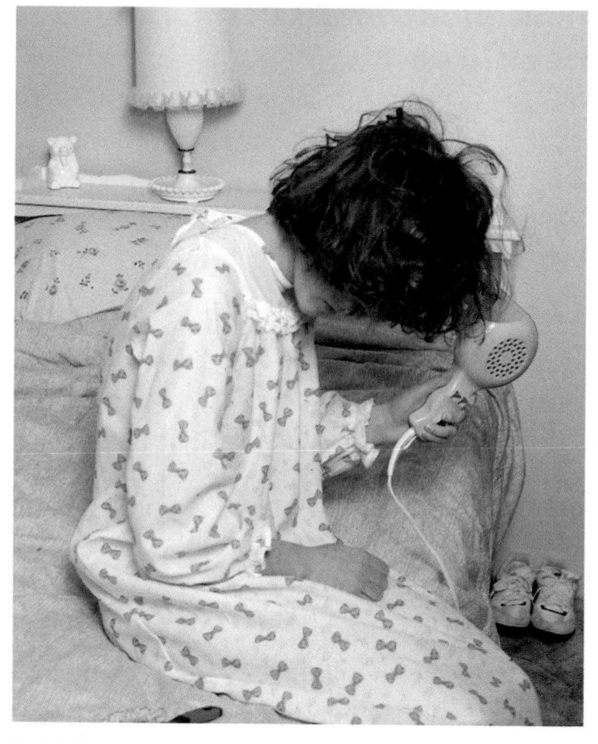

Figure 1

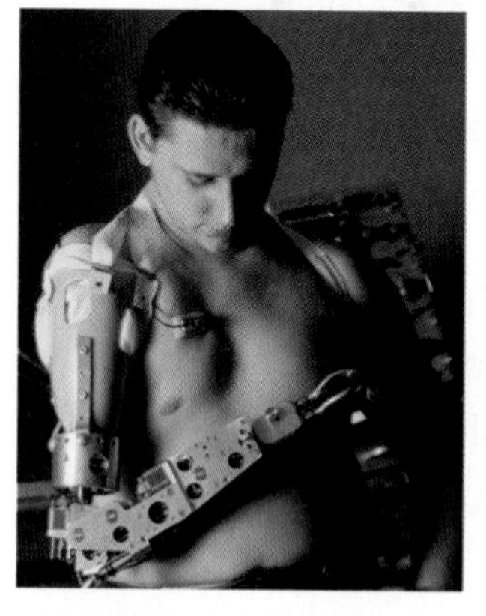

Figure 2

Conductors

Electrons move easily through a conductor. A **conductor** is a substance that carries electrical energy without much resistance, so electrons lose little energy. Copper is a good electrical conductor because the electrons lose very little energy. In other materials, such as the tungsten filaments in a light bulb, the electrons lose much more of their energy. As a result, the electrons' energy is converted into heat energy, and the filament becomes so hot that it glows brightly (**Figure 3**).

In 1827, the German scientist Georg Ohm (**Figure 4**) discovered a relationship, now called Ohm's Law (**Figure 5**), between current and voltage drop.

Ohm's Law states that *the voltage drop between two points on a conductor is proportional (directly related) to the electric current flowing through a conductor.* Stated another way, if the voltage in a circuit remains constant, the current will decrease as the resistance increases.

(b) Define voltage drop. What unit is used to measure voltage drop?

(c) Define current. What unit is used to measure current?

The resistance of the conductor or load relates the voltage drop (V) to the current. This very simple law is used to calculate the resistance of the load when designing many different electrical devices. If the device is supposed to transform electrical energy into another sort of energy, it is designed so that the resistance will be as high as possible.

Solving Problems Using Ohm's Law

The procedure below shows a method for solving problems using the formula for Ohm's Law.

Sample Problem

What is the voltage drop across the filament in a 100 watt light bulb if the resistance of the filament is 150 Ω and a current of 0.80 A is flowing through it?

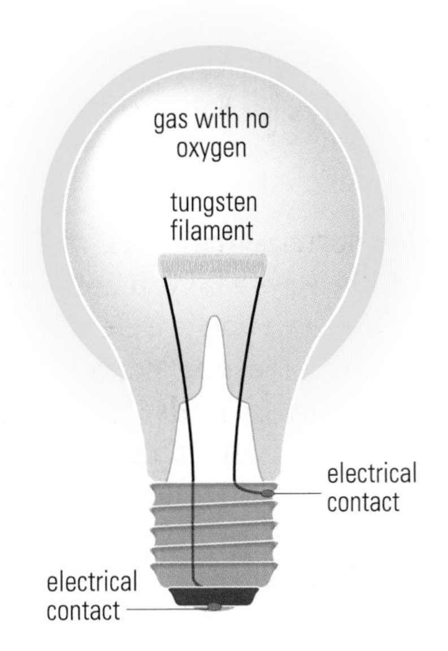

Figure 3

A tungsten filament has a high resistance to electric current. About 90% of energy given off by the tungsten filament is heat energy.

Figure 4

Georg Ohm, 1789–1854

$$V = I \times R$$

or

$$\text{Voltage drop} = \text{Current} \times \text{Resistance}$$

Figure 5

Ohm's Law

Step 1: Read the problem carefully and record all given quantities. Use correct symbols and units.
$I = 0.80\ \text{A}$
$R = 150\ \Omega$
$V = ?\ \text{V}$

Step 2: Write the formula for the problem.
$V = I \times R$

Step 3: Calculate the answer. Record the answer with the correct unit.
$V = 0.80\ \text{A} \times 150\ \Omega$
$V = 120\ \text{V}$

The voltage drop is 120 V.

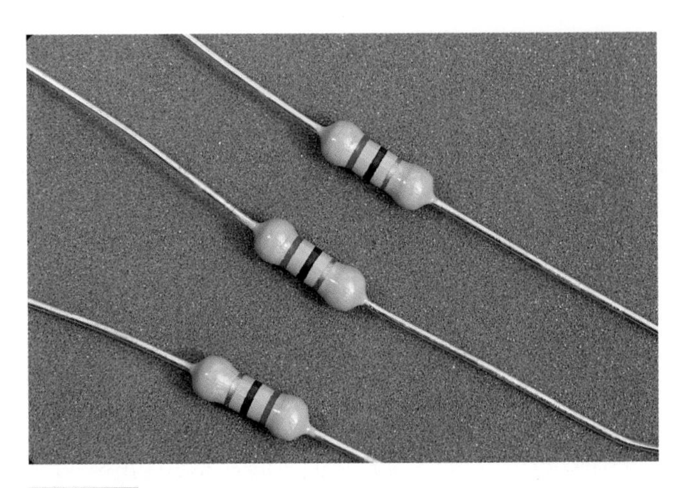

Figure 6

Resistance at Work

Think of a waterwheel that turns when a stream of water is directed at it. The wheel represents resistance in a circuit, while the water represents the flow of electrons. Sometimes, the voltage available from a wall outlet is too high for an electrical device. The high voltage will cause too much current to flow and will "burn" the device. Many electrical devices have resistors that control how much current flows into them (**Figure 6**). Resistors, which come in different shapes and sizes, slow electrical current. The pattern of coloured bands on the resistors is a code that shows the amount of resistance offered by the resistor. The symbol for electrical resistance is *R*, and the SI unit is the ohm (Ω).

(d) How do you think resistance affects electrical devices?

What Is the Pattern?

Table 1 lists the resistance of some electrical loads and the current and voltage drop required to operate them.

Table 1 Resistance of Some Electrical Loads

Ohm's Law V = I × R

Load	Voltage drop (V)	Current (A)	Resistance (Ω)
light bulb (60 W)	120	0.50	240
coffee grinder	120	1.20	100
food dehydrator	120	4.60	26
toaster oven	120	14.0	8.6

(e) Look at **Table 1**. Notice that the voltage drop (F1) does not change. What pattern do you notice for current and resistance?

Look at the circuit shown in **Figure 7**. The resistor in the diagram has a resistance of 10 Ω. The ammeter reads 2 A. The voltage drop is constant.

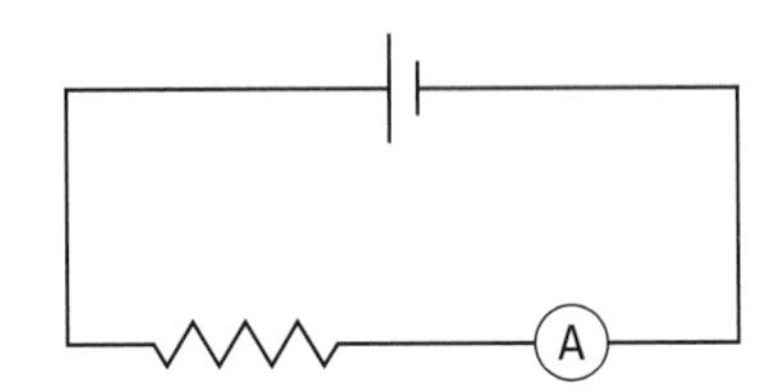

Figure 7

(f) Will the ammeter reading go up or down if a 5-Ω resistor replaces the 10-Ω resistor?

(g) Complete the sentence. If the voltage drop is constant and the resistance increases, the current ____________. If the voltage drop is constant and the resistance decreases, the current ____________.

Try This Activity Are You Resistant?

Most multimeters can measure electrical resistance as well as voltage and current. Set the meter to its resistance scale, and hold one tip of each lead in each hand.

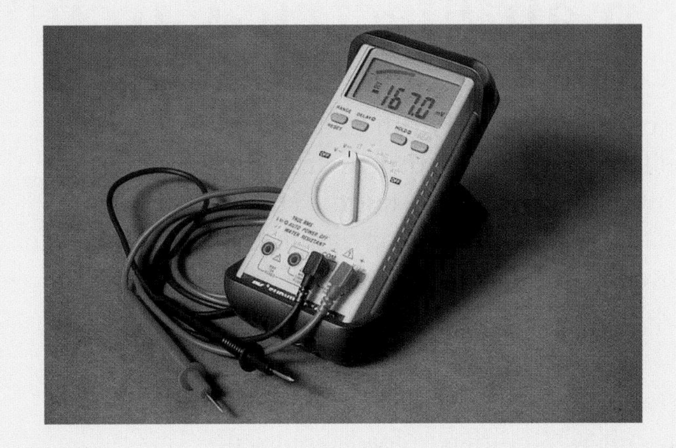

- First, record the resistance of your body with your hands dry. (E2)
- Repeat the measurement after wetting your hands.

(B) What safety issues does this activity identify?

Understanding Concepts

1. Why are electric circuits important?
2. What is electrical resistance?
3. Define, in your own words, a conductor and a resistor.
4. Does the wire in the electrical cord of an electric kettle have a higher or lower resistance than the heating element inside the kettle? Explain your answer.
5. Which one of the following cicuits will have the greatest current? Explain why.

A

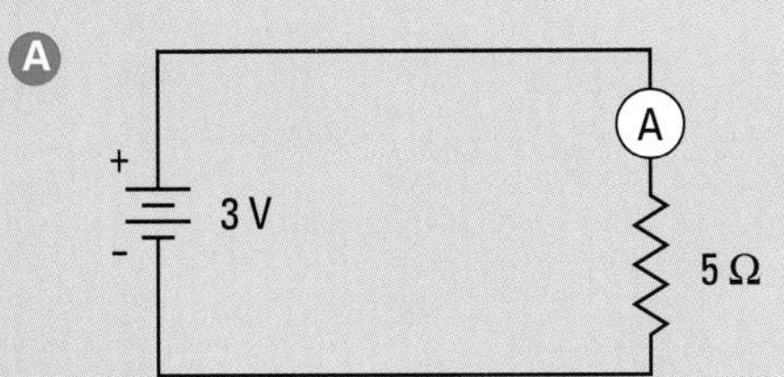

B

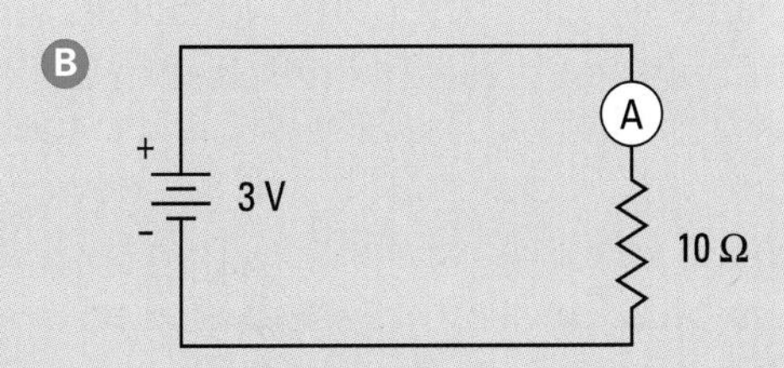

Making Connections

6. Outline four examples each of devices that transform electrical energy into
   (i) heat
   (ii) light
   (iii) sound
   (iv) mechanical energy

Exploring

7. Using several different magazines, make a collage of photos that show the different ways that electrical resistance is used.

Challenge

1 Will you use resistors as part of your electric circuit? Why or why not?

2 Add key points about resistance to your electrical safety pamphlet.

3 Create questions and answers about electrical resistance for your electric game show.

Work the Web

Follow the links through *Science 9: Concepts and Connections*, 3.10, at www.nelson.science.com, and discover the effects on a current of changing voltage and resistance in a simple circuit. Write statements that describe how the current changes when the voltage is changed and when the resistance is changed.

3.11 Investigation

INQUIRY SKILLS
- ○ Questioning
- ○ Hypothesizing
- ● Predicting
- ○ Planning
- ● Conducting
- ● Recording
- ● Analyzing
- ● Concluding
- ● Communicating

The Effect of Resistance on an Electric Circuit

How does the load in an electric circuit affect the voltage drop within that circuit? In this investigation, you will test the effect of different loads on the voltage drop within an electric circuit.

Question

How will voltage readings change in an electric circuit when different loads are used?

Prediction

(a) Predict what will happen to the voltage as you add light bulbs and resistors in series in a circuit. (C2)

Materials

- 1 voltmeter
- 4 D dry cells with holders
- 6 connecting wires
- 3 light bulbs with holders
- 3 resistors (20 Ω, 60 Ω, 100 Ω)
- 1 switch

Procedure

Part 1: Light Bulbs as Electrical Loads

1 Use a table, such as **Table 1**, to record your observations in Parts 1 and 2. (C6)

Table 1

Load	Voltage drop (V) around load
1 light	
2 lights	
3 lights	
resistor 1: 20 Ω	
resistor 2: 60 Ω	
resistor 3: 100 Ω	

2 Construct a circuit based on the circuit drawing in **Figure 1**. Do not close the switch. Ask your teacher to inspect your circuit before continuing. (D3)

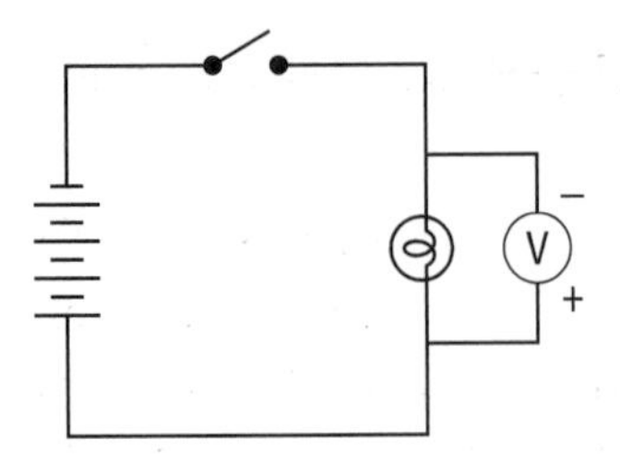

Figure 1

Make sure you connect the terminals correctly to avoid a short circuit. Have your teacher check your circuit before closing the switch.

3 Make sure that the voltmeter (**Figure 2**) is connected according to the negative (–) and positive (+) terminals of the dry cell. The voltage drop (V) will be determined across the light bulb. Close the switch. (D4)

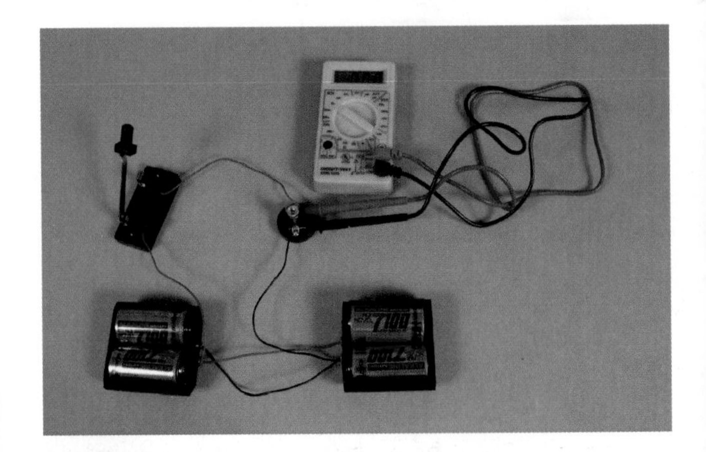

Figure 2

(b) Record your observations in your table.

4 Open the switch. Add another light bulb in series to your circuit. Connect the voltmeter around the two light bulbs. Close the switch.

(c) Record the voltage reading in your table.

5 Add a third light bulb in series. Connect the voltmeter around the three light bulbs. Close the switch.

(d) Record the voltage reading in your table.

Part 2: Resistors as Electrical Loads

6 With the switch open, remove the light bulbs from the circuit and replace them with the lowest-rated resistor. Do not close the switch. Connect the voltmeter around the resistor. Close the switch.

(e) Record the voltage reading in the table.

7 Open the switch and replace the lowest-rated resistor with the next lowest-rated resistor. Close the switch.

(f) Record the voltage reading in the table.

8 Open the switch again and replace the resistor with the highest-rated resistor. Close the switch again.

(g) Record the voltage reading in the table.

Part 3: Resistors in A Parallel Circuit

In Part 2 you measured the voltage drop across a single light bulb and across two light bulbs in a series circuit. Based on your previous investigations, what will happen if you connected the two light bulbs in parallel?

9 Predict the difference in the overall resistance in a parallel circuit as compared to a series circuit. Will it be higher or lower in a parallel circuit?

10 Construct a circuit like the one you made in Step 4 (two bulbs in series). Connect an ammeter in series and connect the voltmeter around the two light bulbs.

(h) Record the voltage and the current.

11 Now construct a circuit with the two light bulbs in parallel. Connect the ammeter in series and the voltmeter around the two bulbs (like **Figure 1b** on page 148).

(i) Record the voltage and the current.

Analysis and Conclusion

12 Analyze your observations by answering the following questions.

(j) What effect did the addition of light bulbs have on the brightness of each bulb?

(k) Use the data in **Table 1** to create a graph in which the *x*-axis represents the number of lights and the *y*-axis represents the voltage drop.

(l) What happened to the voltage readings when you added lights to the series circuit? Was that what you thought would happen? Explain.

(m) Use the data in **Table 1** to create a graph in which the *x*-axis represents the size of resistor and the *y*-axis represents the voltage drop.

(n) Compare your results for the voltage readings of the three resistors in the series circuit. What statement could you make regarding increasing resistance and voltage values in a series circuit?

(o) Why was the switch kept open while you built the series circuit?

(p) Compare the graph lines from the light bulb and the resistor data. Which has the greater voltage drop (V): resistors or light bulbs?

(q) Based on the voltage and current readings in the two circuits in Part 3, do you think the overall resistance in the parallel circuit is greater than the overall resistance in the series circuit? Explain why you think so. (Hint: How did the voltages compare in the two circuits? What happened to the current?)

Understanding Concepts

1. Explain, in your own words, the effect of resistance on voltage drop (V) ratings around a specific load in a simple electric circuit.
2. If the resistance of an electrical load were greater than the voltage available, what would happen to the circuit?
3. How was the voltmeter connected in this investigation? (D4)

Making Connections

4. Most electrical appliances have resistors (**Figure 3**) inside their circuits. Based on how resistors affected the voltage drop in the circuits in this investigation, explain why circuits have resistors.

Figure 3

5. The wires that carried the electric current in this investigation were quite thin. Resistance decreases as the thickness of the wire increases. Why are transmission wires (**Figure 4**) thin?

Figure 4

6. Based on your observations in this investigation, what kind of circuit (series or parallel) would be most suitable to connect all the light bulbs on each of the trees in **Figure 5**?

Figure 5

7. Suppose you had to replace a burned-out bulb in a flashlight. Describe at least two ways that you could determine the correct voltage rating for the bulb.

Temperature can affect the resistance of a conductor. In most conditions, as the temperature increases, the resistance also increases.

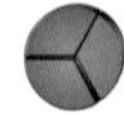

Challenge

1 How could you use resistors when making an electric circuit perform a specific task?

3 What new questions and answers from this investigation could you add to your electric game show?

Home Security Systems Installer

Nikki Abbott

Nikki Abbott has worked in the home security business for the past six years in Sault Ste. Marie, Ontario. She began dispatching crews when alarms were received and then became the technical administrator of the alarms division for ATS (Alarm and Telecom Services Inc.) Nikki needed both interest and training to enter this field.

Nikki holds an Electrician Technician Diploma and a Network Engineering Technician Certificate. After graduating from high school, she completed her training at Sault College of Applied Arts and Technology.

When asked about how electricity relates to her job, Nikki emphasized that "the fundamentals of electricity are definitely an asset to this position!"

As the technical administrator for the alarms division, Nikki has a wide range of responsibilities. She schedules the three technicians in her department and carries out technical research on devices that are used in the field. She programs the monitoring station where the alarms are received and most of the alarm panels.

Nikki gets a lot of satisfaction from working in a field that provides security services to homes and businesses throughout Sault Ste. Marie. She finds the job very challenging since technology is always changing.

"Wireless technology is definitely becoming more popular. Systems that combine intruder alarms and video surveillance are in demand. The average home alarm system can do many things for you, such as linking up your lighting systems and even your entertainment systems. This is called 'X-10' technology, and it's on the rise."

Making Connections

1. (N) (O) Home security is just one area of specialization in electricity. Use newspaper job advertisements, the Internet, or the library to learn about other careers that apply the principles of electricity.
2. (N) (O) Research the technology that is used for electrically monitored security in homes and businesses. What are the basic pieces of equipment needed to provide security surveillance?

Work the Web

Visit www.nelson.science.com and follow the links from *Science 9: Concepts and Connections*, 3.12. From the site information, list questions that you have about a career as a home security systems installer.

Challenge

1 Will your electric circuit be a type of home security device?

The Safe Use of Electricity

You may have never thought about where the electricity in your home comes from, but your home is actually part of one big electric circuit! Let's follow the flow of electricity from the power lines to your home (**Figure 1**) and look at the safety devices that are used.

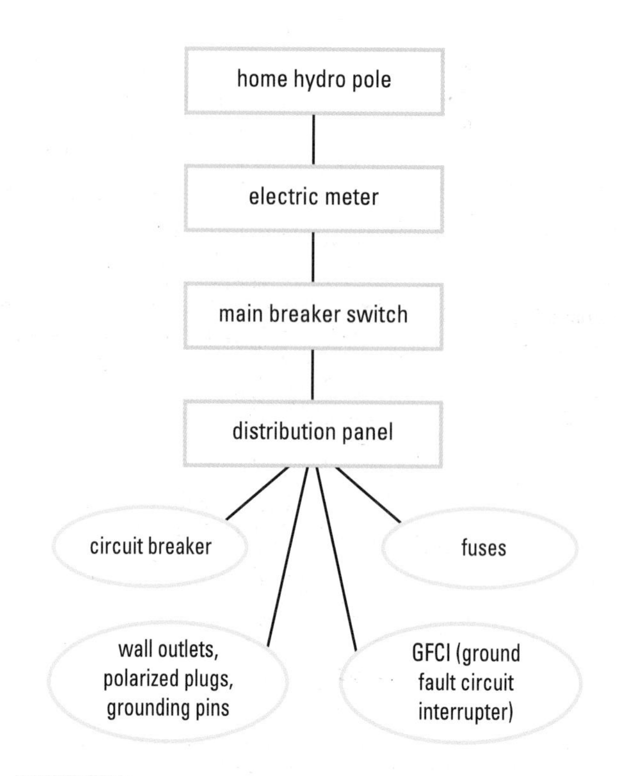

Figure 1

How Electricity Flows to You

Electricity flows into your neighbourhood through high-voltage wires. Transformers safely reduce the amount of power entering homes to 120 V per wire in North America. These wires enter a home (**Figure 2**) from the hydro pole. Three wires connect this pole to the power lines. Two live or hot wires carry power into your home and one neutral wire carries the power back to the hydro pole to complete the circuit.

Figure 2

Electric Meter

The three wires pass from the hydro pole into a protective tube that travels down to the electric meter (**Figure 3**). The **electric meter** measures the total amount of electrical energy used in the building. The three wires then continue through the wall into the main breaker switch.

Figure 3

Main Breaker Switch

The only way to shut off electricity to all the circuits in your home is to shut off the **main breaker switch** (**Figure 4**). For safety, an electrician always turns off the main breaker switch when working on a circuit. The main breaker switch is controlled by a circuit breaker.

Distribution Panel

From the main breaker switch, the three wires pass into a metal box called the **distribution panel** (**Figure 4**). The distribution panel is the place where all circuit breakers (or fuses in some older homes) connect to each circuit. Circuit breakers are reset or fuses are replaced in the distribution panel. Another wire, the **ground**

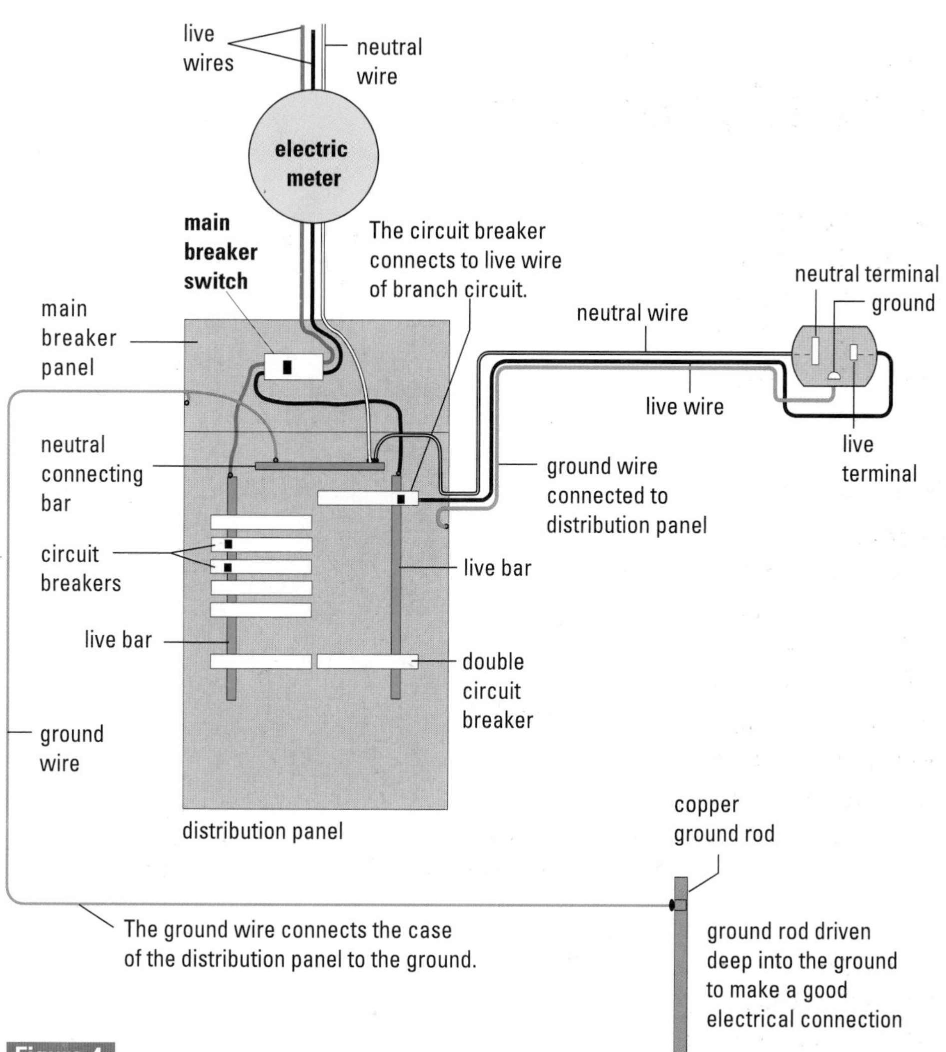

Figure 4

wire, helps protect against electric shock from a short circuit. All ground wires from the outlets in a home connect to the ground bar in the distribution panel. In turn, the ground wire connects the case of the distribution panel to the ground through a metal cold water pipe or grounding rod driven deep into the earth.

Circuit Breakers

All houses have circuit breakers or fuses to control the flow of electricity. A circuit breaker automatically shuts off if there is a current overload. A **circuit breaker** has two different strips of metal back to back that bend when heated. As long as the current in the circuit is less than the maximum current allowed by the circuit breaker, the strip will not bend enough to open or trip the circuit and stop the current flow (**Figure 5**).

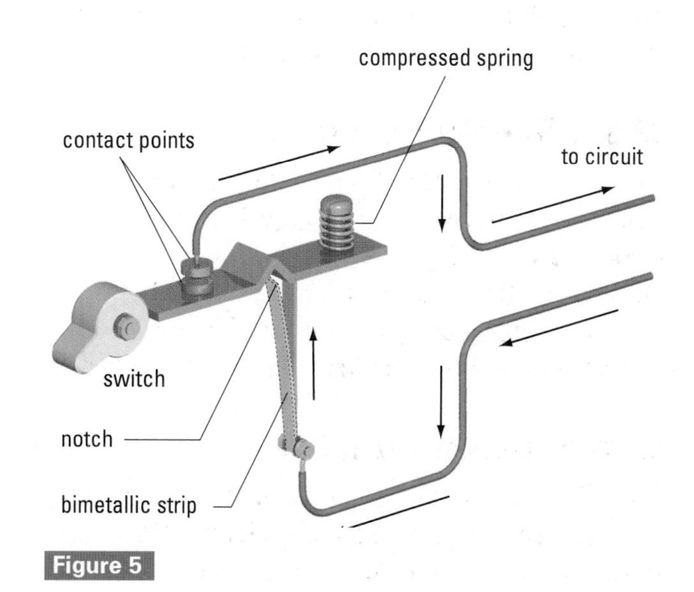

Figure 5

Fuses

Fuses are used in cars, some stoves, and in distribution panels of some older homes (**Figure 6**). **Fuses** are simply pieces of material that will melt when the temperature in the circuit is too high. For example, if the current flowing through a 15-A fuse is higher than 15-A, the fuse melts and the circuit is broken. Fuses are less convenient than circuit breakers because you must replace them every time an overload occurs. A fire can start if you replace a fuse with a fuse of a higher rating.

Figure 6

Fuses are found in some older homes.

Wall Outlets, Polarized Plugs, and Grounding Pins

Electricity has to be safely delivered to various rooms in a building. This is done using **wall outlets** (**Figure 7**). A wall outlet has many safety features. Some of these features include the following:

1. The wall outlet is made of plastic. Plastic does not conduct electricity. Outlets that are cracked should not be used.
2. Most new wall outlets have a third round hole called the **ground terminal** (**Figure 7**). This allows electricity to leave the building and travel to the ground if a short circuit occurs.
3. New wall outlets have **polarized plugs**, as shown in **Figure 7**. If an outlet or plug is not polarized, then the neutral wire might be connected to the electrical device switch. Such a device would still have power even when it is turned off. A polarized outlet and plug prevent this from happening, as the plug fits into the outlet only one way. However, it is still a good idea to unplug a lamp before changing the light bulb.

Figure 7

New wall outlets have polarized plugs. Polarized plugs have one prong that is wider than the other, so there is only one way to insert it into the wall outlet.

Many devices have plugs that have three prongs. The third prong is called the **grounding pin** (**Figure 8**). If the device is damaged, the electrical current will be grounded to the metal pin rather than pass through you. Never remove or bend the grounding pin to connect a three-pin plug to an extension cord or an outlet with only two slots. This removes the grounding.

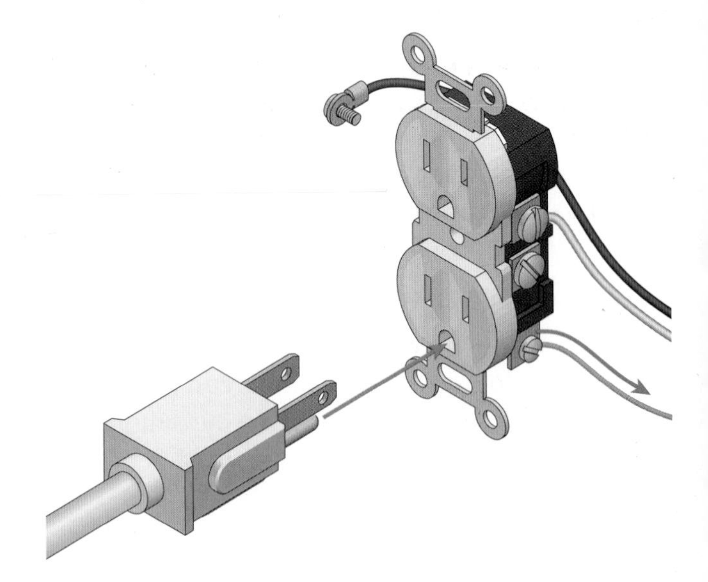

Figure 8

The third prong, or grounding pin, provides protection against a short circuit.

Ground Fault Circuit Interrupter (GFCI)

As you learned earlier (**Figure 5,** section 3.4), very little current is needed to cause serious harm. A circuit has enough electric current flowing in it to kill a person but not enough to cause a circuit breaker to trip, or open. Many fatal accidents occur because of this fact, especially near water. A **ground fault circuit interrupter** (GFCI) is a special kind of outlet. It includes a circuit breaker that responds to very small changes in current and interrupts the flow of electricity. The GFCI outlet can fit into the space of a normal outlet anywhere in a building (**Figure 9**).

The GFCI outlet is specially designed to detect very small differences in the amount of electric current flowing through two wires. If the circuit is operating normally, the current in both the hot and neutral wires will be identical. If the current in the neutral wire is slightly lower than the current in the hot wire, some of the current must be flowing elsewhere. The GFCI outlet detects this difference, and its built-in circuit breaker opens (trips) the circuit very quickly.

GFCI outlets are important in bathrooms, kitchens, laundry rooms, pool areas, outdoor outlets, or any area where water is found.

Figure 9

Ground fault circuit interrupters quickly stop the flow of electricity after sensing very small changes in current.

Challenge

1 Do you need a safety device in your electric circuit?

2 Record how electricity is handled safely. Add this information to your electrical safety pamphlet.

3 Include questions and answers about electrical safety for your electric game show.

Understanding Concepts

1. What are the three different wires found in most homes? What is the purpose of each?
2. Compare the advantages and disadvantages of circuit breakers and fuses.
3. List three safety features related to wall outlets, plugs, and grounding pins, and explain how they provide protection.

Making Connections

4. Complete a GFCI survey of your home. Indicate where GFCIs are installed. Record where GFCIs should be installed if they are not already there.
5. Why would having too many outlets on one circuit cause problems?

Exploring

6. Make a chart and list all the electrically operated safety devices in your home. Are they self-contained, or are they plugged into wall outlets? How do they function? What warning system alerts you if they stop working?

Energy Conservation

The demand for electrical energy continues to rise as industries expand, populations grow, and the use of electrical devices increases (**Figure 1**). What can we do to maximize the efficient use of electricity in the home, the school, and the community?

Figure 1

Computers use a lot of electrical energy.

How Do We Use Electricity?

The first step in making the most of electricity is to explore the ways in which this energy source is used. Electrical devices are used for many reasons. **Table 1** outlines the main functions of these devices.

You should be able to put the devices in your home in at least one of the four areas outlined in **Table 1**. One example is shown in **Figure 2**.

The wise use of electrical energy has two main benefits: It costs less, and it has less impact on the environment.

Common sense is all we need to use electrical energy wisely.

Table 1

Electrical usage	Example
light	table lamp
heat/cooling	water heater
sound	stereo
mechanical energy	washing machine motor

Figure 2

What would our lives be like without electrical energy?

Conserving Electrical Energy

In a typical Canadian province, the use of electrical appliances follows a basic trend (**Table 2**).

Table 2 Top Four Uses of Electricity

Rank	Electricity usage fact
1	heating and cooling homes using heaters and air conditioners
2	heating water for personal uses such as bathing, showering, and washing clothes
3	appliances, such as stoves, refrigerators, toasters, clothes washers and dryers, and dishwashers
4	lighting, which accounts for 15% of home electrical use

Source: Ontario Ministry of Energy, Science and Technology

What recommendations would you make to reduce electricity use in your home? **Table 3** suggests some ways to conserve energy when using different electrical devices.

Table 3 Conserving Electrical Energy

Electrical device	Ways to conserve energy
air conditioner	Seal windows and close blinds or curtains to increase the efficiency of the appliance.
clothes dryer	Clean the lint filter after each load. Lint blocks air flow and increases the drying time for clothes.
dishwasher	Run the dishwasher only when it is full.
electric stove	Cover the pot when boiling water. It will take less time to boil.
lighting	Turn off all unnecessary lights.
television	Have the TV on only when it is being watched.

Making Connections

1. **(a)** What do you think is your family's least efficient use of electricity?

   (b) List three things that you could do to use electricity more efficiently in your home.
2. Do you think we will run out of electricity? Why or why not?

Challenge

1 Is your circuit efficient?
2 Include tips on conserving energy in your electrical safety pamphlet.
3 Add appropriate questions for your electric game show.

Work the Web

Visit www.nelson.science.com and follow the links for *Science 9: Concepts and Connections*, 3.14. Research information on the conservation of electricity in the home and workplace.

Try This Activity

Maximizing the Use of Electricity in the Home

Maximizing the use of electricity requires careful thought and research. Prepare a table that includes a list of appliances and their functions, and recommends ways of conserving electricity (**Table 4**).

1. Refer to **Table 3** and prepare common sense recommendations for using the appliances in an energy-efficient manner.
2. Create an advertisement that suggests ways to use electricity efficiently. (R)
3. Share and compare your findings with those of a classmate.

Table 4

Function	Appliance	Recommendation
heating and cooling	a. b. c. d.	
light	a. b. c. d.	
sound	a. b. c. d.	
mechanical energy	a. b. c. d.	

The Family Energy Audit

In the previous section, you thought about how you could reduce the electricity use in your home. Now, what if you were asked to develop a plan to do this? You might begin by identifying which electrical devices your family uses and categorizing them as essential or nonessential. For example, you could classify the use of a computer or a cell phone (**Figure 1**) as either essential or nonessential, depending on how they are being used.

Electricity is not free. **Table 1** shows the approximate monthly cost of using different appliances. In this activity, you will collect and analyze information about the use of electrical energy in your home (or the home of a friend or a relative), identify the use of the devices as essential or nonessential, and develop a plan to maximize the efficient use of electricity. Saving energy is not only measured in terms of saving money; every time you reduce the amount of energy you use, there are environmental, social, and economic benefits.

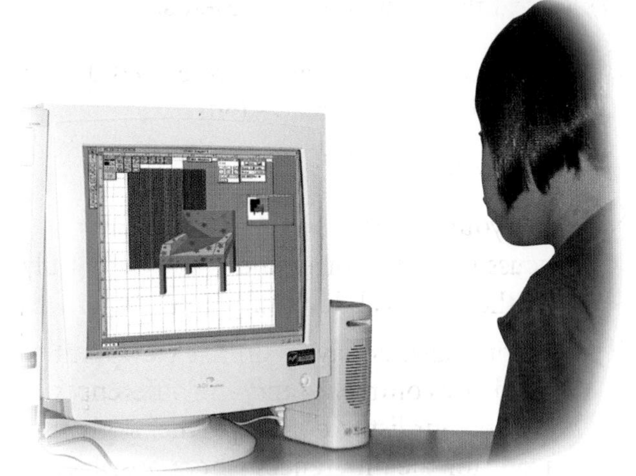

Figure 1

How would you determine which devices are essential or nonessential?

Materials

- data about electrical devices used in the home
- data about patterns of energy use by members of the family in the home

Table 1

Electrical devices	Approximate monthly cost* ($)
air conditioner (room)	7.20–43.30
clothes dryer	4.01–12.03
coffee maker	0.32–2.17
computer (monitor & printer)	0.40–2.89
electric kettle	0.24–1.20
lighting: 60-W incandescent lamp	0.40–2.41
microwave oven	0.40–1.60
television (colour)	0.41–1.20
toaster	0.08–0.41

*Costs are based on $0.082 per kWh.

Procedure

1 Brainstorm a list of electrical appliances used in your home. Identify the use of each device as essential or nonessential.

(E1) **(a)** Record the data in a table (**Table 2**).

Table 2

Electrical appliances	Essential	Nonessential

2 Predict which appliances you think your family will use most and which device will use the most energy during the monitoring period. Use **Table 1** as a reference point.

(b) Record your predictions.

3 Monitor the use of the electrical appliances in your home using **Table 3** to record your observations. Your teacher will tell you the length of the monitoring period.

(c) Record the amount of time (in hours) each electrical appliance is used per day. (E2)

4 During the monitoring period, note your family's activities, the weather, and other events that might affect the amount of electricity used (**Figure 2**).

(E1) (d) Record these activities and events.

5 Calculate which appliances were used the most during the monitoring period.

(E2) (e) Record your answers.

6 Share your observations with a partner. Discuss your categories and how your family used the electrical devices.

(f) Did your partner have a similar viewpoint? Record any comments and any differences between your list and your partner's, giving reasons for the differences where possible.

(g) How did your predictions compare with your actual results? Comment on any differences. (F1)

(h) Review the data in the tables and suggest reasons for any unusual changes in the daily use of electrical appliances by your family.

7 Try to identify ways to conserve energy.

(i) Write a brief proposal to your family, making suggestions about how to reduce the electricity bill. (R)

Challenge

2 Create an electrical safety checklist. Add it to your electrical safety pamphlet.

3 Summarize ways in which we can conserve electricity at home and in the workplace. Use this information for your electric game show.

Table 3

Electrical appliances	Time used per day (h)	Observations (family activities, weather, other events)

Figure 2

Weather affects the amount of electricity used for heating and cooling.

Making Connections

1. How would you convince your family that your plan is reasonable and worth following?
2. What other resources, in addition to electricity, would be conserved if your plan were put into action?

Exploring

3. Visit some environmental websites on the Internet for additional suggestions about conserving electricity. (O)
4. Prepare a questionnaire that asks for the most important electrical appliances in a house by room. Give it to ten people. What were the most popular answers? Do your findings match your own ratings? Why do you think that is?

(E2) Quantitaive Observations
(R) Writing for Specific Audiences
(F1) Interpreting Observation Tables
(O) Internet Research

Efficiency and Electrical Devices

Why does one battery-powered radio (**Figure 1**) continue to play long after another's batteries have run down? The answer is that the efficiency of each radio is different. The **efficiency** refers to how well the electrical energy is changed to useful energy by the electrical device. Assuming that the batteries were the same, we can infer that one radio is more efficient than the other.

The conversions of electrical energy to useful energy are never 100% efficient. Some of the energy is converted into useful energy but some of it is always converted into other forms, in many cases, heat energy. The total energy that comes out of an electrical device is always equal to the energy that goes in. It's just that what comes out may be in more than one form. For example, if we say that an electric motor has an efficiency of 80%, we mean that 80% of the input energy of the energy is converted into useful mechanical energy. The other 20% is probably lost as heat energy. A light bulb has an efficiency of only about 5%. That mean that 5% of the electrical energy that goes into the bulb is converted into useful light energy and 95% of the energy is released as heat energy (**Figure 2**).

The efficiency of an electrical device may be affected by the following factors:

- resistance to current flow
- distance the current flows
- friction
- materials used in the circuit

Figure 1

Not all electrical devices are equally efficient.

Figure 2

A small percentage of the energy put into lighting this boat is actually converted into light. Most is lost as heat.

Determining the Efficiency of an Electrical Device

The energy that goes into an electrical circuit is called the **energy input**. The energy that is produced by the electrical device, as light, sound, heat, or mechanical energy, is called the **useful energy output**. The efficiency of any energy conversion is calculated by using the formula in **Figure 3**. Efficiency is often expressed as a percentage, which is simply the efficiency multiplied by 100%.

$$\text{Efficiency} = \frac{\text{Useful energy output}}{\text{Energy input}}$$

$$\text{Percent efficiency} = \frac{\text{Useful energy output}}{\text{Energy input}} \times 100\%$$

Figure 3

Calculating the percent efficiency of a device

Sample Problem

Determine the percent efficiency of a 60-W fluorescent light bulb that uses 2000 J of electrical energy to produce 400 J of light energy. (A **joule** is the SI unit for measuring energy, and its symbol is **J**.)

Data:

Energy input = 2000 J
Useful energy output = 400 J
Percent efficiency = ?

Equation:

$$\text{Percent efficiency} = \frac{\text{Useful energy output}}{\text{Energy input}} \times 100\%$$

Solution:

$$\text{Percent efficiency} = \frac{400\ \text{J}}{2000\ \text{J}} \times 100\%$$

$$\text{Percent efficiency} = 20\%$$

Statement:

The percent efficiency of a 60-W fluorescent light bulb is 20%.

Understanding Concepts

1. Explain why energy conversions can never be 100% efficient.
2. What is the difference between input energy and useful output energy?
3. Calculate the percent efficiency of an electric motor that uses 15 000 J of energy to produce 11 500 J of useful energy.
4. Calculate the percent efficiency of an incandescent light bulb that produces 2500 J of light energy from 50 000 J of electrical energy.

Making Connections

5. Why do you think it is important to be able to calculate percent efficiency?
6. Provide one example each of useful output energy using **Table 1** as a guide.

Table 1

Useful output energy	Example
sound	
light	
mechanical energy	

Improving Efficiency

Getting the most from the electrical energy that we use is a priority. In Canada, federal law now requires special "EnerGuide" information to be attached to all appliances (**Figure 4**). This label states the amount of electrical energy the appliance uses per year. Consumers can compare the EnerGuide numbers for similar appliances of the same capacity and make an informed decision about which one is the most efficient model.

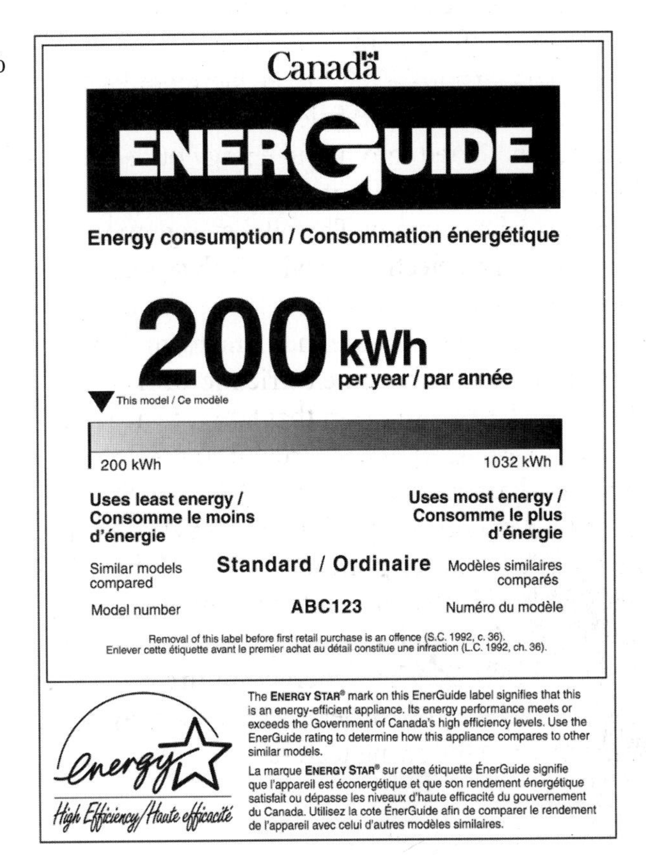

Figure 4

1 Is percent efficiency an issue in your circuit? Why or why not?

3 Develop questions on the percent efficiency of electrical devices for your electric game show. Research factors that produce greater efficiency in electrical devices.

Automobiles and the Fuel Cell

In North America, companies such as Stanley Steamer, Detroit Electric (**Figure 1**), and Ford developed the first steam-, electric-, and gasoline-powered vehicles.

In the early 1900s, the first steam-driven automobile and then the electric-powered car enjoyed popularity over gasoline-powered vehicles. In fact, in 1912, 30 000 electric cars were on the road in the United States, and 6000 new ones were made each year. However, steam and electric cars were not without problems. Steam-powered cars needed water that would not freeze in cold weather. Electric cars could not travel very far before the batteries had to be recharged (**Figure 2**).

(**a**) Why did gas-powered automobiles replace the steam- and electric-powered cars?

(**b**) How are the batteries connected (series or parallel) in **Figure 2**? Explain.

Figure 1
This is a 1914 Detroit Electric Model 46 Roadster.

Figure 2
Four batteries powered the 1914 Detroit Electric Model 46 Roadster.

Electric Cars in the Future

Pollution and a shortage of fossil fuels are two reasons why electric cars may become more popular in the future. The battery for electric cars remains a problem because it would have to be very large to work for long distances. Car manufacturers are looking at other options.

One solution has been to produce a vehicle called the hybrid electric vehicle, or HEV, that uses gasoline and electric power. Many car manufacturers began building these vehicles (**Figure 3**) in the 1990s. However, HEVs use fossil fuels and produce some pollution, so they are not perfect.

(**c**) List the problems that must be fixed to make a practical electric vehicle.

(**d**) What are the advantages of HEVs compared with vehicles with only gas engines?

(**e**) Why are HEVs not a permanent solution to replacing vehicles with gas engines?

Figure 3
Hybrid electric vehicles (HEVs) are highly fuel efficient and combine gas and electric power.

Fuel Cells: A Practical Solution

Governments in Canada and the United States have passed laws that require car manufacturers to increase the efficiency of their vehicles while at the same time reducing pollution.

The future of electric vehicles appears to depend on **fuel cell** technology (**Figure 4**).

Most advances in this field are being made in Canada. For example, Ballard Power Systems is working with car manufacturers to make a fuel cell (**Figure 5**). Fuel cells produce electricity from hydrogen and oxygen. The only other products created by the fuel cell are heat and water. A fuel cell-powered car would not create any harmful byproducts and could run on clean and renewable fuels such as hydrogen or methanol. The fuel cells would be connected in series to power the vehicle. The number of fuel cells needed would depend on the size and performance of the vehicle.

(f) What have governments done to encourage manufacturers to produce electric cars?

(g) Why are fuel cells the most likely power source for the vehicles that we will drive in the future?

Figure 4

A city bus powered by a fuel cell

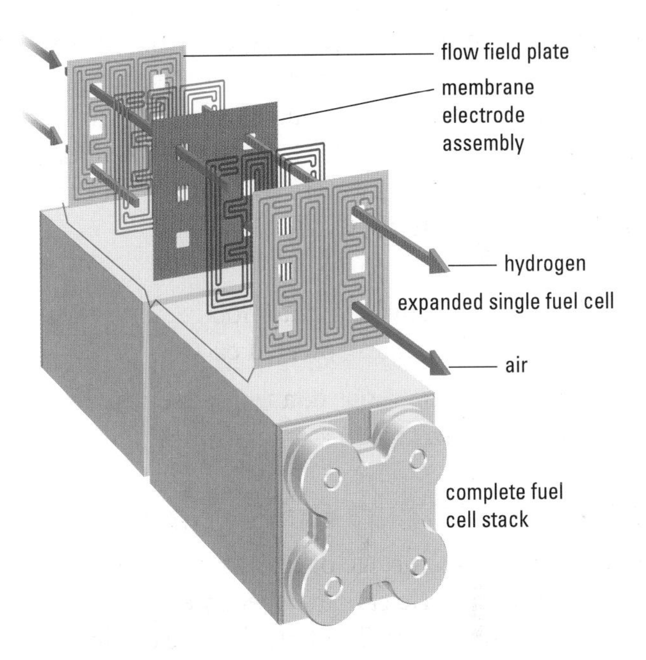

Figure 5

Hydrogen interacts with the membrane in a fuel cell to produce electricity, water, and heat.

A Versatile Energy Source

Fuel cells, and their ability to produce electricity from hydrogen, open up many applications. Imagine electrical generators the size of refrigerators that could provide electrical energy for isolated buildings! This innovation would be invaluable in remote communities. Fuel cells can even run on the methane produced in sewage treatment plants and landfill sites around the world.

(h) Why would it be helpful if fuel cells ran on the methane produced from sewage treatment plants and landfill sites?

Making Connections

1. Brainstorm a list of other practical applications for fuel cell technology.
2. What other raw materials could provide the hydrogen needed to run a fuel cell? How could this be useful for farms in isolated areas?

Exploring

3. Research on the Internet and in the library to find out more about fuel cells. Will the future energy source that powers vehicles be fuel cells? What are the advantages and disadvantages of this technology? (N) (O)

Comparing Electrical Energy Production

Everyone uses electrical energy daily. Electricity can be moved easily from one place to another and can be converted into other forms of energy that we need in our everyday lives. The need for electricity is growing so we must find ways of producing the additional energy needed by our modern society.

Figure 1

Renewable sources of electricity are those that will not run out over time.

Renewable and Nonrenewable Energy Sources

If we had a choice, all of our energy would come from renewable sources. A **renewable energy source** is one that does not get used up in producing energy. Renewable sources of electrical energy presently in use are wind, tides, solar, and hydro (water) power (**Figure 1**). However, these sources cannot presently supply all of the energy needed so we have to use nonrenewable sources. **Nonrenewable sources** are those that cannot be replaced in a short period of time, such as a human lifetime. Nonrenewable sources included fossil fuels such as coal, oil, and gas, and nuclear fuels.

About 89% of the energy produced in Canada is produced from nonrenewable sources. The remaining 11% is produced from hydro and other renewable sources. **Table 1** shows the sources of electrical energy in Ontario.

Table 1 Electricity Sources

Electricity sources	Ontario's electricity consumption (%)
nuclear energy	37
natural gas	7
coal or oil	29
hydro (water power)	26
alternative sources: solar, wind, biomass, waste	1

Source: The combination of electricity sources used to generate the electricity consumed in Ontario in 2000 (Statistics Canada Catalogue No. 57-003-XPB for 2000 and the Ministry of Energy, Science and Technology).

Comparing Energy Production Technologies

The main factors to consider when comparing electrical energy production methods are the following:

- cost
- efficiency
- geography
- storage
- impact on the environment

Cost

Some sources of energy are more expensive than others. **Table 2** shows the cost of producing electricity from wind, tides, hydro, solar cells, and from nonrenewable sources. The costs shown in this table are compared to the cost of hydroelectricity. For example, producing electricity with solar cells is 22 times more expensive than producing electricity with hydro power. Currently, solar cells (shown in **Figure 1**) cost too much to be used on a large scale. As costs decrease and efficiency increases, solar cells will become more important.

Efficiency

Electrical energy production methods vary in efficiency. **Table 3** shows the efficiency of each source. Almost all sources of nonrenewable energy produce the same amount of electricity available. Wind energy is the most efficient method of producing electricity but as you can see in **Table 2** it is 4 to 5 times more expensive than hydro power.

Geography

Most alternative energy sources depend on the conditions of a specific location. For example, hydroelectricity can only be generated in a location that has large amounts of running water (**Figure 2**). Each renewable energy source depends on a specific geographic location. However, usually at least one of these sources can be used in any location.

Nonrenewable energy sources, such as fossil fuels, also depend on specific locations. Oil and natural gas are usually pumped out of the ground. Coal is mined. Presently, fossil fuels are easy to obtain and transport.

Nuclear generating stations must be built near a body of water to operate. However, once built, they do not need other geographic conditions to operate.

Table 2 **Relative Cost of Energy Sources**

Energy source	Relative cost (compared with hydroelectric)
hydroelectric **(Figure 2)**	1
tidal	1
wind	4–5
solar cells	22
nuclear	2–5
fossil fuels	2–5

Table 3 **Electrical Conversions from Energy Sources**

Source of energy	Percentage of energy converted to electricity
solar cells	17.0
wind generators	30.0
hydroelectric generators	20.0
nuclear	20.0
oil	20.2
coal	22.2

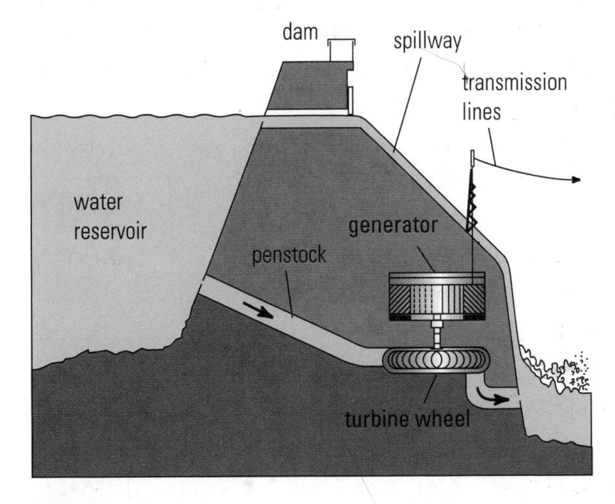

Figure 2

How hydroelectric energy is produced

Storage and Transmission

It is difficult to actually store electrical energy once it is produced. When you recharge a battery, you are converting electrical energy from another source into chemical energy in the battery. Electrical energy itself is not stored in a battery. An electrical device called a **capacitor** can store small amounts of electrical energy. However, no practical way exists to store the huge amounts of electrical energy produced at a generating station. As electrical energy is produced, it must be used. One problem with alternative energy sources is that they may not produce electricity when it is needed.

Electrical energy must be moved from the location where it is produced to where it will be used. A network of wires and transmission towers carries the energy from the generating stations to homes where it is used. Since the conducting wires offer resistance to the flow of electricity, some of the energy is lost as heat when it is carried over long distances.

Impact on the Environment

All types of energy sources have an impact on the environment. Windmills take up large spaces, and birds can be hurt or killed if they fly into the blades. Generation of electricity using hydro power often requires flooding the land and sometimes blocks fish migration routes. Burning fossil fuels produces air pollution and contributes to global warming. Acid rain, caused by rain falling through polluted air, can harm lakes, vegetation, and buildings and statues made of stone. Nuclear energy produces nuclear waste, which is very harmful to living things and the environment. Because of this, long-term storage of nuclear waste material is a big concern. In addition, the construction of large transmission lines requires large amounts of land that might otherwise be used for farming or forestry.

As we use energy daily, we all need to be aware of the risks and benefits associated with the production and use of electrical energy.

Try This Activity: Risk–Benefit Analysis of Sources of Energy

Prepare a table with the following headings:

Risks				Benefits			
Possible result	Rank of that result (scale of 1 to 5)	Probability of that result happening	Probability value (rank × probability)	**Possible result**	Rank of that result (scale of 1 to 5)	Probability of that result happening	Probability value (rank × probability)
Total probability values (risks)				**Total probability value (benefits)**			

(K) Conduct a risk–benefit analysis on the various sources of electricity. As you study the ways of generating electricity with renewable and nonrenewable energy sources, add this information to your table. Use the Internet, the library, or other resources to help you make a recommendation based on your analysis. Refer to section K of the Skills Handbook for a summary of risk–benefit analysis. Which energy source would you recommend? Why?

SKILLS HANDBOOK: (K) Analyzing the Issue

Understanding Concepts

1. Electricity sources were compared using cost, efficiency, geography, storage, and impact on the environment.
   - **(a)** Which energy sources were the most cost effective? the least cost effective?
   - **(b)** Which energy sources were the most efficient? the least efficient?
   - **(c)** Could at least one of the energy sources be used in Canada, no matter the location? Explain.
   - **(d)** Why is storage of electricity a problem?
   - **(e)** Which energy sources are the most environmentally friendly?
2. Why is most electricity produced from nonrenewable sources in Canada?
3. Why is the location of the place where the electricity is generated so important for most kinds of renewable energy sources?

Making Connections

4. Rate the four renewable energy sources from this section from best to worst. Decide on the criteria you will use to determine which is best and which is worst. Explain your reasoning.
5. Rate the three nonrenewable energy sources from this section from best to worse. Decide on the criteria you will use to determine which is best and which is worst. Explain your reasoning.

Exploring

6. (N) (O) Research using the Internet and other resources to assess the different methods of producing electricity using renewable energy resources. What methods seem to be favoured? Why? (See **Figure 3**.)

Figure 3

Reflecting

7. How do you think the electrical energy needs of your area might be met 50 years from now?

Work the Web

How can solar cells be combined with other energy sources? How many people in North America are using solar cells in their homes? Visit www.nelson.science.com and follow the links from *Science 9: Concepts and Connections*, 3.18.

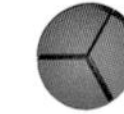

Challenge

3 Develop questions on renewable and nonrenewable energy resources for your electric game show.

3.19 Explore an Issue

DECISION-MAKING SKILLS

- ● Define the Issue
- ○ Identify Alternatives
- ● Research
- ● Analyze the Issue
- ● Defend a Decision
- ○ Evaluate

Choosing an Energy Source

One fact is certain in Canada. Because our population and our economy are growing (**Figure 1**), we will need to produce more electrical energy. In section 3.18, you learned about the different types of renewable and nonrenewable energy sources. As our need for electrical energy grows, we are faced with a difficult choice. What sources should we use to produce the extra electrical energy?

Figure 1

Today's Energy Sources

At present, most of our electrical energy is produced by fossil fuels, nuclear fuel (**Figure 2**), and hydroelectric generating stations (**Table 1**). It is unlikely that we will be able to provide more energy by using new hydroelectric resources. This is because most of the hydroelectric sites that can produce large amounts of electricity are already being used. Without considering the renewable sources, which of the two nonrenewable energy sources would be preferable to meet future energy needs?

Both options have problems, as you saw in section 3.18.

Table 1 Electricity Sources in Ontario

Energy source	Number of stations	% of total electricity supply
fossil fuels	6	17
nuclear energy	5	48
hydroelectric	69	25
independent producers or purchased from other utilities	—	10

Renewable Energy Sources

Several renewable energy sources—geothermal, solar cells, biomass, wind, and fuel cells—could be used to provide electrical energy in Ontario. However, the cost of using any of these alternatives is currently too high for them to be workable options.

Right now, none of the renewable energy technologies can be manufactured and operated cheaply enough to make them an acceptable alternative in the near future.

Based on the fact that they are the cheapest sources, let's discuss only the two nonrenewable sources—fossil fuels and nuclear power.

Figure 2

Debate: Fossil Fuels or Nuclear Power?

Statement

Future electricity needs should be met using additional nuclear generating stations (**Figure 2**) rather than fossil fuel stations.

For

- Nuclear energy can supply a great deal of electrical energy using only a few generating stations. This is good for the environment.
- The nuclear industry in Canada is good for the economy. It has created more than 30 000 jobs for people across the country.

Against

- Nuclear energy produces dangerous nuclear waste, and currently we do not have a safe way to dispose of it.
- An explosion involving radioactive nuclear material would have a disastrous impact on the environment and the communities around the station.

What do you think?

- In your group, discuss the statement and points for and against. Write down any additional points that your group considers.
- (N) (O) Search for additional information using the library, CD-ROMS, or the Internet.
- (K) From your research, prepare a summary of additional information on the topic, and use this to form your own opinion on the statement.
- Share the information with your group members, and then decide whether your group agrees or disagrees with the statement.
- Write a summary of your group's final position.
- (L) (S) Prepare to defend your group's position in a class discussion.

Understanding Concepts

1. Why are most renewable energy sources not considered a possible solution to our future electrical energy needs?

Exploring

2. (N) (O) Places such as New Zealand use geothermal energy to generate electricity (**Figure 3**). Geothermal energy is the heat energy taken from beneath the Earth's surface. Is this an option for Canadians? Why or why not? Use Internet or library resources to find out.

Figure 3

Work the Web

Follow the links for www.nelson.science.com to *Science 9: Concepts and Connections*, 3.19, to find out how many years we have left until our nonrenewable energy sources run out.

Challenge

3 What information about fossil fuels and nuclear energy could you add to your questions and answers for your electric game show?

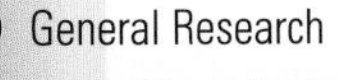

Building a Home Wiring Model

In this activity, you will put the knowledge and skills you have learned in this unit to use! You will build a wiring model of your home. For your model, you will create a floor plan outlining the main electrical devices used in your home, draw a circuit diagram based on your floor plan, and build your model using available materials.

Figure 1

Materials

- Popsicle sticks
- 2 1.5-v batteries
- 9 light-emitting diodes (LEDs)
- black insulated wire (thin gauge)
- white insulated wire (thin gauge)
- wire stripper
- 10 switches
- soldering materials (optional)
- glue and tape
- other materials as required

Only a licensed electrician should do any home wiring.

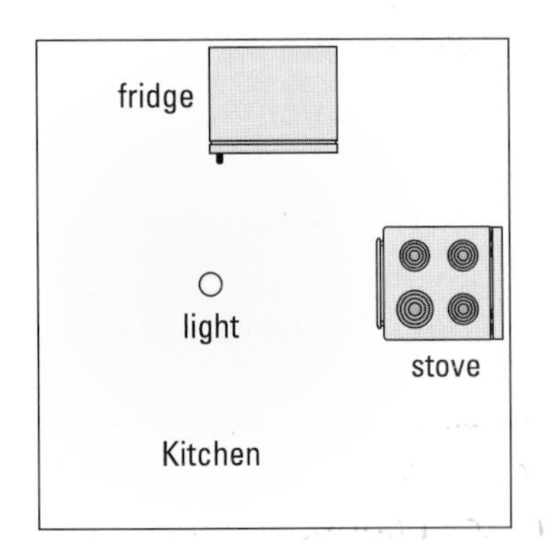

Figure 2

Procedure

1 Work with a partner. Decide what living area you want to base your wiring model on (**Figure 1**). The living area should have a maximum of three rooms, with three devices per room.

2 To begin, draw a floor plan.

(a) Sketch the living area and label the main electrical loads that are present in each room. **Figure 2** is a sample drawing of a kitchen.

3 Before an electric circuit is wired in a home, a circuit diagram is created. In a real home, it provides a plan for the electrician to follow.

(b) Draw a circuit diagram using cells as the D3 energy source. Outline the devices used in the circuit diagram. **Figure 3** is a circuit diagram that matches the kitchen in **Figure 2**.

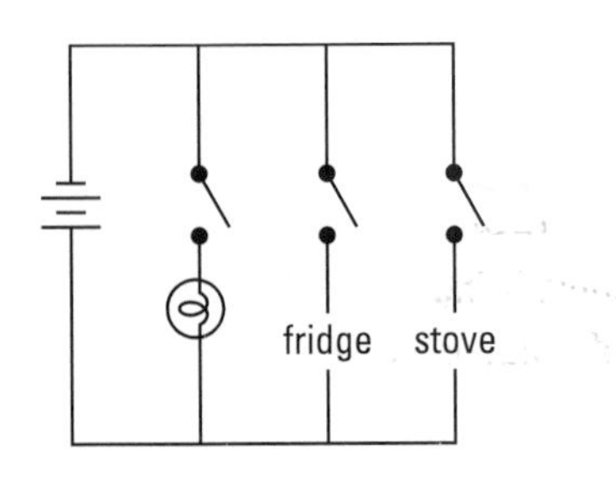

Figure 3

Homes are wired in parallel. This is done so each device can be shut off without affecting the other devices in the circuit.

 SKILLS HANDBOOK: D3 Drawing and Constructing Cicuits

4 Now build your model.

(c) Use Popsicle sticks, glue, and a platform for building on, such as a cardboard box, and build the framework for the home wiring model. Each room will be wired with a maximum of three electrical devices. The appliances will be represented by the LEDs. **Figure 4** is an example of a home wiring model.

5 It is now time to wire your model. You must follow some basic rules when wiring your model:

- LEDs are polarized. They have a positive and a negative wire. The long wire is positive and the short wire is negative.
- Black wire represents a positive charge.
- White wire represents a negative charge.
- Positive is connected to positive and negative is connected to negative.

(d) How do you think you could test your model to make sure it is wired in parallel? A completed home wiring model is shown in **Figure 5**.

6 Summarize the home wiring model activity in Ⓢ a presentation to the class. Your presentation should include (i) a completed floor plan, (ii) a circuit diagram of the floor plan in parallel circuit form, and (iii) a working home wiring model.

Figure 4

The framework for the home wiring model is made out of Popsicle sticks.

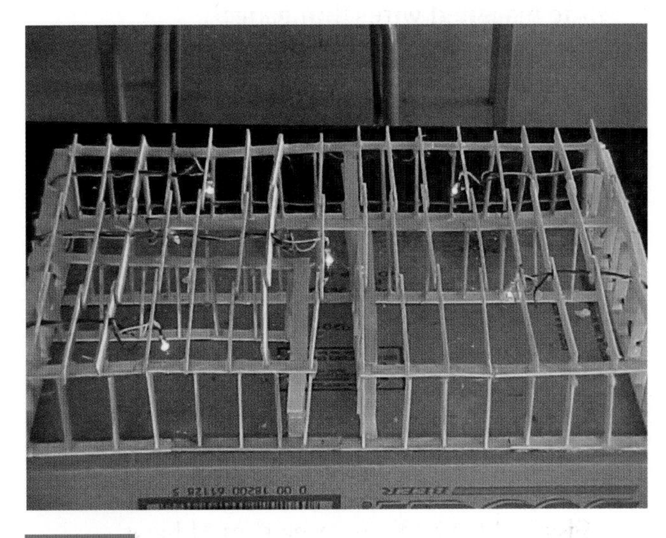

Figure 5

Challenge

2 What areas of the home represent the greatest danger from electricity to the people living there? Explain your reasoning. Include this information in your electrical safety pamphlet.

Making Connections

1. Why would an electrician require a circuit diagram? What other jobs might benefit from having a circuit diagram?
2. Summarize the main points to remember when wiring an electric circuit in parallel.
3. List three things that an electrical inspector would check if this were a real home.

Unit 3 Summary

Key Expectations

Throughout this unit, you have had opportunities to do the following:

- Explain common electrostatic phenomena (3.1, 3.2, 3.3)
- Compare qualitatively static and current electricity (3.2)
- Describe the concepts of electric current, voltage, and resistance with the help of a water analogy (3.4)
- Explain how electric current, voltage, and resistance are measured using an ammeter and a voltmeter (3.8, 3.9, 3.10, 3.11)
- Describe qualitatively the effects of varying electrical resistance and voltage on electric current in an electric circuit (3.10, 3.11)
- Apply the relationship

  voltage drop = resistance × current

  to simple series circuits (3.10)
- Determine quantitatively the percent efficiency of an electrical device that converts electrical energy to other forms of energy, using the relationship

  $$\text{Percent efficiency} = \frac{\text{energy output}}{\text{energy input}} \times 100$$

  (3.16)
- Demonstrate knowledge of electrical safety procedures when planning and carrying out investigations and choosing and using materials, tools, and equipment (3.1, 3.3, 3.5, 3.7, 3.8, 3.9, 3.11, 3.13, 3.20)
- Identify an authentic practical challenge or problem relating to the use of electricity (3.14, 3.15, 3.16, 3.17, 3.18, 3.19, 3.20)
- Formulate questions about the problem or issue (3.14, 3.15, 3.16, 3.19)
- Demonstrate the skills required to plan and conduct an inquiry into the use of electricity, using instruments, tools, and apparatus safely, accurately, and effectively (3.1, 3.3, 3.5, 3.7, 3.9, 3.11, 3.20)
- Select and integrate information from various sources, including electronic and print resources, community resources, and personally collected data, to answer the questions chosen (3.1–3.20)
- Organize, record, and analyze the information gathered (3.1, 3.3, 3.5, 3.7, 3.9. 3.11, 3.14, 3.16, 3.17, 3.18, 3.19, 3.20)
- Communicate scientific ideas, procedures, results, and conclusions using appropriate SI units, language, and formats (3.1, 3.3, 3.5, 3.7, 3.9, 3.10, 3.11, 3.13, 3.20)
- Design, draw, and construct series and parallel circuits that perform a specific function (3.6, 3.7, 3.9, 3.11, 3.20)
- Use appropriate instruments to collect and graph data, and determine the relationship between voltage and current in a simple series circuit with a single resistor (3.11)
- Describe and explain household wiring and its typical components (3.4, 3.5, 3.6, 3.7, 3.11, 3.13)
- Develop a solution to a practical problem related to the use of electricity in the home, school, or community (3.14, 3.15, 3.16)
- Compare electrical energy production technologies, including risks and benefits (3.18, 3.19)
- Explain how some common household electrical appliances operate (3.4, 3.5, 3.6, 3.7, 3.15, 3.20)
- Describe careers that involve electrical technologies, and use employability assessment programs, newspaper job advertisements, and/or appropriate Internet sources to identify the knowledge and skills requirements of such careers (3.12)

Key Terms

ammeter
capacitor
charging by contact
circuit breaker
circuit diagram
conductor
current electricity
distribution panel
efficiency
electric circuit
electric current
electric meter
electrical load
electrostatics

energy input
fuel cell
fuses
ground fault circuit interrupter (GFCI)
ground terminal
ground wire
grounding pin
hot or live wire
joule (J)
law of electric charges
main breaker switch
negative charge
neutral wire
parallel
polarized plug
positive charge
potential difference
renewable energy source
resistance
resistor
series
static electricity
useful energy output
voltage
voltmeter
wall outlet

What HAVE YOU *Learned?*

Review your answers to the What Do You Already Know? questions on page 127 in Getting Started.

- Have any of your answers changed?
- What new questions do you have?

Unit Concept Map

Use the concept map to review the major concepts in Unit 3. This map can help you begin to organize the information that you have learned. You can copy the map, and then add more links. You can also add more information in each box.

You can use a concept map to review a large topic on a general level, or you can use it to examine a very specific topic in detail. Select one concept from this unit that you need to study in detail and make a detailed concept map for it.

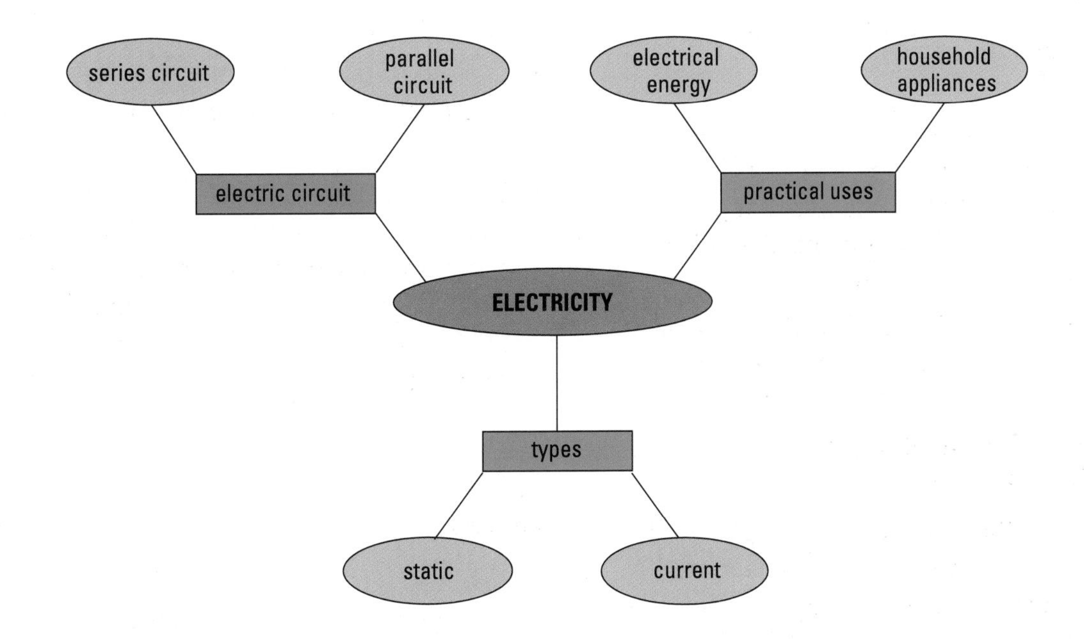

Unit 3 Review

Understanding Concepts

1. Match the terms *static electricity* and *current electricity* to the following definitions and characteristics:
   (a) Electrons do not move along a path
   (b) Stationary buildup of electric charge on a substance
   (c) Electrons move along a path
   (d) Lightning is an example
   (e) Household wiring is an example
2. What statements make up the law of electric charges?
   (a) Like charges repel one another
   (b) Like charges attract one another
   (c) Unlike charges repel one another
   (d) Unlike charges attract one another
3. What particles are lost and gained when working with static electricity?
4. Why do two different substances build up static charges when rubbed together?
5. In your notebook, write the word(s) needed to complete each statement below.
   (a) A positively charged object has a(n) ____________ of electrons.
   (b) The law of electric charges states that unlike charges ____________
   (c) When a positively charged object touches an uncharged object, the uncharged object becomes ____________.
   (d) When cells are connected in ____________ the voltage of the battery increases.
   (e) When additional loads are connected in a parallel circuit, the total current ____________.
   (f) The amount of electrical ____________ use depends on the time the appliance is operating.
   (g) As the number of loads in a parallel circuit increases, the effective resistance ____________.
   (h) Nuclear fuels and fossil fuels are both ____________ sources of energy.
   (i) The primary purpose of a circuit breaker is to ____________.
   (j) The third, rounded pin on the plug of an appliance connects to the ____________ wire when it is plugged into a socket.
6. Indicate whether each of the statements is true or false. If you think the statement is false, rewrite it to make it true.
   (a) Neutral objects are attracted to charged objects.
   (b) Like charges repel one another and unlike charges attract one another.
   (c) To connect cells in series, the positive terminal of one cell is connected to the negative terminal of the next cell.
   (d) When the filament of a bulb in a series circuit (**Figure 1**) burns out, all the other bulbs remain lit.

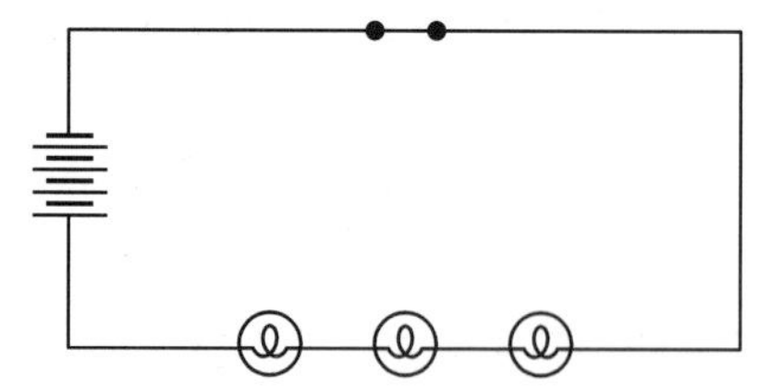

Figure 1

   (e) The amount of electrical energy in a battery depends on the electrical load connected to it.
   (f) Most of the electrical energy produced in Ontario is produced by nonrenewable energy sources.
   (g) The primary purpose of the circuit breaker and the fuse is the same.
   (h) A GFCI measures the current flowing in the ground wires.
   (i) The pressure in a water line can be compared with the voltage in an electric circuit.

(j) A large amount of current is needed to kill a person.

(k) Many energy conversions are 100% efficient.

(l) Electric cars were more popular than gas-powered cars at the turn of the twentieth century.

(m) Producing electricity by wind power does not depend on geographic location.

(n) The difference in the amount of voltage flowing after the electrons have flowed through a conductor is called voltage drop.

7. Match the terms *electric current, voltage,* and *resistance* to the following definitions:

(a) The force that moves electric charge in a circuit

(b) Electric charges that are moving from one place to another

(c) The force that tries to stop or slow the electric charge in a circuit

8. An electric circuit has four main parts. Identify them and explain how they work together in the circuit.

9. Describe the similarities and/or differences between each pair of terms listed below:

(a) positive charge; negative charge

(b) dry cell; battery

(c) open circuit; closed circuit

(d) series circuit; parallel circuit

(e) ammeter; voltmeter

(f) renewable energy source; nonrenewable energy source

(g) neutral wire; live or hot wire

(h) circuit breaker; fuse

(i) input energy; useful output energy

10. Match the terms *series* and *parallel* to the following characteristics:

(a) electrical energy not shared

(b) one path

(c) more than one path

(d) electrical devices can be on or off within the circuit

(e) electrical devices must all be on or all be off

(f) electrical energy is shared

11. Draw circuit diagrams for the following electrical circuits:

(a) one dry cell, one closed switch, one light, wired in series

(b) two dry cells, one open switch, two motors, wired in parallel

12. Homes are wired in parallel and not in series. List two advantages and two disadvantages of this method.

13. Explain how electric current, voltage, and resistance in an electric circuit can be compared with water flowing in a plumbing system. Use a diagram in your answer.

14. What electrical devices are used to measure the

(a) amount of electric current flowing in a circuit?

(b) electrical pressure in a circuit?

15. A voltmeter and ammeter are connected in electric circuits differently. Explain how the two electrical measurement devices are connected in an electric circuit to measure potential difference and current.

16. What is the difference between a conductor and a resistor?

17. Explain how a fuse controls the flow of electric current in a circuit.

18. Why should a burnt-out fuse never be replaced by one with a higher current rating, or with a piece of aluminum foil?

19. What does a load do to the flow of electrical current in a series circuit (**Figure 2**)?

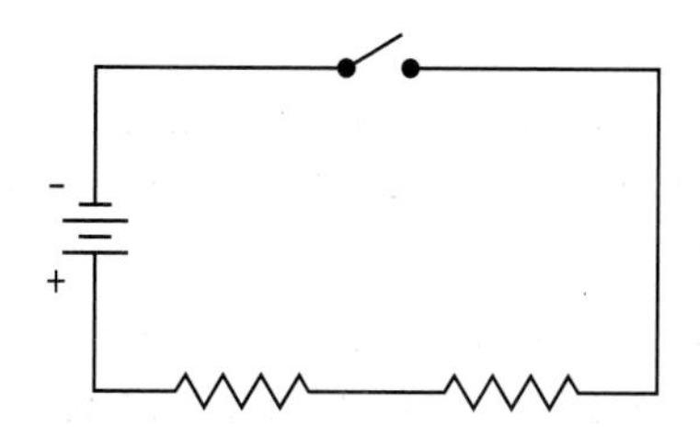

Figure 2

20. A simple series circuit has a current of 2 A and a resistance of 0.5 Ω. Calculate the voltage drop across this circuit.

21. A red-hot stove element has a current of 12 A and a potential difference of 240 V. Calculate the resistance of the element.

22. (a) Why aren't electrical devices 100% efficient?

    (b) Where is the energy lost in this process?

    (c) Can the efficiency of an electrical device be greater than 100%? Explain your answer.

23. What is the formula for calculating percent efficiency?

24. Calculate the percent efficiency of an electrical motor that uses 8000 J of electrical energy to produce 6200 J of energy output.

25. How does a fuel cell work?

26. What electrical energy source is the most commonly used in Canada? Why?

27. How much current is needed to kill a person?

Applying Skills

28. (a) Describe how to charge a pith ball by contact.

    (b) Draw a series of diagrams to show the movement of electrons as the pith ball in part (a) is charged.

29. Design and carry out an experiment that shows the law of electric charges using an ebonite rod, a Lucite rod, fur, polyester, string, and a retort stand.

30. Draw a labelled diagram showing how a typical circuit is connected from the distribution panel to the circuit breaker and then to three wall outlets.

31. **Figure 3** shows two circuit diagrams.

    (a) In which circuit will the ammeter reading be the greatest? Explain why.

    (b) Calculate the current in the circuit in **Figure 3a**.

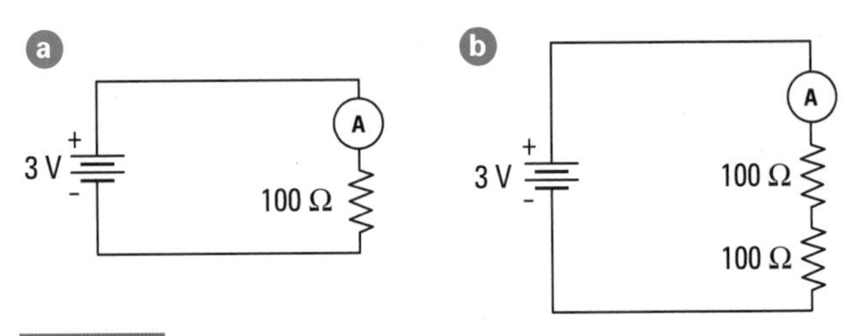

Figure 3

32. Design and draw a circuit that includes four dry cells, four light bulbs each with their own switch, a clock, and a motor that is protected by a fuse.

33. Describe and explain what would happen in the circuit diagram shown in **Figure 4** if

    (a) the switch is closed;

    (b) the switch is closed and light bulb 1 is unscrewed;

    (c) the switch is closed and light bulb 3 is unscrewed;

    (d) the switch is closed, and light bulb 6 is removed and replaced by a copper wire.

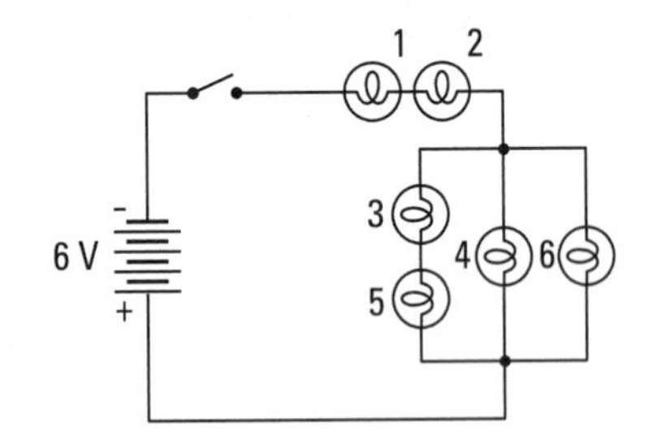

Figure 4

34. Determine the values of voltage indicated by the meter needle positions A and B shows in **Figure 5**.

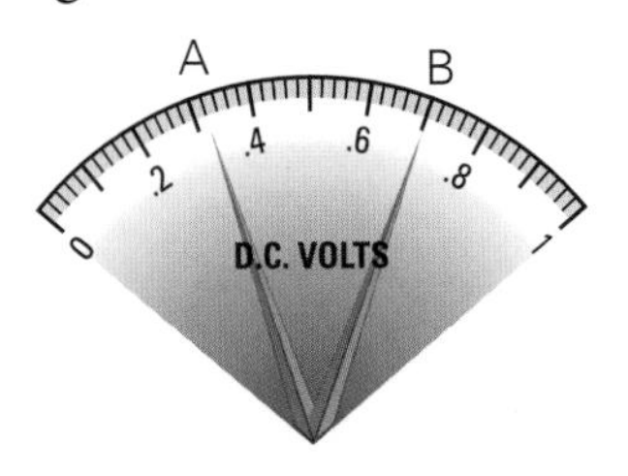

Figure 5

35. Determine the values of current indicated by the meter needle positions A and B shows in **Figure 6**.

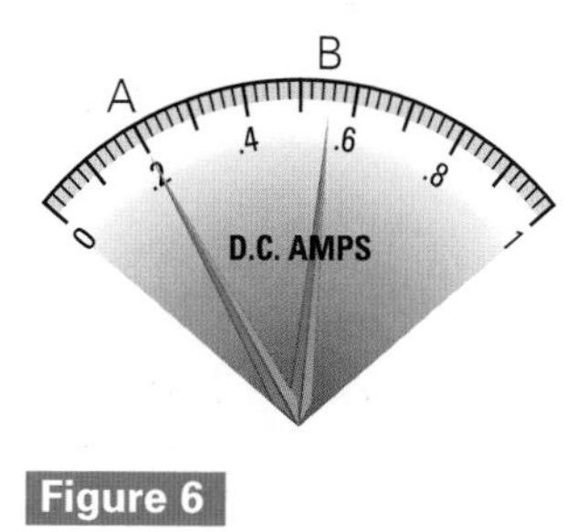

Figure 6

Making Connections

36. List three ways to control static electricity.
37. (a) What effect does an electric current have on human muscles?
    (b) Explain what is meant by the term "let-go threshold"?
38. Why are vehicles operated by fuel cells considered to be less polluting than those operated by gas? How could you convince someone to use such a vehicle?
39. Why are generators so important in rural areas?
40. If the demand for electricity were greater than the amount being generated, what might happen?
41. Why is the efficiency of the energy conversion process so important when considering different ways to produce electrical energy from both renewable and nonrenewable energy sources?
42. Identify five ways that electrical energy could be conserved in your home and at school.
43. Describe how you would use the Energuide labels attached to large appliances to decide which dishwasher you would buy.
44. How can GFCIs be used to upgrade the safety of bathrooms and other hazardous areas in a home?
45. Why is a GFCI considered better than a circuit breaker as a safety device?
46. What kinds of lights are used in most office buildings and schools? Explain why.
47. What safety precautions must be followed
    (a) when wiring a circuit?
    (b) around water and electricity?
48. Describe two types of jobs for someone with an electrician's diploma.

Space Exploration

Scientists are learning more about planet Earth and its solar system every day, thanks to the men and women involved in space exploration. By using telescopes, space probes, and piloted missions, we learn about what is happening in space. By increasing our knowledge and understanding of space, we satisfy our natural curiosity and desire to explore, and we come closer to answering the question "How do we fit into the universe?"

Unit 4 Overview

Overall Expectations

In this unit, you will be able to

- learn about our solar system and the universe
- design and conduct investigations into the motion of objects in the sky
- identify how our interest in space has helped us to understand outer space, Earth, and living things

The View from Earth

Many myths and legends were created to explain what people saw in the sky. The patterns in the sky helped people to find their way and to create a calendar.

Specific Expectations

In this unit, you will be able to

- recognize natural objects in the sky
- describe and compare different parts of the solar system
- connect the beliefs of various cultures about the various objects that appear in the sky to parts of their civilization
- carry out experiments about the motion of visible celestial objects, and gather, record, and communicate the outcome

Understanding Space

The universe is constantly changing. Its objects are always moving. Models and simulations help us to understand what is actually happening to the many objects in the universe.

Specific Expectations

In this unit, you will be able to

- describe the Sun and its effects on Earth
- identify problems that occur when observing celestial objects, and describe ways these problems can be solved
- identify and describe major components of the universe
- outline the current theory of the origin of the universe and the evidence that supports it
- select and integrate information from a variety of sources

Space Research and Exploration

Space exploration and the related technology increase our understanding of the universe and improve many areas of life on Earth.

Specific Expectations

In this unit, you will be able to

- describe and explain the effects of space on organisms and materials
- describe problems that may occur when observing celestial objects, and describe their solutions
- outline questions about a problem or issue in space exploration
- plan and conduct an investigation about space exploration
- identify and assess the impact of developments in space research and technology on other fields
- describe Canada's role in space research and development

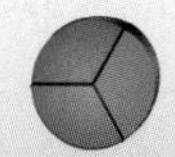

Challenge

Space—What Is Next?

As you learn about space, think about the challenges that scientists have overcome. In all cases, the key is communication. You will be able to show your learning by completing one of the following Challenges. For more information on the Challenges, see the following page.

1 A Tour Book

The universe is huge with no simple road map available. You have been hired to create a tour book of the universe. You must include directions to all locations and interesting information about each tour site. Your teacher will let you know how many sites you must include in your tour book.

2 A Space Colony

One goal of the space program is to discover a planet that will support human life. This requires water and oxygen at the very least. Design a space colony suitable for humans to occupy permanently.

3 A Space Technology Information Package

You will organize information about space exploration and the study of astronomy, highlighting its influence on our lives, and communicate it effectively to the public.

Record your ideas for your Challenge as you progress through this unit, and when you see

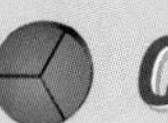

Challenge

Unit 4 Challenge

Space—What Is Next?

People are interested in space exploration. We want to know what else is in space and whether it is possible to live anywhere else in the universe. Through education, research, and communication, we will find out.

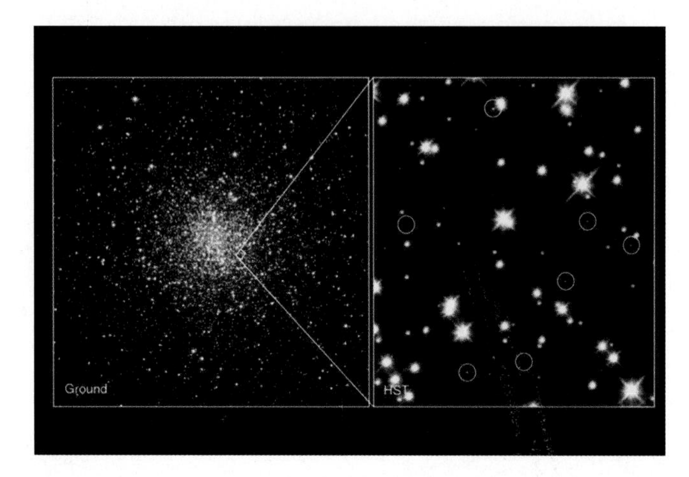

1 A Tour Book

One way that people learn about a location is by experiencing it firsthand. People plan trips by looking through tour books and deciding what interests them most. You have been hired to create a tour book of the universe. You must include interesting information and directions to all the important sites. Your teacher will let you know how many sites you must include. Create a tour book of the universe that will make people want to visit.

Your tour book should include

- a creative title
- at least seven sites (your teacher will let you know how many sites you are responsible for)
- directions to the sites
- basic information about each site, such as how it was formed, how old it is, when and how it was discovered
- an explanation of why each site is included in this tour book and what makes it special
- a photo or picture of each site
- colour and computers graphics or layout

2 A Space Colony

One long-term goal of the space program is to discover a planet that will support human life, or to create a colony using technology. Assume that all the technology needed has been developed, such as a device that can generate oxygen. Design a space colony in which humans could live permanently. What and who would you take with you?

Your plan should include

- the name of the colony
- a model or detailed outline of the colony
- a list of the number, sex, and age of the colonists and their skills
- the number of spacecraft needed to get to the colony
- an outline of how the colonists will be able to manage on their own in the short and the long term
- a description or diagram of the technology that has been developed for this program

3 A Space Technology Information Package

Researchers have developed a lot of important technology in the past 50 years. In 1952, the first satellite was launched. Today, the world has come together to build the International Space Station (ISS). Research two technologies that have helped us learn more about space, and create an information package. The package can be in written form or poster form or both. You will have to decide at which age group your space technology information package will be aimed.

Your information package should include

- a title page and table of contents, if it's written
- a title and clear subheadings, if it's a poster
- a description of the development of each technology
- a picture or diagram of each technology
- an explanation of the purpose of each technology
- an account of Canada's role in each technology
- a summary of what we have learned from using the technologies
- an explanation of how these technologies have affected our daily lives (for example, Velcro was originally developed for use in space)

When preparing to use any test or carry out an experiment, have your teacher approve your plan before you begin.

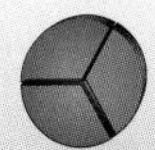

Assessment

Your completed Challenge may be assessed according to the following:

Process

- Understand the specific challenge.
- Develop a plan.
- Choose and safely use appropriate tools, equipment, materials, print resources, and computer software.
- Analyze the results.

Communication

- Prepare an appropriate presentation for the task.
- Use correct terms, symbols, and SI units.
- Use information technology.

Product

- Meet established criteria.
- Show understanding of the science concepts.
- Show effective use of materials.
- Address the identified situation or problem.

Unit 4 — Getting Started

Going on a Trip?

Curiosity is a part of human nature. It is what helped to develop vaccines and what took us to the Moon.

Curiosity has been the driving force behind learning more about outer space and perhaps finding a new location to colonize. The International Space Station (ISS) is the start of a worldwide effort to colonize outer space.

The year 2001 saw a tourist visit the International Space Station for the first time. This is one of the many events associated with the space program that make the world pay attention to what is happening in the universe.

The space program is an international project. People from many countries are involved in projects such as the ISS and the missions to Mars.

One long-term goal of the space program is to create an environment that would support life somewhere other than on Earth. So far, the efforts have produced six successful missions to the Moon, many satellites orbiting the Earth, the ISS, and several space probes to Mars and beyond. Mars seems to be the most promising location to colonize.

The astronauts who live on the ISS are colonists (**Figure 1**). Each astronaut was chosen for his or her specific skills. All had to pass tough physical tests. Each one had to make decisions about what to bring for the stay on the ISS.

Technology may advance so far that visiting the ISS or Mars will be possible for your children or grandchildren.

Figure 1

Astronauts out for a space walk outside the ISS

What DO YOU ALREADY *Know?*

1. List five things that are necessary for survival. Begin your list with *warmth*.
2. When people are camping, or when there is a blackout, certain things make life easier, such as matches and candles. List four other items that you can use without electricity.
3. Everyone needs an area that is all his or her own. Think about what things truly make you feel at home. List four of these items.

Throughout this unit, note any changes in your ideas as you learn new concepts and develop your skills.

Try This Activity A Packing List

You are going to visit the ISS for three months. You will be expected to help around the space station. You must make a list of everything you want to bring, so that ISS officials can approve it.

Nutritious food will be provided for you; however, it will be freeze-dried (just add water) and will not include specialty items like ice cream or chocolate. Your part of the refrigerator is half the size of a shoebox. Fresh fruit and vegetables are available once a month when shipments come from Earth.

Your personal space on the ISS is about half the size of your locker. As for clothing, the ISS provides everything. You can take a two-and-a-half minute shower once every three weeks. A limited amount of water is available for washing twice a day.

Answer the following questions honestly as you try to decide what to take with you.

(a) Think about a typical day in your life here on Earth. List five things you do and what you need to do them.

(b) From your list in (a), choose two items that you could live without for three months in space. Cross them out.

(c) What would you need in space that you would not need on Earth? Add three items to your list in (a).

(d) Since you will be in space for three months, how much of each item in (a) will you need? Add the quantities to your list.

(e) Compare your list with those of your classmates. Make changes to your list if you want to.

Challenge

2 Keep the packing list in a folder. This is the beginning of your preparations to go to the colony. You can add to the list or change it as you work through the unit.

What Can We See in the Sky?

What are some of the objects you can see when you look up at the sky? You can see the Sun, the Moon, airplanes, satellites, and, on a clear night, stars and planets. The night sky has very distinct patterns and motions. Astronomers can predict, with accuracy, when we can see certain objects in the sky. The study of what is in space beyond Earth is called **astronomy**.

Sky watching is a pastime for many people. People who sky watch are called **astronomers**. Using only your eyes, you can be an amateur astronomer.

Star Constellations

People have been stargazing for thousands of years. Some groups of stars seem to form the shapes of animals, ancient heroes or gods, and everyday objects. Names have been given to some of these patterns. Groups of stars that seem to form shapes or patterns are called **constellations** (**Figure 1**). Since the stars are moving, what looked like a bird thousands of years ago may look like a fish today. For this reason, books show the shapes of the constellations by joining the stars together with lines (**Figure 2**).

Constellations, because of their predictable motion, have been used as calendars, timekeepers, and direction finders by people travelling on land and by sea. Different constellations were, and still are, used for these purposes in some parts of the world.

Objects in the Solar System

The **solar system** is made up of the Sun and all the objects that travel around it. The planets are shown in **Figure 3**. The planets are **nonluminous**. Light must reflect off them for them to be seen. Objects that give off their own light, like the Sun and other stars, are **luminous**.

Figure 1

The Big Dipper is part of the constellation Ursa Major, also called the Great Bear. It is probably the best-known group of stars in the Northern Hemisphere.

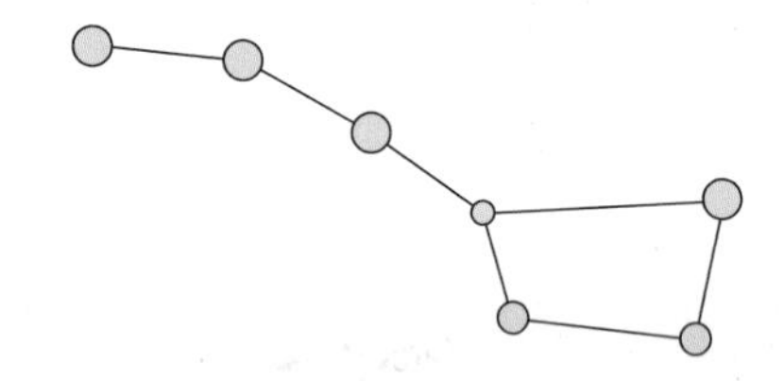

Figure 2

A common way of drawing the Big Dipper

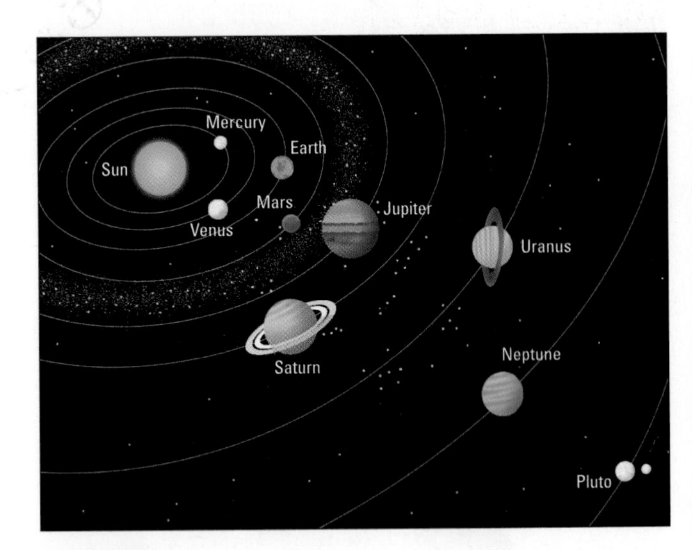

Figure 3

This drawing shows the planets of our solar system (not drawn to scale).

Observing Stars and Planets

A **star** gives off huge amounts of heat and light. A **planet** is a sphere-shaped object that follows a certain path around a star. Looking up at the night sky, how can you tell a planet from a star? We can only see five planets without binoculars or telescopes: Mercury, Venus, Mars, Jupiter, and Saturn. These planets are closer to the Earth than the stars are, so they appear larger than the stars. **Table 1** lists the many differences between stars and planets. If you have been stargazing and have seen a star that looks like something from **Figure 4**, chances are you were looking at a planet.

Venus

Jupiter

Mars (bright when close to Sun)

Sirius

Saturn

Mars (faint when far from Earth)

Figure 4

The brightest star in the night sky is Sirius. It is brighter than Saturn but not nearly as bright as Jupiter or Venus.

Table 1 Comparing Planets and Stars

Feature	Planet	Star
location	in the solar system	far beyond the solar system
distance from Earth	fairly near	very far
real size	smaller than most stars	usually larger than the planets
reason we see object	reflects light from the Sun	gives off its own light
surface temperature	usually cool or very cold	very hot
what object is made of	usually rocks or gases	gases under high pressure and temperature
observable feature	has a steady light	appears to twinkle
long-term observable feature	very slowly wanders through some constellations	appears to move as part of a constellation

Understanding Concepts

1. Give one reason why
   (a) it is very difficult to see Neptune and Pluto without a telescope.
   (b) we see Jupiter in the night sky more often than we see Mercury.
2. In your own words, explain four differences between a star and a planet for someone who knows little about astronomy.

Making Connections

3. People once used constellations to help them find their way when they travelled. Why do you think this was possible?
4. Describe how constellations could help a traveller today.
5. Do you think that astronomy is an important area of study? Why or why not?

Challenge

1 Will you choose one of the objects in our solar system as a site to visit in your tour book?

Planets on the Move

The Effect of Earth's Rotation

If you watch the sky at night for at least an hour, you will notice that the positions of the stars and planets slowly change. Like the Sun and the Moon, most stars seem to rise in the east, travel across the sky, and set in the west. This makes sense, if you think about Earth's motion.

A day on Earth is 24 hours long. This is the time it takes for Earth to complete one rotation. A **rotation** is one spin on Earth's axis. Due to the spinning motion, the Sun, the Moon, and most stars seem to rise in the east and set in the west. The Earth's axis is the imaginary straight line from the North Pole to the South Pole. **Polaris**, or the **North Star**, which is a star in the Little Dipper, is on this imaginary axis, right above the North Pole (see **Figure 1a**). For this reason, Polaris can be seen at all times of the year from countries in the Northern Hemisphere, including Canada. Polaris is never seen from countries in the Southern Hemisphere, such as Australia, because Earth blocks the view.

Polaris seems fixed in the night sky, because it is in line with Earth's axis. All other stars seem to revolve around Polaris. This can be demonstrated, as shown in **Figure 1b**.

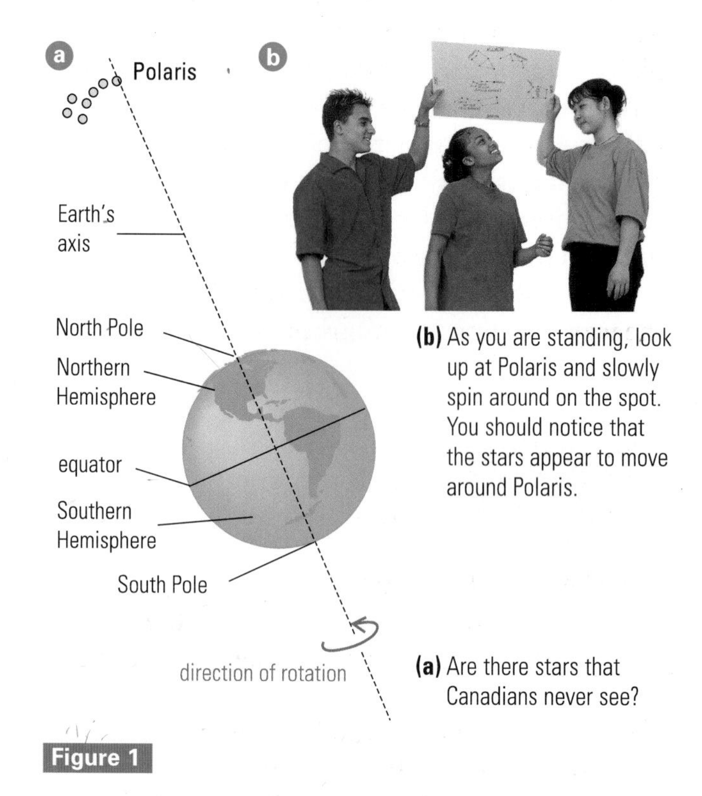

(b) As you are standing, look up at Polaris and slowly spin around on the spot. You should notice that the stars appear to move around Polaris.

(a) Are there stars that Canadians never see?

Figure 1

The Effect of Earth's Revolution

Another type of motion is **revolution**, which is the movement of one object around another. It takes Earth one year to revolve around the Sun. This motion lets us see different constellations during different seasons, as shown in **Figure 2**.

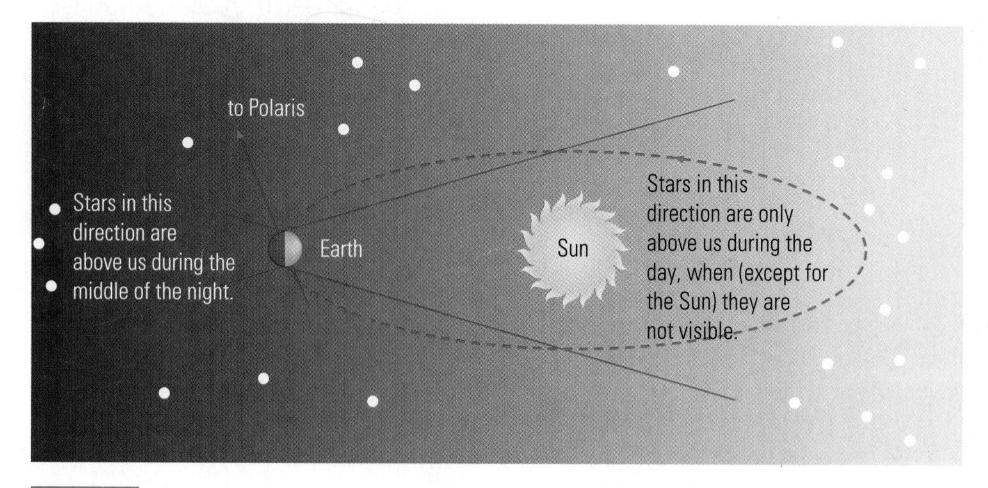

Figure 2

In the Northern Hemisphere, the only constellations we can see in all seasons are those that appear close to Polaris. The rest we see only when they are not in the sky at the same time as the Sun.

The Effect of Planets' Motion

In most cities, if you were to look up at the night sky, you would not see very much. The city lights make it hard to see the stars. In ancient times, before city lights, people were more aware of the night sky. Not only did they track constellations, but they also took note of five bright points of light that moved differently from the way the constellations did. These were called "wandering stars" or planets, taken from the Greek word *planetes*, meaning "wanderers."

The reasons the planets move differently from the way stars do are that the planets are all travelling around the Sun and that they are much closer to us than the stars. If you observe the planets over weeks or months, you will see that they do not travel across the sky in a night. They wander through the constellations.

The planets seem to move through many constellations that are shaped like animals. The Greek word for "animal sign" is *zodion*, so these constellations were called the **zodiac constellations**. You may have been asked the question "What is your sign?" Your zodiac sign depends on your birth date.

Try This Activity: You Are Earth

- Use your classroom to stargaze.
- Place diagrams of the Big Dipper, the Little Dipper, and Cassiopeia on the ceiling of the classroom.
- Place a diagram of Aquila (summer) on the east wall of the classroom, Pegasus (autumn) on the north wall, Orion (winter) on the west wall, and Leo (spring) on the south wall.
- Choose one person to be the Sun in the middle of the room. Pretend you are the Earth.
- Stand east of the model of the Sun and slowly turn counterclockwise. When your back is to the Sun, it is night on Earth. What can you see? As you spin around to day, there may be constellations on the horizon, but it is too bright for you to see them clearly.
- Try this at locations north, south, and west of the model of the Sun.

Challenge

1 Polaris has had a huge effect on Earth's history, as some explorers from ancient times used it to find their way to North America. Would you consider Polaris as a site for your tour book?

2 Is Polaris a possibility for a space colony? Why or why not?

Understanding Concepts

1. Describe Earth's rotation. What time period does it define for Earth?
2. Describe Earth's revolution. What time period does it define for Earth?
3. Why does a constellation appear to change position during the night?
4. In **Figure 3**, the stars appear to revolve around a single spot. What was the camera pointing at?

Figure 3

Night sky as observed in the Northern Hemisphere

Making Connections

5. **(a)** Describe two problems that using stars to navigate while travelling could cause.

   (b) Would one star be more useful for navigation than any of the others? Explain your answer.

Different Views of the Sky

People have not only looked around them, but they have also looked up. Many cultures recorded their history along with their observations of the night sky. Sometimes legends, myths, or religious beliefs were developed to explain the natural world. An explanation of the unknown allowed people to put aside their fears.

Ideas about the Night Sky

Records of astronomy have been found in caves, in tombs, and on pottery, all of which are thousands of years old. This makes astronomy one of the oldest sciences (**Figure 1**).

The stories about the objects in the sky vary from culture to culture. Canadian First Nations have many legends that describe **celestial** or space events. Here are a few of them.

- The Menomini of the Great Lakes region tell this legend about meteors: When a star falls from the sky it leaves a fiery tail. It does not die; rather, its shade goes back to shine again.
- The Kwakiutl of the west coast region believed that a lunar eclipse (Earth casting a shadow on the Moon) occurred when the sky monster tried to swallow the Moon. By dancing around a smoky fire, they hoped to force the monster to sneeze and cough up the Moon.
- The Tsimshian, also of the west coast, believed that each night the Sun went to sleep in his house but allowed the light from his face to shine through the smoke hole in the roof. The stars were sparks that flew out of the Sun's mouth. The full Moon was the Sun's brother who rose in the east when the Sun fell asleep.

(**a**) What did each group actually see happening in the sky?

(**b**) Why would each of these celestial events have scared people?

(**c**) Why do you think people were unafraid when a legend explained these events?

Figure 1

This ancient Aztec carving show that Aztecs studied the sky.

Developing Skills

Seven thousand years ago, people began settling on farms and in towns. The cycles of the Sun and Moon helped create a calendar. The calendar helped people to decide when to plant and harvest crops, as well as when to hold feasts.

- More than 3000 years ago, the Chinese developed a very accurate 365-day calendar.
- Greek and Chinese astronomers created star maps more than 2000 years ago. These maps, created using only their eyes and instruments to measure angles, showed the brightness and position of more than 800 stars.
- The Maya are an Aboriginal people of Central America. More than 1000 years ago, they developed ways to keep track of the motion of the planets. An ancient Maya book shows that they were able to predict the appearance of Venus. In 500 years, their predictions were off by only two hours.

(d) What skills did Chinese and Greek astronomers need to make their calendars?

(e) List three ways in which a calendar helped people 3000 years ago.

(f) List three ways in which a calendar helps you today.

(g) Three thousand years ago, Greek and Chinese astronomers mapped 800 stars. We know now that there are billions of stars. Why did these astronomers map only 800 stars?

Evidence of Applications

Many structures that are thousands of years old were built to show celestial events.

- In Egypt, a temple was built with a long, narrow entrance through which the Sun's rays hit a statue of the pharaoh on two days of the year: one in February, the other in October.
- Stonehenge (**Figure 2**) was built in England 5000 years ago. It is believed to show the longest day of the year and predict eclipses.

(h) If the Egyptian statue was in sunlight for only two days of the year, the Sun must be in that position only on those days. How did astronomers from ancient times discover this?

Figure 2

The huge stones at Stonehenge are arranged in a well-organized pattern. Scientists think that the people who built Stonehenge used the pattern to predict cyclic events.

Challenge

3 How did legends and myths lead to modern astronomy? How do these stories affect our modern point of view?

Work the Web

Find out more about the legends and myths of different cultures. Start your search by visiting www.science.nelson.com and following the links from Science 9: Concepts and Connections, 4.3. Summarize your favourite legend or myth.

Making Connections

1. In Canada, the Sun is highest in the sky on the first day of summer and lowest on the first day of winter. Use this information to design a calendar that will tell you when school begins, when the new year starts, when summer holidays begin, and when it is your birthday, according to the position of the Sun.
2. State two similarities between what people in ancient times used to view the sky and what you use to view it today.
3. State two differences between what people in ancient times used to view the sky and what you use to view it today.

Exploring

4. The reason we have a day and night cycle is that the Earth rotates on its axis once every 24 hours. Create a legend or myth to explain day and night.

A Seasonal Star Map

Reading a map is a useful skill. Like a street map, a star map is a tool that helps you to identify what is around you. In this activity, you will build your own seasonal star map, and you will learn to use it in the classroom. With your parent's or guardian's permission and under adult supervision, you can also use the map outside at night.

Part 1: Building a Seasonal Star Map

Materials

- copy of seasonal star map (Skills Handbook)
- cardboard (8.5″ × 11″)
- acetate (8.5″ × 11″)
- scissors
- glue
- copy of star map window frame (Skills Handbook)
- paper fastener or thumbtack
- phosphorescent paints (optional)

Be careful when using scissors. If you use any phosphorescent paints, be sure to follow the instructions on the container.

Procedure

1 Use **Figure 1** as a reference. Carefully cut out the seasonal star map and a piece of cardboard the same size.

2 Glue the map onto the cardboard.

3 Make a hole in the centre of the cardboard and the star map for the paper fastener.

4 If possible, use one colour of phosphorescent paint for the constellations and another for the labels on the window frame and the star map. (The paint will help you see the map in the dark after it has been in bright light.)

5 Cut out the star map window frame. Cut out the window area from the star map window frame without cutting the frame itself.

6 Glue the window frame to a piece of acetate. Cut the outer part of the acetate, but *do not cut* the window out of the acetate.

7 In the centre of the circle, make a hole for the paper fastener.

8 Use the paper fastener to fasten the acetate to the star map. Be sure the window frame rotates freely.

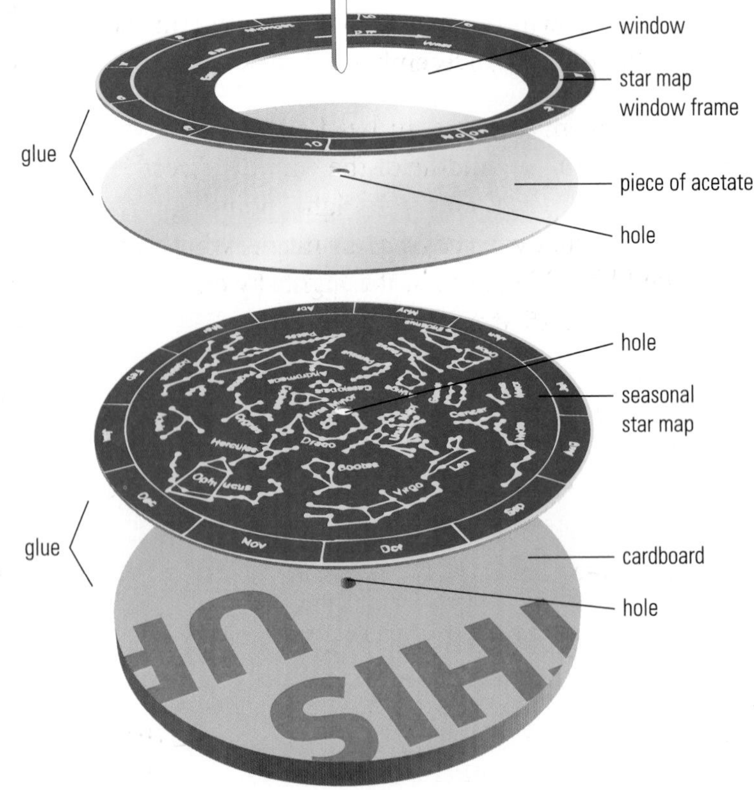

Figure 1
Putting together the seasonal star map

A planisphere is a seasonal star map available in specialty stores. If you can buy or borrow one, compare the planisphere to the seasonal star map you made. If you buy a planisphere, make sure it is for your location. For example, a planisphere for use in northern Canada is different from one for use in southern Canada.

Part 2: Using the Seasonal Star Map

Materials

Seasonal star map made in Part 1

Procedure

9 While in the classroom, pretend you are outdoors. Put an N on the north wall, a W on the west wall, an S on the south wall, and an E on the east wall of the classroom.

10 To use the map at midnight on December 15, rotate the window frame so that "midnight" is in the middle of December.

11 (F3) Hold the map over your head, with the map facing you, the middle of the window directly above your eyes, and "midnight" pointing north. What you see on your map is what you would be able to see in the night sky on December 15 at midnight.

(a) (C2) Which constellations would you be able to see near the northern horizon? the western horizon? the eastern horizon? the southern horizon?

(b) Which constellations would you see directly overhead?

12 Compare your predictions with those of another student.

(c) After comparing your results from (a) and (b) with those of another student, do you want to change your predictions? If so, make the changes in a different-coloured pen.

13 Repeat steps 10, 11, and 12 for midnight on May 15.

(C2) **(d)** Record your predictions.

14 If you look at the sky before midnight, the stars you expect to see will be rising in the east. If you look at the sky after midnight, the stars you expect to see will be setting in the west. Prove this to yourself by setting your star map to 8:00 P.M. on December 15 and then to 4:00 A.M. on December 15.

(e) Describe what you discovered.

Work the Web

Find out what is visible in your region at night at this time of the year, other than the stars on your star map. Are any planets expected to be visible, or are any comets or meteors predicted to pass through your region? Start your search by visiting www.science.nelson.com and following the links from Science 9: Concepts and Connections, 4.4.

Understanding Concepts

1. As you rotate the window frame, which constellations can you see no matter what month is shown?
2. Name the constellations that are visible nearer the horizon at midnight in (a) July and (b) March.

Making Connections

3. In societies where calendars are not common, the seasonal positions of the constellations are used to indicate when various festivals or activities should take place. If a crop has to be planted in October, what constellations should a North American farmer look for?

Exploring

4. Describe how you would try to find a planet in the sky if you were observing it tonight.

2 How would a seasonal star map help in a space colony? Would it have to be modified? Why or why not?

The Planets in the Solar System

Included in our solar system are the Sun, the nine known planets, and the moons of those planets. All the planets revolve around or **orbit** the Sun. The time it takes a planet to complete its revolution or orbit is called one **orbital period**. A planet's orbital period is considered a year for that planet.

All the planets also rotate on their axes. The rotation period is the time it takes the planet to spin once on its axis. This defines a day for that planet and creates its day and night cycle.

Each planet has a gravitational pull. If the planet has a greater mass than Earth, its force of gravity is stronger than that of Earth. If it has a smaller mass than Earth, its force of gravity is weaker.

The planets can be split into two groups: the **inner planets** and the **outer planets**. Pluto is in a category by itself, because it has features that fit into both groups.

Unpiloted space probes, such as *Voyager 2* (**Figure 1**), have flown close to each planet, except Pluto, and have sent back pictures and information to Earth. This is safer and faster then sending humans out into space for long periods.

Figure 1

Once the unpiloted space probe *Voyager 2* leaves our solar system, it will take about 20 000 years to travel as far as the closest star other than the Sun.

The Inner Planets

The planets closest to the Sun are Mercury, Venus, Earth, and Mars. They are also known as the **terrestrial planets,** because they are similar to Earth. They are all small, rocky planets.

Mercury

- Mercury is the closest planet to the Sun at about 59 million km away.
- Mercury has the shortest orbital period or year: 88 Earth days.
- One "day" (rotation) on Mercury takes 59 Earth days.
- Surface temperatures can reach 400°C during the day and –180°C at night.
- Mercury has no moons, rings, or atmosphere.
- Mercury's craters were caused by chunks of rock colliding with the planet (**Figure 2**).
- Because Mercury is so close to the Sun, it is not usually visible at night. If it is visible, it usually right after sunset or right before sunrise.

Figure 2

The surface of Mercury

Venus

- Venus is the second closest planet to the Sun at a distance of about 108 million km.
- The orbital period or year (225 Earth days) on Venus is shorter than its rotation period or day (243 Earth days).
- Venus rotates in the opposite direction from most other planets.
- Venus is the hottest planet. Its surface temperature is 470°C.
- Venus has no moons or rings.
- It has huge volcanoes (**Figure 3**).
- Venus is the brightest object in the sky after the Sun and Moon. It is known as both the morning star and the evening star, even though it is a planet.

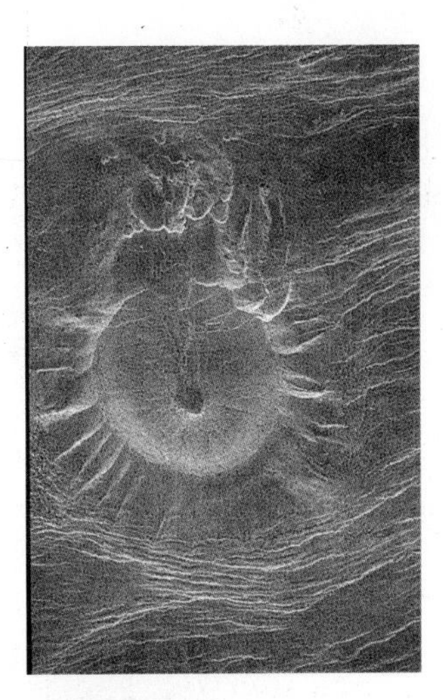

Figure 3

Volcanoes of Venus

Earth

- Earth is third closest to the Sun at a distance of about 150 million km.
- Surface temperatures on Earth can vary from −85°C to 65°C.
- Earth has one moon but no rings.
- Earth supports life as we know it. Water covers 70% of Earth's surface (**Figure 4**).
- The ozone in the atmosphere protects Earth from the damaging light rays of the Sun, but it lets enough light through to support life.
- The soil on Earth is important for plant growth.
- Earth has some active volcanoes, but for the most part, it is a stable planet.

Figure 4

Earth

Mars

- Mars is fourth closest to the Sun. It is about 228 million km away.
- An orbital period or year on Mars is 687 Earth days. This is nearly twice an Earth year.
- A day on Mars is 24.65 Earth hours.
- The surface temperatures on Mars range from –120°C to 30°C.
- Mars has two moons and no rings.
- Mars is known as the red planet because of the reddish colour of its rusty soil.
- In 1997 the probe *Pathfinder*, along with the *Sojourner* rover, landed on Mars and sent back all kinds of data (**Figure 5**).

Figure 5

Scientists believe that Mars is a possible location for a space colony.

The Outer Planets

The planets further from the Sun are Jupiter, Saturn, Uranus, and Neptune, also known as the **gas giants**. They do not seem to have a solid surface, but rather they have a small dense core of liquid or solid elements.

Jupiter

- Jupiter is fifth closest to the Sun, at a distance of 778 million km.
- Its orbital period or year takes 11.86 Earth years.
- Jupiter has the shortest day: 9.85 Earth hours.
- Its average surface temperature is –160°C.
- Jupiter has 39 moons and a thin, almost invisible ring around it.
- Jupiter has many coloured bands around it, which are different cloud belts of gases spinning around the surface.
- Jupiter is the largest planet (**Figure 6**).
- It has a Great Red Spot that is actually a huge hurricane larger than two Earths.
- Galileo Galilei discovered Jupiter's first four moons (the Galilean moons) by using the first telescope. Using a telescope, Io, Europa, Ganymede, and Callisto (the moons) are visible from Earth.

Saturn

- Saturn is sixth closest to the Sun, at a distance of 1.5 billion km.
- An orbital period or year can take 30 Earth years.
- A day on Saturn is 10.65 Earth hours.
- Its average surface temperature is –180°C.
- Saturn has 30 moons and the most rings (more than 1000) of all the gas giants (**Figure 7**). The rings are made up of chunks of ice and rock.
- Saturn is the second largest planet in the solar system.
- Saturn could float on water because it is much less dense than water.

A probe launched today would take 30 years to reach Neptune. The *Voyager 2* (**Figure 1**) was launched in 1977. It took only 12 years to reach Neptune because Jupiter, Saturn, Uranus, and Neptune were lined up in a way that allowed their forces of gravity to increase the speed of the probe. This alignment happens only once every 176 years.

Figure 6

Jupiter and one of its moons

Figure 7

Saturn's rings

Uranus

- Uranus is the seventh closest planet to the Sun. It is 2.9 billion km away.
- Its orbital period or year is 84.1 Earth years.
- Uranus rotates in the opposite direction from most other planets.
- Its day takes 17.3 Earth hours.
- Its average surface temperature is –210°C.
- Uranus has 20 moons. Its rings are almost invisible from Earth (**Figure 8**).
- It is the third largest planet in the solar system.
- Its atmosphere is mostly hydrogen.

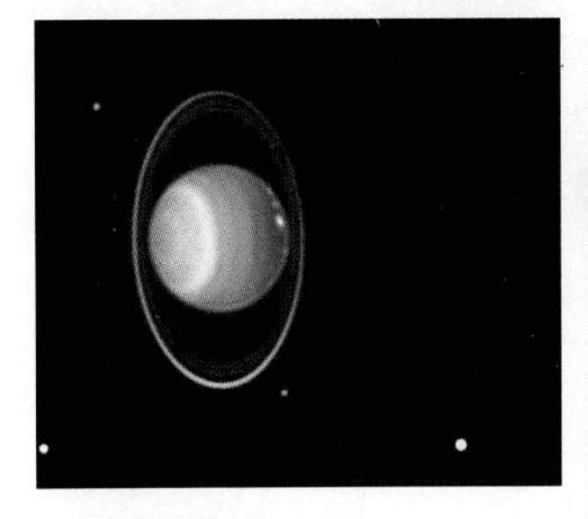

Figure 8
Uranus

Neptune

- Neptune is the eighth planet from the Sun. It is 4.5 billion km away.
- Its orbital period or year is 165 Earth years.
- A day on Neptune is 15.7 Earth hours.
- Its average surface temperature is –220°C.
- Neptune has eight moons and several rings.
- Neptune appears blue because of methane gas.
- Neptune sometimes has a Great Dark Spot, which appears to be a storm (**Figure 9**).

Figure 9
Neptune

In a Class by Itself

Pluto

- Pluto is the farthest planet from the Sun. It is 5.9 billion km away.
- It has a long orbital period or year (248 Earth years), like the outer planets.
- A day on Pluto is 6.7 Earth days.
- Pluto is cold. Its surface temperature is –220°C.
- Pluto has one moon (**Figure 10**) and no rings or atmosphere.
- Pluto is the smallest planet. It is a mixture of rock and ice like the inner planets.
- Pluto was discovered in 1930.

Figure 10
Pluto and its moon, Charon

Challenge

1 Should a planet from our solar system be included in your tour book?

2 Would the space colony be located somewhere in our solar system? Give reasons for your choice. How will the planet conditions need to be adjusted for humans to survive?

3 What technologies have been developed to gather information about the objects in our solar system?

Understanding Concepts

1. Create a mnemonic sentence to help you remember, in order, the names of the planets.
2. Why are the inner planets also called the terrestrial planets?
3. **(a)** List two features of Earth that make it unique among the planets.
   (b) List two features of Earth that make it similar to the other planets.

Making Connections

4. List five ways that humans have had an impact on Earth. Explain how each way has had positive and negative effects on Earth.

Work the Web

Find out more information about your favourite planet. Is there a probe headed there right now? When was the last time a probe reached the planet? Have scientists found out anything new about the planet lately? Start by visiting www.science.nelson.com and following the links from Science 9: Concepts and Connections, 4.5.

4.6 Investigation

INQUIRY SKILLS
- ○ Questioning
- ○ Hypothesizing
- ○ Predicting
- ○ Planning
- ○ Conducting
- ● Recording
- ● Analyzing
- ● Concluding
- ● Communicating

Planets and Retrograde Motion

People in ancient times thought that Earth was the centre of the universe. They believed that everything, including the Sun and the stars, revolved around Earth. In the early seventeenth century, this idea was replaced by the theory that the planets revolve around the Sun.

Seen from Earth, the stars, the Moon, and the Sun rise in the east and set in the west. The planets do not. The planets appear in the night sky in the east, but they wander slowly through the constellations over a few months. In this way planets are different from stars.

We can only see five planets without binoculars or telescopes: Mercury, Venus, Mars, Jupiter, and Saturn. We can see Mercury and Venus only at dusk or dawn. Mars, Jupiter, and Saturn appear at different times during the night and year. Once a year, these planets seem to loop backward in the night sky (**Figure 1**). This apparent backward loop is called **retrograde motion**. The Sun-centred model explains retrograde motion. The Earth-centred model did not explain this motion.

The planets are sometimes in retrograde motion because of the path of their orbits (**Figure 2**) and their speeds in those orbits. Earth catches up to, and then passes, Mars in orbit. This occurs because Earth orbits the Sun faster than Mars does. It takes Earth 12 months, while it takes Mars 22.5 months.

Figure 1

Jupiter's path across the night sky while in retrograde motion

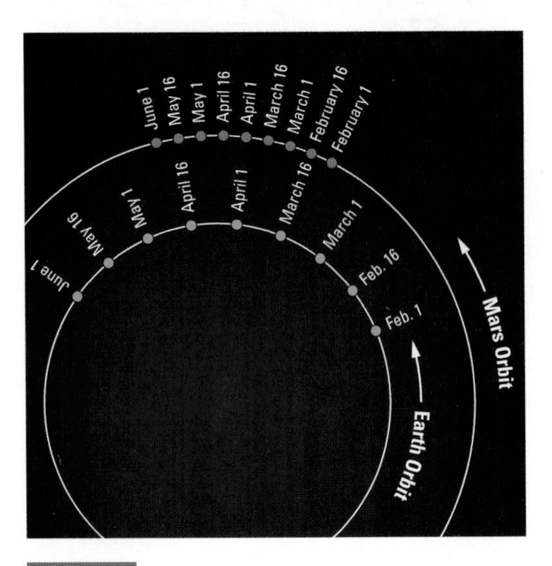

Figure 2

The orbits of Earth and Mars

Question

When does Mars show retrograde motion?

Materials

- large classroom chalkboard
- 2 "planets" (polystyrene balls, one coloured blue, one red) on 1.5 m sticks
- chalk
- measuring tape
- 22 cm × 36 cm paper
- graph paper
- ruler

Procedure

1 Work in a clear floor space.

2 Your teacher will mark Earth's and Mars's orbits on the floor using masking tape and a marker, as shown in **Figure 2**.

3 Mark a point "O" at the right edge of the chalkboard. Label it 0 cm.

4 Work in groups of three. One person is Earth and takes the blue planet. One person is Mars and takes the red planet. The third person is the recorder.

5 Choose a date. The Earth person stands at that date on the Earth orbit. The Mars person stands at that date on the Mars orbit. The recorder is at the chalkboard.

6 The Earth person places Earth on the spot, stands behind Earth, covers one eye, and lines up Earth and Mars. The Earth person tells the recorder where to move and mark the apparent position of Mars with an X (**Figure 3**).

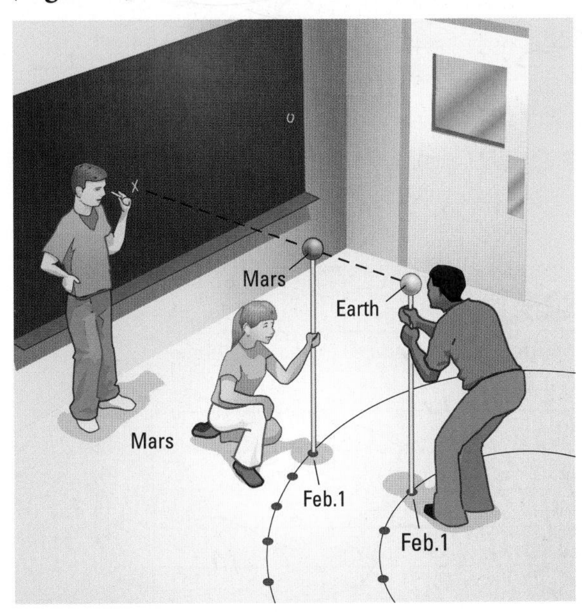

Figure 3

7 (E1) The recorder labels the position with the appropriate date.

8 Steps 4 to 7 are repeated by all groups in the class until all the dates are marked on the board.

9 (E2) The recorder and the Mars person measure the distance from O to each X. Record this distance under the date on the chalkboard.

Analysis and Conclusion

(a) Copy the information on the chalkboard onto the 22 cm × 36 cm paper. Title it "Mars's Motion."

(b) Copy and fill in **Table 1**.

Table 1 **Mars's Motion**

Date	Distance from "O" (cm)
February 1	
February 15	

(c) (W2) Plot the points on a graph. Put the date on the horizontal axis (x-axis) and the distance on the vertical axis (y-axis). Join the points in a smooth curve.

(d) (F4) Using your table, when was Mars in retrograde motion?

(e) Using your graph, when was Mars in retrograde motion?

(f) Was it easier to figure out when Mars was in retrograde using the table or the graph? Why?

Understanding Concepts

1. Which planet, Earth or Mars, takes longer to orbit the Sun?
2. Why do Venus and Mercury never seem to be in retrograde motion?
3. Why do Mars, Jupiter, and Saturn seem to loop backward?

Making Connections

4. If you looked at Uranus, Neptune, and Pluto using a telescope, would you ever see them in retrograde motion? Why or why not?

Challenge

3 How has technology played a role in improving human understanding of retrograde motion?

Work the Web

Are any planets currently in retrograde motion? Start your search by visiting www.science.nelson.com and following the links from Science 9: Concepts and Connections, 4.6. Record your answers.

Other Objects in the Solar System

Figure 1

The *Apollo 11* astronauts brought many samples to Earth for study and collected much data on the Moon.

About 60 million years ago, the dinosaurs died out very quickly. Why?

Scientists have found evidence that a fast-moving object from outer space crashed into Earth, sending material flying into the atmosphere. This material reduced the amount of sunlight reaching Earth. The lack of sunlight caused the climate to change, and any organism that could not adapt to this change died.

Scientists hope that as they learn more about space, they will find clues to how life may have formed on Earth.

Planets and Their Moons

Objects that revolve around planets are called **satellites.** Satellites can be natural or artificial. The natural satellites are called **moons.** Several planets have more than one moon.

Earth has one moon, and astronauts visited it six times between 1969 and 1972 (**Figure 1**). The first successful mission was *Apollo 11*, which landed on the Moon on July 20, 1969. Scientists at NASA (National Aeronautics and Space Agency) are still analyzing the data from the *Apollo* missions.

The moons of other planets were discovered after the invention of the telescope. **Table 1** lists the number of known moons within our solar system. Some of the planetary moons have been investigated by space probes. We have gathered information about the two moons of Mars, Phobos and Deimos, and many of the moons of Jupiter, Saturn, Uranus, and Neptune (**Figure 2**).

Table 1

Planet	Number of known moons
Mercury	0
Venus	0
Earth	1
Mars	2
Jupiter	39
Saturn	30
Uranus	20
Neptune	8
Pluto	1

a

b

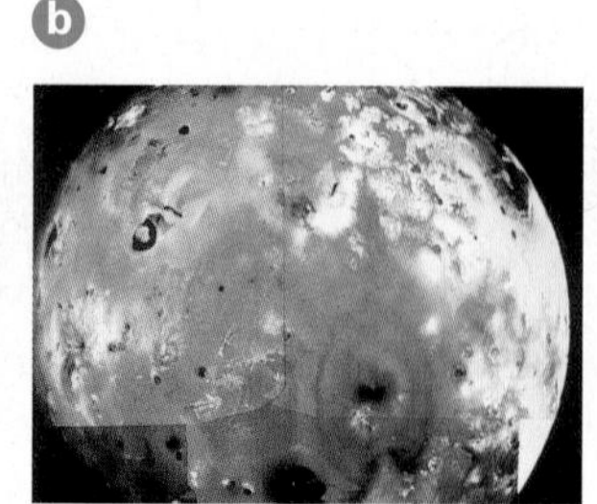

c

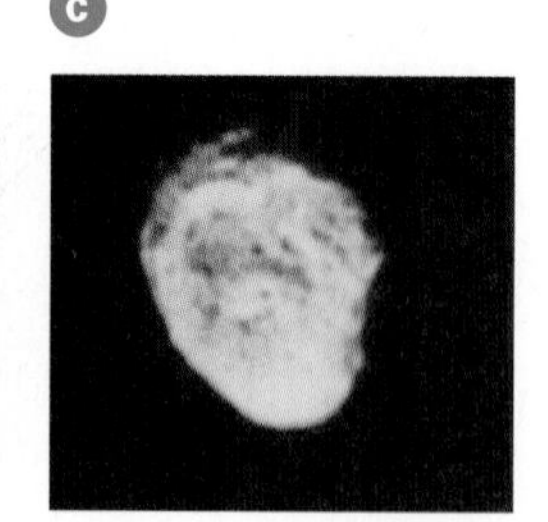

Figure 2

(a) Jupiter's moon Europa has an icy surface with very few craters. It is nearly as big as Earth's Moon.
(b) Io, the closest moon to Jupiter, is the only moon in the solar system to have active volcanoes, more violent than any on Earth.
(c) Saturn's moon Hyperion is only 360 km across and has an irregular shape, possibly the result of repeated collisions with large space rocks.

Asteroids

Refer to the model of the solar system from section 4.1, page 192. Our solar system has a large gap between the orbits of Mars and Jupiter. In that gap is a ring of thousands of small, rocky objects called **asteroids**. They are materials that never formed into planets. This ring of asteroids, known as the **asteroid belt**, was investigated by several space probes.

Asteroids also share Jupiter's orbit, and others travel in paths that may take them closer to the Sun or Earth. In 1937 an asteroid called Hermes came within 800 000 km of Earth—that is really close for an asteroid! **Figure 3** shows the orbits of some asteroids and their names.

In February 2001 the *NEAR* spacecraft landed on Eros, an asteroid. This is a promising achievement as asteroids contain many minerals that we may someday mine. The largest asteroid is only 1000 km in diameter, so astronauts must be careful when navigating through the asteroid belt to avoid any collisions.

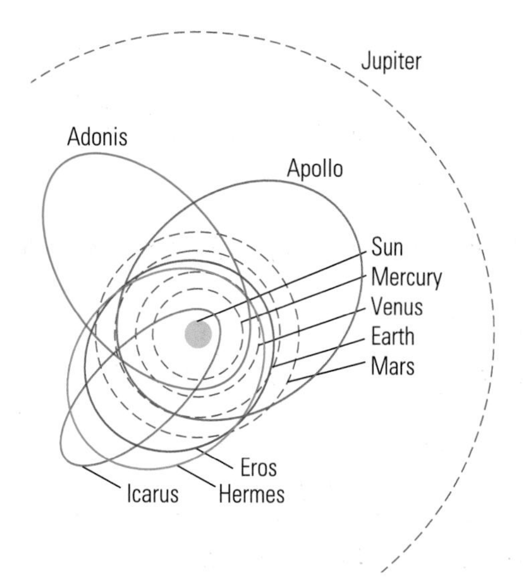

Figure 3

Orbits of several asteroids (not drawn to scale)

Meteors and Meteorites

A **meteoroid** is a lump of rock or metal that is pulled through Earth's atmosphere. As it falls, it rubs particles of the air, causing heat. As a result, the meteoroid burns up. This fireball shows up as a bright streak in the sky, called a **meteor**. Meteors are often mistakenly called shooting stars because of the light given off by the burning meteoroid.

If a meteor actually hits Earth before it completely burns up, then it is known as a **meteorite**. Meteorites can cause a lot of damage, because they are travelling very fast when they hit Earth (**Figure 4**).

Figure 4

The Barringer Crater in Arizona was formed 20 000 to 30 000 years ago by a large meteorite. The crater is 1.2 km across and 120 m deep. Why are most craters this shape?

Have you heard of shooting stars or falling stars? They are not stars at all: they are meteors. If you look at the sky on a clear night, you might see a meteor. If you look at the sky during a meteor shower, your chances of seeing a meteor are high. The three most active meteor showers are the Perseid shower (August 12), the Geminid shower (December 14), and the Quandrantid shower (January 3 or 4).

Comets

One evening, an amateur astronomer from Japan was using binoculars to sky watch. He noticed an object that he had not seen before. He recorded what he saw, and then reported it to professional astronomers. To everyone's excitement, he had discovered a **comet**. A comet is a chunk of frozen ice and dust that travels in a very long orbit around the Sun. The comet was called Comet Hyakutake, after the man who first observed it (**Figure 5**).

The nucleus of a comet is only a few kilometres wide. However, its tail can be up to a million kilometres long. As a comet travels close to the Sun, it is warmed by the Sun. The frozen gases that make up the comet melt and then evaporate. These gases are pushed outward by the Sun's wind and rays. They form a bright glowing tail visible for several months as the comet passes close to the Sun.

Comets have fixed orbits, and astronomers can predict when we will see certain comets. The most famous comet, Halley's comet, was last seen in 1986. It has a revolution of 76 years.

Figure 5

Comet Hyakutake, discovered in 1996, was the brightest comet seen in 20 years.

Exploring Minor Bodies

Scientists are trying to figure out what formed Earth and other objects of the solar system. Scientists think that asteroids and comets are made of the same elements as planets and moons. Asteroids and comets are known as the **minor bodies** of the solar system. Unlike Earth, the minor bodies have not changed much. Scientists send probes to the minor bodies to learn more about them in hopes of learning about the origins of Earth and the solar system.

A probe called *Deep Space 1* was launched in 1998 (**Figure 6**). Its goal is to explore as many minor bodies as possible. The probe is different from other probes as it is lightweight, uses an efficient fuel, and has an onboard computer that will guide the probe to a safe landing on a minor body. This form of artificial intelligence has improved the chances of success on these unpiloted missions.

Figure 6

Deep Space 1, 1998

Try This Activity Debate: Extraterrestrial Mining

The human race has a long history of exploration. We set out to find resources and then decide how to best use them. Earth's resources are being used up. Scientists are turning their eyes toward outer space as a new supplier of natural resources.

Many of the objects within our solar system have valuable resources. For example, Mars is rich in iron, and Jupiter's moon Ganymede contains hydrogen peroxide.

Some people have suggested that mines (**Figure 7**) and colonies could be built on these planets or moons. This raises the question of who owns the solar system and, more important, of who is responsible for the well-being of the solar system.

"We should learn to manage Earth's resources better, before trying to use extraterrestrial resources."

You have been asked to participate in a panel discussion, either supporting or opposing the statement above. You may be an investor, a miner, an astronomer, an environmentalist, a social worker, or someone (I) who simply has an opinion. List three of your (L) own ideas that support your position.

Support

- Earth has many unused resources. If we used them properly and recycled, we would not need space mining.
- The cost of technology is too high. The money could be used on Earth.
- We do not own space.

Oppose

- Using resources from space reduces our dependency on Earth.
- If we set up mining colonies, further space exploration would be simple.
- If we don't own space, who does?

Understanding Concepts

1. **(a)** What is an asteroid?
   (b) Where is the asteroid belt?
2. **(a)** Explain the difference between a meteoroid and a meteorite.
   (b) Why are meteorites less common than meteors?
3. When will Halley's comet be seen next?

Making Connections

4. Describe what might happen if a giant meteorite crashed into Earth's surface (a) on land and (b) on water. Use a map to show where such a meteorite would have the least impact on human life.

Reflecting

5. Space exploration is costly. Do you think that sending probes to explore the minor bodies is necessary? Give two reasons.

Figure 7
An artist's version of an extraterrestrial mining facility

Challenge

1 Could the asteroid field be a tour stop? What about a comet or even a meteor shower?

3 What technologies have made it possible for us to see objects that are in space?

4.8 Case Study

Telescopes

The **universe** comprises everything that exists everywhere. That is a lot to learn about. One way we can learn about the universe is by looking at images of planets, stars, and other objects in space.

The technologies behind computers and telescopes improve quickly. Images that we see today are much sharper, much clearer, more colourful, and more accurate than images from even a few years ago.

Using Telescopes

Telescopes cost anywhere from a few hundred dollars to hundreds of millions of dollars. From Galileo's first telescope to the most sophisticated instruments now orbiting Earth, the main purpose of a **telescope** has not changed. Its purpose is to gather light. The light forms an image that can be seen or recorded using cameras or other devices.

The design of a telescope has not changed much since Galileo's time (**Figure 1**). However, telescopes are now being built that are larger and more powerful and they are being built in many different locations. Light from large cities is a problem when using a telescope. Many observatories are built on the tops of mountains, away from the bright lights of cities, in locations where the sky is clear and dark.

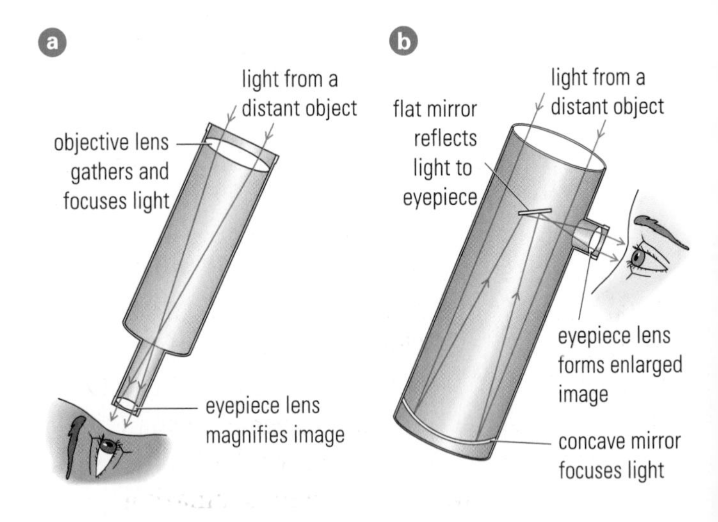

Figure 1

(a) In a **refracting telescope**, light rays bend as they pass through a light-gathering lens. The lens can be at most 1 m across. If the lens if any larger, then it is too heavy and sags. This sagging distorts or deforms the image. This type of telescope was first built by Galileo in the first decade of the seventeenth century.
(b) A **reflecting telescope**, which was first built by Newton in 1668, uses a mirror to gather light. The support for the mirror is underneath, so the mirror can be huge without distorting the image.

Cameras can be programmed to gather more light over an extended period. They use electronic detectors instead of film. The result is a much sharper image (**Figure 2**).

Figure 2

These views are of the same object. From left to right, these images show light-gathering times of 1, 5, 30, and 45 minutes.

The Hubble Space Telescope

Stars give off light. Some light is absorbed while passing through the atmosphere and causes fuzzy images from space.

In 1946, Lyman Spitzer, Jr. (1914–97), a theoretical astrophysicist, first came up with the idea of a telescope in space. The problem of city lights and the atmosphere would be solved by orbiting a telescope above the Earth's atmosphere.

In 1990, the Hubble Space Telescope (HST) was launched (**Figure 3**). The HST is a reflecting telescope whose purpose is to gather light from distant objects and send the images through a computer to Earth. The HST was named in honour of Edwin Hubble (1889–1953). In 1929, Hubble discovered that the universe is expanding and changed the way astronomers look at the sky.

Figure 4 shows one of the most amazing images ever taken of outer space. It is called the *Hubble Deep Field.* You are looking at light that is up to eight billion years old.

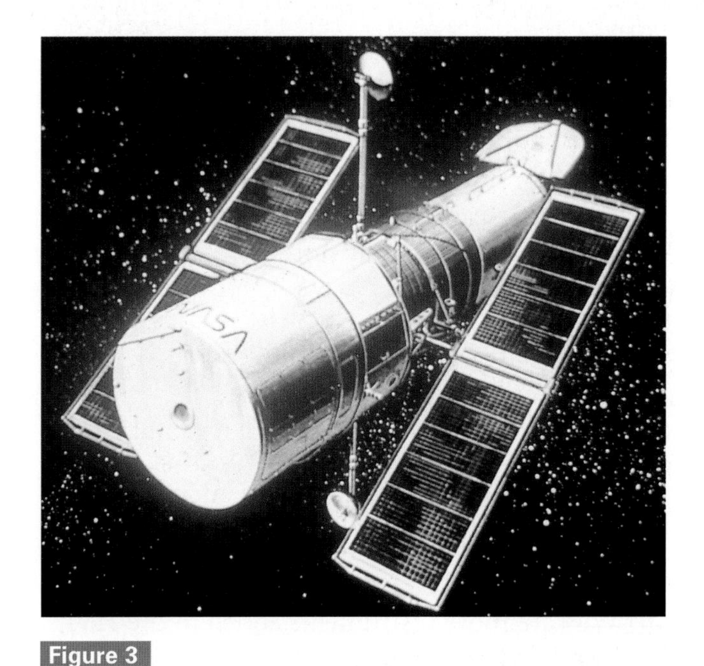

Figure 3

The Hubble Space Telescope

Figure 4

This image was captured by the HST when it was aimed toward a part of the sky above the Northern Hemisphere. The stars are the objects that appear to have spikes. Everything else is a galaxy.

The image from **Figure 4** was obtained by HST over 100 h.

(a) Do you think the image would have been as clear if a time-lapse photo of 100 h had been taken by a camera attached to a telescope here on Earth? Why or why not?

The "optical" portion of the HST includes a large mirror that was specially built. Unfortunately, a flaw in the mirror was found after the launch of the HST in 1990. This error was corrected by astronauts in 1993. Since then, many other instruments have been added to the HST, further improving its performance (**Figure 5**).

(b) Why are more instruments added to the HST rather than being sent into orbit on their own?

(c) List two advantages to sending instruments into orbit separate from the HST.

Advanced technology is needed to keep a telescope, travelling at almost 30 000 km/h while orbiting Earth, aimed at one spot for an extended period.

(d) Why wouldn't the telescope naturally stay pointed in the same direction?

Figure 4 sees deeper into space than any other image. That is why it is called the *Hubble Deep Field.* The farthest objects, which are galaxies, are up to eight billion years old. We are seeing what the galaxies looked like eight billion years ago.

(e) How could information from these galaxies help scientists understand more about our galaxy and solar system?

Challenge

2 Would telescopes be a necessary part of the space colony? Why?

3 The development of telescopes has improved our ability to see what is in space. The Hubble Space Telescope is a technology that has increased our knowledge. Investigate at least two different types of telescopes (including the Next Generation Space Telescope), where they are located, how they work, and for what purpose they were developed.

Figure 5

Astronauts working on the HST

Understanding Concepts

1. **(a)** List two problems that might make it difficult to use telescopes to see the sky from Earth.
   (b) Where would you build an observatory to avoid these problems? Explain your choice.
2. How does the image in **Figure 4** allow scientists to "look back in time"?

Making Connections

3. How have improvements in technology changed our view of the universe?

Work the Web

Investigate Canada's role in the development of the Next Generation Space Telescope. Start your search by visiting www.science.nelson.com and following the links from Science 9: Concepts and Connections, 4.8. Record your information in point form.

Space Artist

Paul Fjeld

Paul Fjeld: "I want to paint a picture as if you are actually there."

Paul Fjeld gives a thumbs up sign as he steps out of the high performance jet in which he has just pulled 6 *g*'s. His call sign is "Fingers," and he flies in jet fighters whenever he can as part of his research: Paul Fjeld is a Canadian artist interested in space.

From 1971 to 1973 the *Montreal Star* sent a young artist to cover the last three launches of the *Apollo* missions. He soon convinced NASA to include him in their official art program. Soon Fjeld got to go where even the press could not. His greatest thrill was working in mission control during the 1975 *Apollo-Soyuz* mission.

Fjeld documents history through art. His first painting was of a crippled Skylab. He worked with the engineers to determine what the damage might look like to prepare a visual image for the press and public.

Fjeld's work has appeared in *National Geographic* and on the cover of *Aviation Week.* Actor Tom Hanks hired Fjeld to recreate the moment of the *Apollo 11* lunar landing on a huge set for the TV series *From the Earth to the Moon* (**Figure 1**).

What is next for Fjeld? another painting? a movie script? Whatever it is, it will almost certainly involve his passion: space.

Figure 1

Paul Fjeld's painting of the *Apollo 11* lunar landing

Making Connections

(N) **1.** What skills and attitudes does an artist need to (O) capture what a camera cannot?

2. If you were going to follow a similar career path, (N) what courses would you take in high school and (O) college or university?

Work the Web

Find out more about the NASA Art Program. Start your search by visiting www.science.nelson.com and following the links from Science 9: Concepts and Connections, 4.9.

Challenge

2 Would a person with artistic skills be important for the space colony? Explain your answer.

4.10 Case Study

The Sun: An Important Star

The most important star for Earth is the one at the centre of our solar system: the Sun. It provides the energy needed by all plants and animals, and its gravitational pull keeps us in a steady orbit. By studying the Sun, we also learn about other stars. Actually, we compare all other stars to it, since the Sun is the star we know the most about.

(a) Why is the Sun the most important star? Give two reasons.

(b) Why do we want to learn more about the Sun?

Since the Sun is the closest star to Earth, for us it is the brightest. It is so bright that you cannot see any other stars while the Sun is in the sky.

The Sun and all stars produce huge amounts of heat and light energy through a process called **nuclear fusion**. Under extremely high pressures and temperatures, two hydrogen nuclei combine to form a helium nucleus (**Figure 1**). Every second, the Sun makes more energy than people have used throughout history.

(c) Describe the process of nuclear fusion. Include a diagram.

Scientists have calculated that the Sun has been producing energy for about five billion years and has only used up 25% of its hydrogen. They estimate that the Sun will continue to produce energy for at least another five billion years before it runs out of hydrogen for nuclear fusion.

(d) What might happen to living things on Earth when the Sun runs out of fuel?

(e) When might this happen?

Some people take advantage of the Sun's energy by using solar panels (**Figure 2**).

Ⓝ **(f)** Do some research and explain why everyone Ⓞ does not use solar panels.

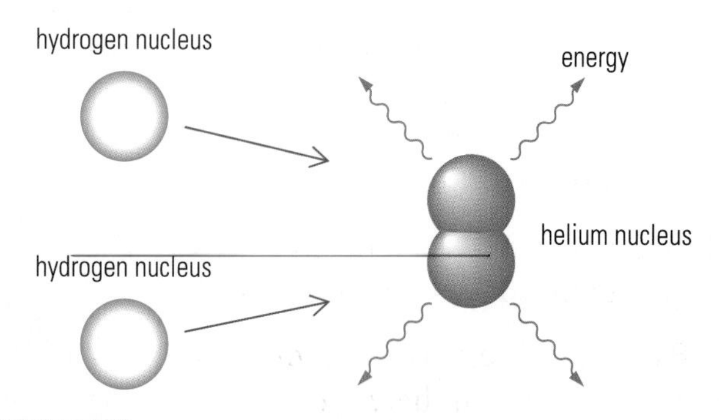

Figure 1

During nuclear fusion, two substances combine to form a new substance, producing energy.

Never look directly at the Sun. It can damage your eyes.

Figure 2

A home that uses solar panels for energy, rather than oil or natural gas

A Close Look at the Sun

It is very difficult to study the poles of the Sun from Earth. The probe *Ulysses* was launched in 1990 to study the Sun's poles. Another probe, *SOHO*, has 12 instruments on board for observing the Sun.

(g) Why do we send unpiloted probes to study the Sun?

Based on observations from the probes and from here on Earth, we can draw a model showing the various layers of the Sun (**Figure 3**).

(h) What is the temperature of the photosphere? the corona? the core?

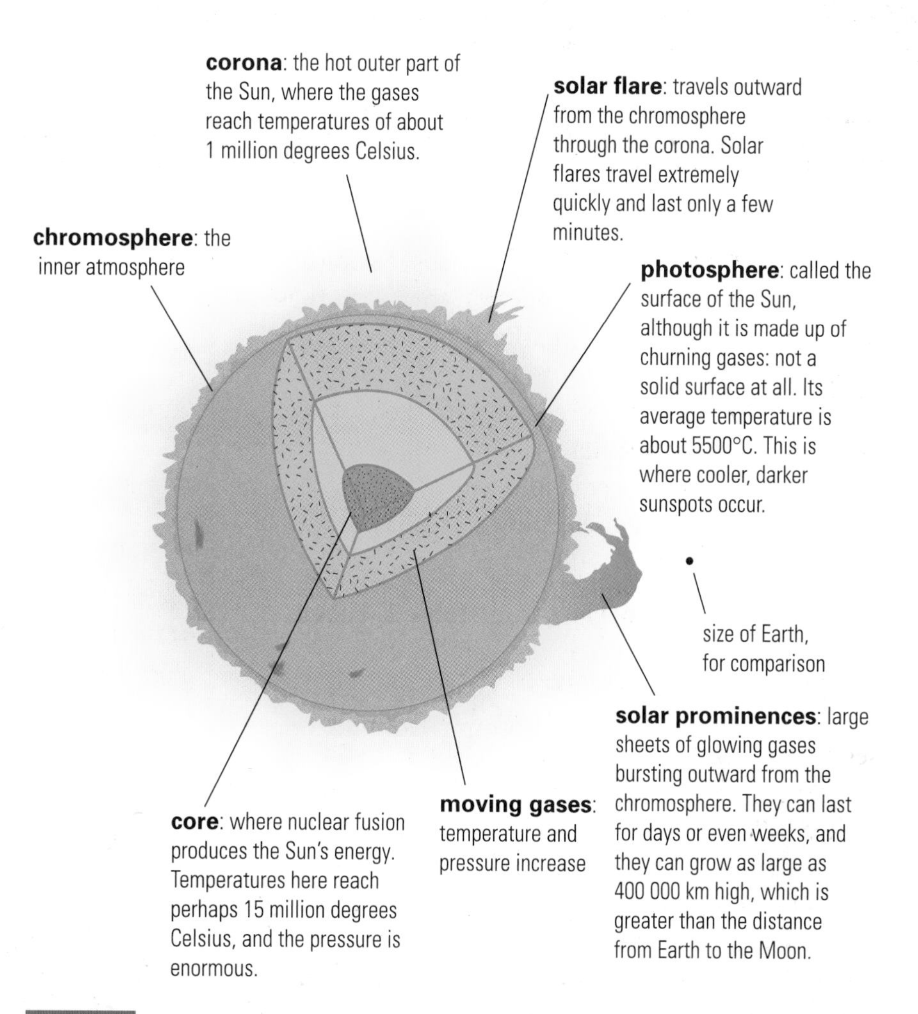

Figure 3

The Sun's Effects on Earth

Although astronomers can see **solar flares** as they happen, other effects only become apparent a few days afterward. Solar flares give off charged particles, which travel slowly toward Earth. These particles interfere with Earth's magnetic fields at the north and south poles. The resulting electrical effects in the atmosphere interfere with radio communications in the area and can damage satellites.

These same charged particles produce beautiful auroras, or shifting lights, over the North Pole (Aurora Borealis or the Northern Lights) (**Figure 4**) and the South Pole (Aurora Australis or the Southern Lights).

(i) List two differences between a solar flare and a solar prominence.

Figure 4
The Aurora Borealis

Sunspots

Approximately every 11 years, scientists observe problems that occur with our radio and other communication systems. Problems also occur with the distribution of electricity.

Scientists have discovered that these problems occur just after violent (magnetic) storms on the Sun's surface. These storms appear to occur when there are many dark regions on the Sun, called **sunspots** (**Figure 5**). These are the darker, cooler regions of the photosphere.

Sunspot activity seems to occur in cycles. Astronomers have been keeping a record of sunspot activity since 1700.

(j) Graph the sunspot data from **Table 1**. Label the vertical axis (*y*-axis) as number of sunspots. Label the horizontal axis (*x*-axis) as year. Write a title for your graph. Connect the points. (W2)

Table 1 **Sunspot Data**

Year	Number of sunspots
1979	155
1980	155
1981	141
1982	116
1983	67
1984	46
1985	18
1986	13
1987	29
1988	100
1989	146
1990	142
1991	156
1992	95
1993	55
1994	30
1995	18

(F2) **(k)** In what years was sunspot activity at a peak?

(F2) **(l)** In what years was sunspot activity the least?

(F2) **(m)** How many years is one cycle?

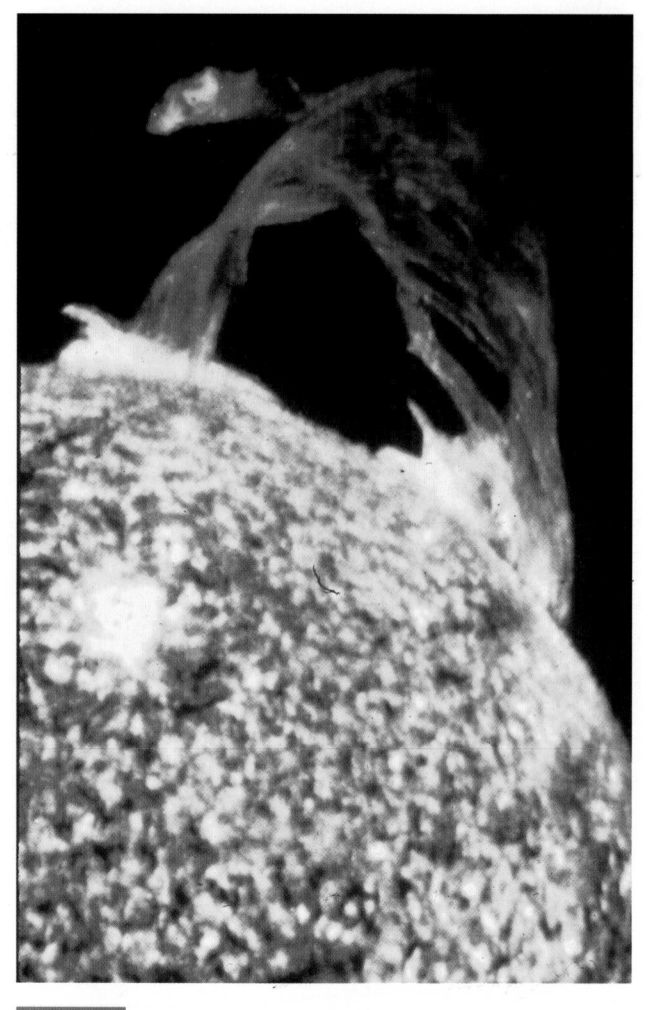

Figure 5

The small, dark regions in this photograph of the Sun are called sunspots.

Sunspots are huge cooler areas in the Sun's photosphere. Even the smallest sunspots observed are larger than Earth.

Challenge

1 The Sun is our most important star. Could this be a tour stop?
2 Would the space colony need to be close to, but not too close, to a star?
3 The Sun is vital to our survival. What technologies have helped us learn more about the Sun and, by extension, how stars are formed?

Try This Activity The Brightness of Stars

Question: How does the brightness of stars change with distance?

You will need

- lamp with 100-W or 200-W nonfrosted bulb
- metre stick
- sheet of black paper with 1 cm^2 cut from centre
- sheet of graph paper
- two 20 cm × 25 cm sheets of cardboard
- scissors
- tape or glue

Be careful when using scissors.

- Cut out a hole with a 5-cm diameter in the centre of one sheet of cardboard.
- Centre and glue the circle on the black paper, and glue both onto the cardboard. Make sure the opening in the black paper is over the opening in the cardboard.
- Glue the graph paper onto the other piece of cardboard.
- Turn on the lamp. Turn off the classroom lights.
- Place the cardboard with the single square opening 10 cm from the bulb.
- Make sure the cardboard is perpendicular to the bulb's filament and that the filament is pointed toward the opening.
- Place the graph paper and cardboard behind the black paper and cardboard, and line up one of the graph paper squares with the square opening of the black paper.
- Notice the light from the bulb lights 1 cm^2 on the graph paper at a distance of 10 cm.

a) Move the graph paper and cardboard to 20 cm, 30 cm, 40 cm, and 50 cm away from the black paper and cardboard (see **Figure 6**). What happened? Were more or fewer squares lit? Was the light brighter or fainter?

b) What is the significance of a star's light being quite faint at night?

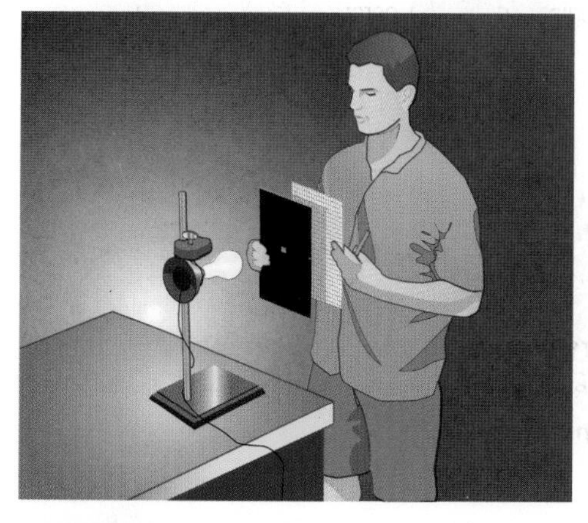

Figure 6

Understanding Concepts

1. Why is the Sun so bright? Why are other stars so faint?
2. Describe the differences between a solar flare and a solar prominence. Which one affects us and how?
3. Describe the process that occurs in the Sun's core to produce so much energy.
4. When is the next year of maximum sunspot activity predicted to be?

Making Connections

5. Is the possible "death" of the Sun in five billion years a problem that we should be worrying about? Why or why not?

Reflecting

6. Why do we consider the Sun our most important star?

Work the Web

The number of sunspots per year increases and decreases in a regular cycle. Find out where we are in the cycle. Start your search by visiting www.science.nelson.com and following the links from Science 9: Concepts and Connections, 4.10.

Galaxies and Star Clusters

When you look at a map of your province, you see cities, towns, and villages. These are places where people cluster or group together. The spaces between these regions are quite large and are called rural areas. Similarly, when you look at a map of the universe, you see different-sized clusters of stars with different features.

Galaxies

A **galaxy** is a huge collection of gas, dust, and hundreds of billions of stars. These stars are attracted to each other through the force of gravity, and they are always in motion.

Earth and the solar system are part of the Milky Way Galaxy. You may be able to see the Milky Way in the summer and winter. It looks like a trail of spilt milk across the sky. The Milky Way is the combined light of the 400 billion distant stars in our galaxy. The Milky Way is disk-shaped, with the Sun near the outer part of the disk (**Figure 1**). The central bulge contains the most stars. The spiral arms move around the central bulge.

The Milky Way Galaxy is called a **spiral galaxy** because of its shape. **Figure 2** shows photographs of three other spiral galaxies.

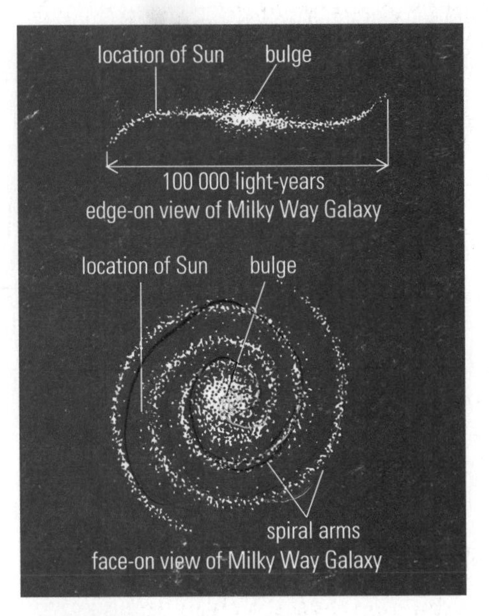

Figure 1

The spiral arms of the Milky Way Galaxy have huge concentrations of stars. The Sun is on one of the outer spirals.

The universe has billions of galaxies. We call galaxies that gather in groups galaxy clusters or superclusters. The Milky Way Galaxy belongs to a group called the Local Group and to the supercluster called the Virgo Cluster.

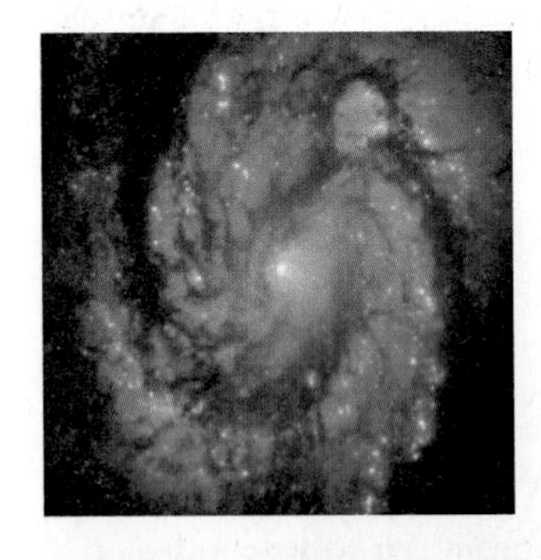

(a) This giant spiral galaxy resembles the Milky Way Galaxy.

(b) This spiral galaxy is coloured to show young giant stars (blue).

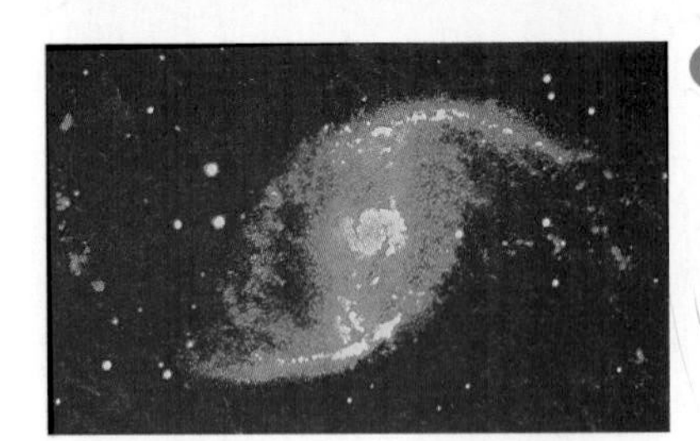

(c) An example of a barred-spiral galaxy, coloured to show the central bar

Figure 2

Types of Galaxies

There are three types of galaxies. **Figures 1** and **2** were examples of spiral galaxies. There are also **elliptical galaxies**, which are shaped like a football (**Figure 3**), and **irregular galaxies**, which have no familiar shape at all (**Figure 4**).

Figure 3
An elliptical galaxy

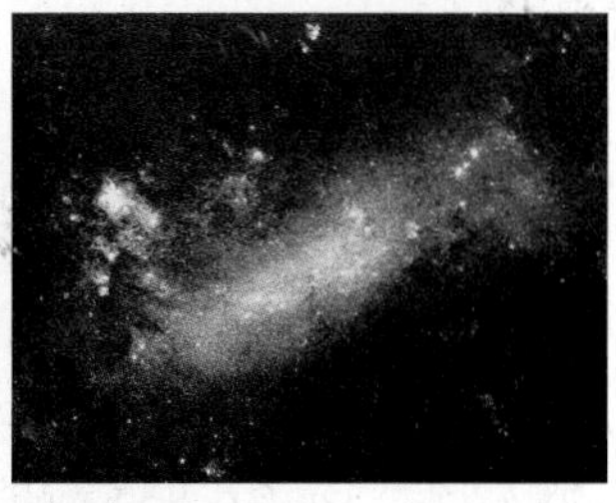

Figure 4
An irregular galaxy

Unusual Galaxies

The further we see, the more discoveries we make, such as the following:

- Some galaxies appear to be in the process of colliding, recombining, and tearing away stars from one another.
- Smaller galaxies are "eaten" by larger ones.
- **Quasars** are very distant objects that look like faint stars, but they give off up to 100 times more energy than the Milky Way galaxy.

Star Clusters

Groups of stars that are close together and that travel together are called **star clusters**. A single cluster can have anywhere from 10 to a million stars. Clusters are part of galaxies.

A familiar star cluster is the Pleiades, from the constellation of Taurus, shown in **Figure 5**.

Figure 5
The Pleiades

Try This Activity

Model a Spiral Galaxy

Make your own spiral galaxy model. (C2) One way is to use a plastic container, water, a stir stick, and a drop of food colouring. Place the water in the container, add a drop of food colouring, and stir. Make a prediction first.

(E1) What will happen? Record your (E3) prediction and observations. Do not forget to draw a diagram.

Show your model to your teacher.

Challenge

1 Some galaxies, such as the Andromeda Galaxy, are beautiful. Some people may enjoy a tour stop at a galaxy.

3 Would a space colony in another galaxy, so far away from home, be desirable?

Understanding Concepts

1. How are galaxies classified? Draw and label an example of each.
2. Arrange the following in order of size, starting with the largest: star cluster, galaxy, universe, star, planet.
3. How do you think the Milky Way galaxy got its name?

Work the Web

Which galaxy is your favourite? Find out more and summarize the information in a paragraph. Start your search by visiting www.science.nelson.com and following the links from Science 9: Concepts and Connections, 4.11.

INQUIRY SKILLS
- ○ Questioning
- ○ Hypothesizing
- ○ Predicting
- ○ Planning
- ○ Conducting
- ● Recording
- ● Analyzing
- ● Concluding
- ● Communicating

Flame Tests

When each element on Earth burns, it produces a different colour. For example, sodium burns bright yellow, oxygen burns white, and lithium burns red. Any compound containing these elements will burn the corresponding colour.

A star's colour can indicate two things: its temperature (**Table 1**) and, through spectroscopy, its chemical composition (**Figure 1**). The colour of a star mostly indicates its temperature, blue stars being the hottest and red stars the coolest. However, in **spectroscopy**, which is the spectral analysis of a flame, scientists use a device called a **spectroscope** to look at the patterns of light emitted and identify which elements are present in a particular star. These patterns are unique for each element.

You will be performing flame tests to determine the identity of the unknown substance. These tests are too general to analyze a star.

Some of these substances are poisonous. If you get any of these substances on your skin, in your eyes, or on your clothing, wash the area immediately with cold water and report it to your teacher. Inform your teacher of any spills.

Table 1

Colour	Temperature range (°C)	Example(s)
blue	25 000–50 000	Zeta Orionis
bluish-white	11 000–25 000	Rigel, Spica
white	7 500–11 000	Vega, Sirius
yellowish-white	6 000–7 500	Polaris, Procyon
yellow	5 000–6 000	Sun, Alpha Centauri
orange	3 500–5 000	Aldebaran
red	2 000–3 500	Betelgeuse

Figure 1
(a) Sunlight produces part of the spectrum of light.
(b) An element such as hydrogen produces only certain parts of the spectrum that can be viewed with a spectroscope.

Question

What is the identity of an unknown substance?

Materials

- safety goggles
- apron
- Bunsen burner or candle
- matches or flint lighter
- 7 labelled 50-mL beakers, each containing one splint
- splints previously soaked in 0.5 mol/L solutions of one of the following chlorides: lithium, sodium, potassium, copper (II), barium, and one unknown
- one splint that has not been soaked; it is the control
- beaker containing water for extinguishing splints

Procedure

C6 **1** Copy **Table 2** into your notebook.

Table 2

Substance	Flame colour
wooden splint	
lithium	
sodium	
potassium	
barium	
copper (II)	
unknown	

2 Review the safety procedures for lighting a burner. Put on your safety goggles and apron.

3 Obtain from your teacher seven labelled beakers (per group) containing the wooden splints. Six of the splints have been soaked in solutions of the following chlorides: lithium, sodium, potassium, copper (II), barium, and unknown. The seventh splint has not been soaked. It is a control. Handle the splints by the dry end only.

4 Carefully light the burner and adjust it to produce a blue flame.

5 Burn the dry wooden splint in the flame. (E1) Record the colour of the flame in your notebook.

6 One at a time, holding the dry end of the splint, put the soaked end of the splint in the flame until a colour is observed. Do not let the splints burn. This should only take a few seconds.

7 Put out the flame by dunking the splint in the beaker of water.

(E1) **8** Record the flame colour.

9 Repeat steps 6 to 8 for the remaining five substances.

10 Clean up your work station and wash your hands. Dispose of your solutions as directed by your teacher.

Analysis and Conclusion

11 Answer the following questions using full sentences.

(F4) **(a)** What was the identity of the unknown (G) substance? How did you decide?

Figure 2

Another way to do flame tests is to dip a loop of platinum wire into a solution and then into the flame.

Understanding Concepts

1. A sample of an unknown white solid is burned. The flame is a deep red. Is the sample table salt (sodium chloride)? Why or why not?

Making Connections

2. How do flame tests relate to the study of the stars?

Challenge

3 How has the understanding of flame tests and spectroscopy helped scientists to understand the meaning of the colours of stars?

Evidence of an Expanding Universe

Imagine a duck bobbing up and down on the surface of a pond. What would you see in the water? You would see ripples spreading from the duck in circles that have the same centre. The distance from wave crest to wave crest is called a **wavelength** (**Figure 1**).

Now picture yourself standing on the shore. Your friend is standing on the opposite shore, and the duck is between you. The duck begins to swim toward you. What would you see? What would your friend see?

As the duck swims toward you, the ripples get closer together. The wavelength gets shorter. The duck is swimming away from your friend. Your friend sees wavelengths that are getting longer. **Figure 2** shows this.

Wavelengths get shorter as an object moves toward you. Wavelengths get longer as an object moves away from you. How does this relate to astronomy?

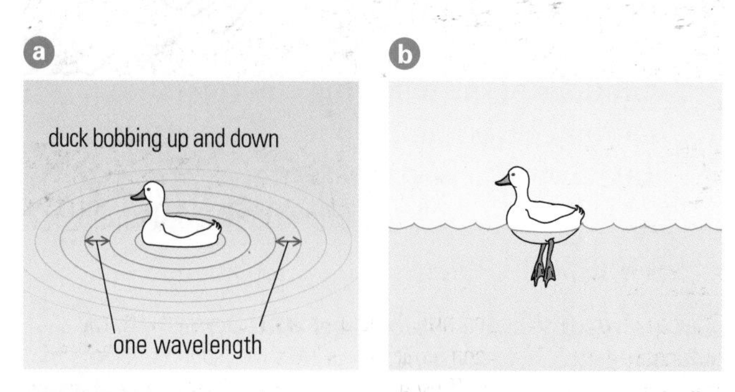

Figure 1

(a) The duck is sending out ripples of constant wavelength.
(b) Side view of the waves

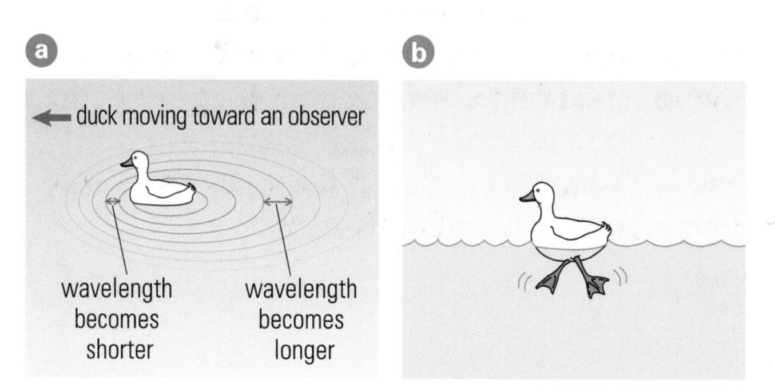

Figure 2

(a) The wavelength becomes shorter if the duck moves toward you and longer if the duck moves away from you.
(b) Side view of the waves

The Visible Spectrum

In the **visible spectrum** (**Figure 3**), the colours are always in the same order: red, orange, yellow, green, blue, violet (ROYGBV). Red has the longest wavelength, and violet has the shortest wavelength. The light from a source such as a galaxy can be divided into its lines of colours or wavelengths to produce a spectrum.

Figure 3

The visible spectrum

Some people believe that a rainbow has seven colours: red, orange, yellow, green, blue, indigo, violet. Indigo is really a transition from blue to violet, not a separate colour. The number six is often associated with evil, and people in ancient times did not want something as beautiful and natural as a rainbow to have six parts. They included indigo to make it seven colours.

Red Shift and Blue Shift

Astronomers want to learn as much as possible about stars and galaxies. So, they keep a record of the patterns of light for different stars and galaxies.

When Edwin Hubble recognized a pattern of light for a galaxy, he then noticed that the colours of the lines looked different. It had shifted toward the red end of the spectrum (**Figure 4**). This meant that the wavelength was getting larger. The galaxy was moving away from Earth.

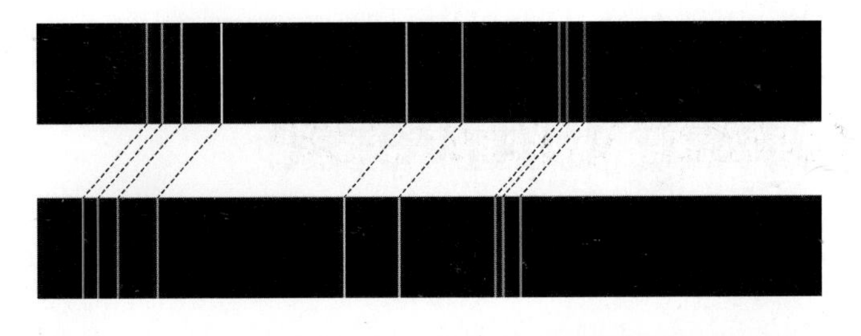

Figure 4

The top diagram shows the nine lines of a spectrum given off by a stationary galaxy. The second diagram shows the same nine lines of the same spectrum given off by a moving galaxy. The bottom spectrum is shown for reference.

This movement toward the red end of the spectrum is called **red shift.** Red shift supports the idea that the universe is expanding.

If the pattern of light shifts toward the blue end of the spectrum, the galaxy or star is moving toward us. This is called **blue shift.**

Try This Activity The Duck

Use a basin of water and an object ("the duck") that floats to illustrate red shift and blue shift.

(a) What happened to the water when you dropped the duck into the basin?

(E1) **(b)** (E3) Push the duck along the surface of the water in a straight line. Sketch what happens to the waves. Record any observations.

(c) What happened to the wavelength behind the duck?

(d) What happened to the wavelength in front of the duck?

(e) Rewrite the sentences using the correct word: The wavelength gets shorter as an object moves toward/away from the observer. The wavelength gets longer as an object moves toward/away from the observer.

Understanding Concepts

1. Copy the pattern of ripples from **Figure 5** into your notebook. Label the ripples that have shorter wavelengths and those that have longer wavelengths. Indicate the direction of movement of the object causing the waves.
2. What does "red shift" mean?
3. If astronomers were to observe a "blue shift" for a certain star, what could they conclude? Why?
4. Sketch a diagram of the ripples that an object travelling from the top to the bottom of this page would make.

Figure 5

Work the Web

Where is the edge of the universe? Can you find any photos of the edge of the universe? Start your search by visiting www.science.nelson.com and following the links from Science 9: Concepts and Connections, 4.13.

Challenge

1 Would tourists be interested in the edge of the universe?

3 How has technology helped us to look at and understand what we are seeing in space?

A Model of the Expanding Universe

As you learned in section 4.12, scientists use a spectroscope to look at the light from galaxies and stars. The light is split into a pattern of separate lines of colour. The colour associated with each substance as it is heated or burned is specific to that substance (**Figure 1**). The spectroscope can show the absorption spectra as either missing lines in the spectrum (**Figure 1a**), or as thin lines of colour that match the absorbed colours (**Figure 1b, Figure 1c**)

A spectroscope can be attached to a telescope and the spectrum of objects analyzed. The patterns of light can tell us what elements are present in a star or galaxy, how much of each element is present, its temperature, and whether the star or galaxy is moving toward or away from Earth.

Most galaxies and some stars that scientists have observed seem to be experiencing red shift. This supports the idea that the universe is expanding.

What is really expanding? Only the space between galaxies is increasing. In this activity, you will use a model to demonstrate the idea of an expanding universe.

Question

Can an expanding universe be modelled?

Materials

- 1 round balloon
- black, fine point, felt-tip pen
- tape measure

Figure 1
(a) light from the sun
(b) sodium light
(c) hydrogen light

Try This Activity Spectroscopy

Your teacher may have simple, handheld spectroscopes that you can use.

You will need

- spectroscope
- white light bulb (light source)
- sealed container of translucent (clear) potassium permanganate solution

Plug in and turn on the light. Turn out the classroom lights. Try to eliminate other light sources.

Look through the spectroscope. Aim the slit of the spectroscope toward the white light source. What do you see? You should see the visible spectrum—one right in front of you and a secondary spectrum to the side.

Place the potassium permanganate solution in front of the white light source. Look through the spectroscope while aiming the slit at the solution. What do you see? You should see the visible spectrum with some lines missing in front of you and those missing lines in the secondary spectrum to the side. This pattern of light is unique to potassium permanganate.

Repeat the process for other translucent solutions.

Procedure

(C6) **1** Copy **Table 1** into your notebook.

Table 1

	Measured distances (cm)			Calculated distances (cm)	
	A to B	B to C	C to D	A to C	B to D
orange stage					
basketball stage					

2 Blow up the balloon to the size of an orange.

Do not over inflate the balloon.

3 (E2) While your partner keeps the balloon at that size by pinching the opening, use the felt-tip pen and tape measure to mark the balloon with four dots, each separated by 1 cm. Label the dots in order A, B, C, D. Determine the exact distances between points using the tape measure and record them (**Figure 2**).

4 (E2) Blow up the balloon until it is about the size of a basketball. Measure and record the distances between the dots.

5 Calculate and record the distances A to C and B to D for both stages.

Analysis and Conclusion

6 Look at how much the distances A to B, B to C, and C to D increased when you inflated the balloon to the second stage. Compare those increases.

7 Imagine you are standing on A while the balloon inflates.

(a) (F1) Which dot would appear to be moving away from you the most quickly?

(b) (F1) Which dot would appear to be moving away from you the most slowly?

(c) (F1) Which dot would appear to be moving toward you?

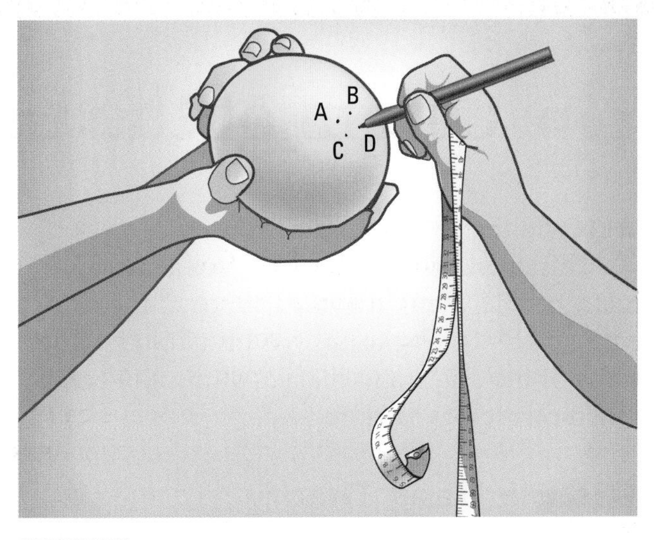

Figure 2

8 Imagine that the dots are galaxies of stars. Complete this sentence by choosing the correct word. "The galaxies that are closer to/further from Earth seem to be moving away from Earth *faster* than galaxies that are closer to/further from us."

Understanding Concepts

1. **(a)** What instrument does an astronomer use to determine the patterns of light of a star or galaxy?
   (b) What can the patterns of light tell a scientist about a star or a galaxy?
   (c) Why is using the instrument from (a) better than using only a telescope?

Reflecting

2. Although the balloon model was useful to illustrate an expanding universe, it had a major limitation. What was that limitation? (Hint: What about the galaxies that are inside the balloon?)
3. Describe a model of the expanding universe that would eliminate the limitation mentioned in question 2. Make a sketch if necessary.

Challenge

2 Is an expanding universe an issue for the people planning transportation and communication to and from the space colony?

The Origin of the Universe

"How did the universe begin?"

This question has been asked often and by many people. Many astronomers spend their entire careers trying to answer this question. The study of the origin and changes of the universe is called **cosmology** (**Figure 1**).

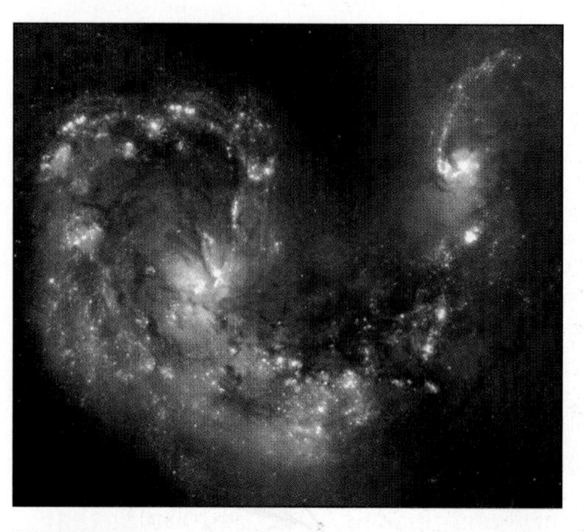

Figure 1

The Antennae galaxies are a pair of galaxies whose collision has resulted in a lot of star birth activity.

The Big Bang Theory

If galaxies are moving apart, then they used to be closer together. "Time zero" is when the galaxies were next to each other.

Scientists have estimated that time zero was 10 billion to 15 billion years ago. At that time, all the matter of the entire universe was packed together into one small, extremely dense, hot mass under enormous pressure. The event where this mass began to move apart is called the Big Bang (**Figure 2**). Scientists use the **Big Bang Theory** to describe the beginning of the universe that we know.

Red shift supports the Big Bang. Scientists have collected data that suggest that the universe is still "glowing" from that initial bang, similar to a match that continues to glow after the flame has been put out.

The Big Bang Theory can be used to explain why the universe is expanding, but we still have a few questions.

Scientists know what the universe looked like a millionth of a second after the Big Bang, but what did it look like a millionth of a second before the Big Bang?

For this reason, the theories about the origin of the universe are likely to change.

Big Bang

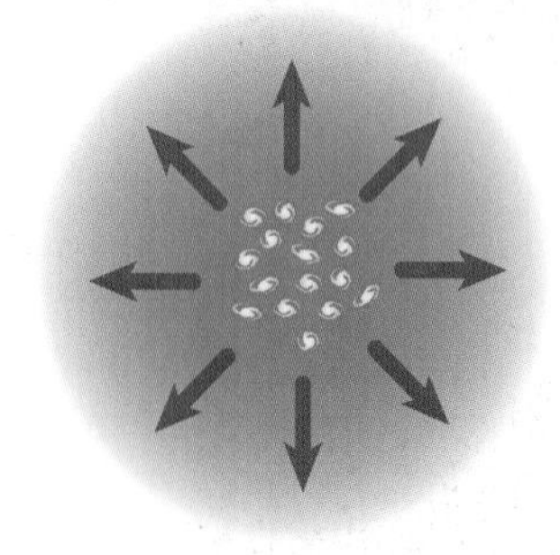

galaxies forming

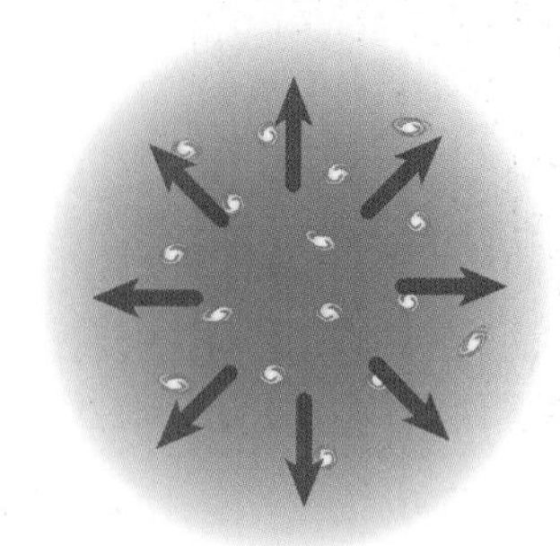

the present:
galaxies dispersing

Figure 2

These diagrams illustrate the Big Bang Theory of the formation of the universe.

The Origin of the Solar System

If the Big Bang explains how the universe began, what explains the formation of the solar system? In its early stages, our solar system was part of a nebula. A **nebula** is a spinning mass of gas and dust. All stars are born from nebulas.

Gravity causes the matter of the nebula, the clouds of gas and dust, to spin together. The gas and dust in the centre become a star and the rest of the matter is pulled together by gravity to form the planets, as shown in **Figure 3**.

(a) Gravity caused the matter of a nebula to spin together.

(b) As the process continued, a bulge formed toward the centre. This bulge became the Sun. Some of the cooler material clumped together.

(c) These chunks eventually formed the planets.

Figure 3

Astronomers think there were three main stages in the formation of the solar system.

Try This Activity

Stirring Things Up

1. Use a stick to stir a bucket of water and sand. What happens to the sand? Draw a diagram. (E3)
2. What happens if you try to stir the water the opposite way? Draw a diagram. (E3)
3. How does this model support the theory of the origin of the solar system?

Understanding Concepts

1. What is the estimated age of the universe?
2. How does the Big Bang Theory explain an expanding universe?
3. How could you use all the students in your class to model the expanding universe?
4. What force is responsible for bringing together particles in space?
5. How is the formation of a star linked to the formation of the solar system?

Reflecting

6. How does a scientific theory, such as the Big Bang Theory, differ from a belief?

Challenge

1 Would tourists like to visit a nebula?

3 Scientists in CERN, Switzerland, managed to create a mini Big Bang. Why are scientists so interested in how the universe began?

Work the Web

People in all parts of the world have had, and still have, other ideas about the origin of the universe. Find out about two of those ideas. Start your search by visiting www.science.nelson.com and following the links from Science 9: Concepts and Connections, 4.15.

SKILLS HANDBOOK: (E3) Scientific Drawings

4.16 Case Study

Satellites

A **satellite** is any object that revolves around another object. In this section, we look at humanmade satellites that orbit Earth.

Many people think that all satellites orbit thousands of kilometres above the surface of Earth. This is not true. Even the International Space Station is orbiting only 400 to 500 km above Earth's surface (**Figure 1**).

Hundreds of satellites orbit Earth. They are used for communication, weather forecasting, military purposes, and science.

Some satellites detect naturally occurring radiation, as well as their own radiation. Computers analyze the reflected signals and convert them into images.

A Canadian satellite system called RADARSAT constructs radio wave images of Earth's surface.

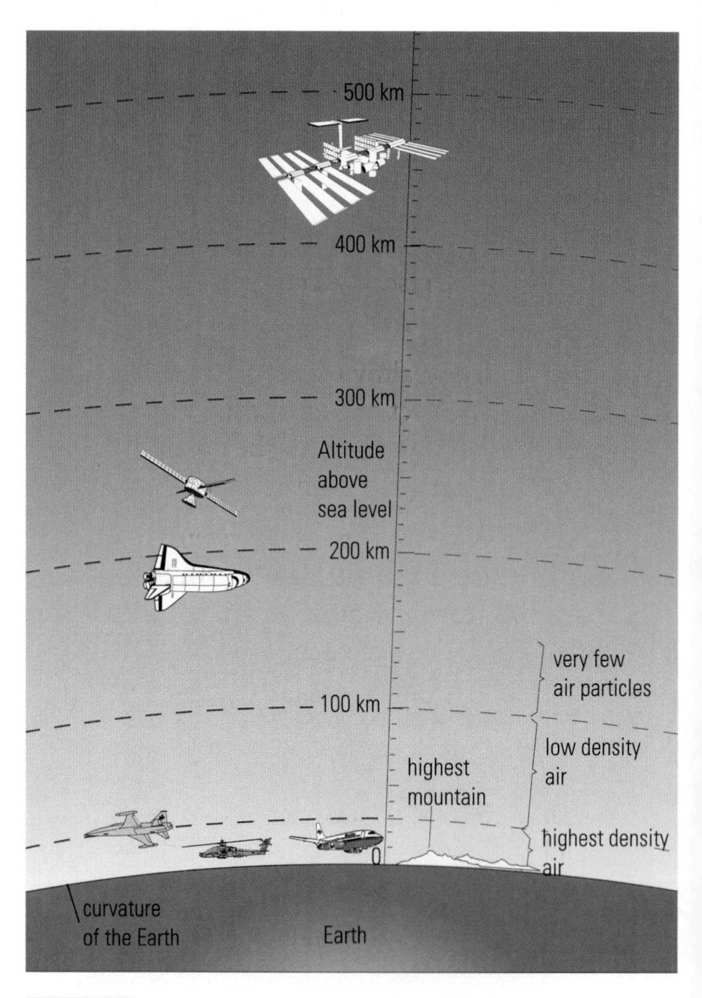

Figure 1

Vehicles in and beyond Earth's atmosphere

RADARSAT

In the spring of 1997, Manitoba experienced one of the worst floods in its history. Soon after the flood began, satellite images helped emergency crews plan disaster control and relief (**Figure 2**).

The word **radar** is short for "radio detection and ranging." A radar device gives off bursts of radio waves, picks up their reflections off objects, and figures out how far away those objects are. Radio waves can travel through clouds, so they can be used in all types of weather, as well as at night.

(a) List two advantages of using radio waves rather than visible light.

(b) What are two uses of radar other than for emergency relief?

RADARSAT is a satellite system that uses radar. It looks at features on land and on water using radio waves. In addition to floods, RADARSAT helps in large-scale emergencies, such as earthquakes, mudslides, ice storms, ice jams in melting rivers, and oil spills.

(c) How might RADARSAT help in your region?

Figure 2

This radar image of part of Manitoba has been computer improved. The red line shows the normal course of the river. The blue area shows the area under floodwaters in 1997.

Many industries benefit from RADARSAT images. Resources such as oil, natural gas, water, and minerals are found underground. Often, surface features of the ground help to determine where these resources are located. These features are easier to find using satellite images. Satellite images also let us monitor crop conditions, forests, soil humidity, how fast and how high a river flows, the number of fish, and shipping conditions.

(d) Why might it be better to search for underground resources using RADARSAT images?

To provide a healthy planet for the future, we must learn to protect our environment. Satellites (**Figure 3**) help us to monitor the environment and make informed decisions about our actions.

(e) RADARSAT is an expensive system, but it provides great benefits. Is it worth the cost? Give two reasons why or why not, along with a brief explanation of each.

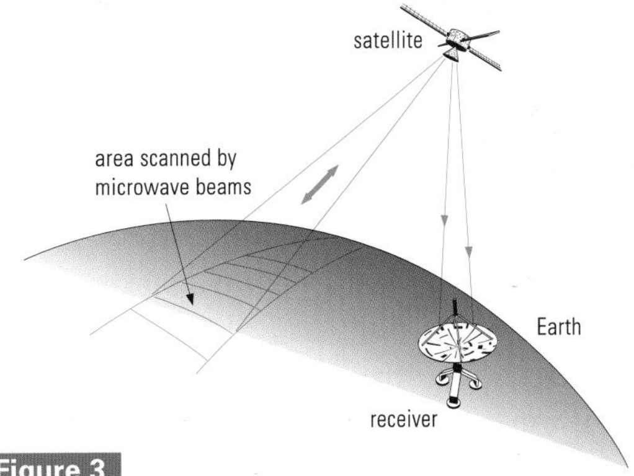

Figure 3

This satellite sends out bursts of microwaves. The satellite receives the reflected waves, converts the information to digital format, and beams it back to Earth.

Work the Web

Find out more about the satellites orbiting Earth. Record your information in point form. Start your search by visiting www.science.nelson.com and following the links from Science 9: Concepts and Connections, 4.16.

Understanding Concepts

1. List five uses for satellites.
2. What are two benefits of using RADARSAT in Canada?

Making Connections

3. Radar is related to bats. Explain how.
4. (L) Who should be responsible for cleaning up the pieces of space junk that crash to Earth? Give reasons for your answer. (Remember that some of these falling pieces contain nuclear waste.)

Challenge

3 Satellites have improved our knowledge of space. How?

Try This Activity Tracking Satellites

You will need
- Internet access
- seasonal star map or planisphere
- observation sheets (you design these)
- binoculars (optional)

1. (N) (O) Research which large satellites will be passing over your region in the next few weeks. Interpret the information to determine the satellites' paths.
2. (F3) On your star map, trace and label the paths of two of the satellites.
3. Choose a clear night and a time to view the night sky.

   Always obtain your parent's or guardian's permission and supervision to go out at night to make these observations. Dress appropriately for the weather.

4. (E1) Try to observe the satellites you have researched. Using binoculars may help. Record whether you observed the satellites. Give reasons why you could or could not see them.

The International Space Station

To survive in a space colony, we must learn more about how the human body can survive in space. Much of the research can be done on space stations that are orbiting Earth.

The first space station was Salyut, followed by Skylab. The first continuously occupied space station was the Soviet Union's Mir. The research done on Mir helped scientists to plan their next big step: construction of the International Space Station (ISS).

The ISS is the biggest technological project ever. Sixteen nations in six countries (Brazil, Canada, Europe, Japan, Russia, and the United States) have combined their resources to build this space station. When it is complete (which it is expected to be by 2004), it will have four research modules, a service module, a habitation module, remote robotic controls, a cargo block, a docking station for shuttle crafts, and huge solar panels, all connected to a central truss more than 100 m long (**Figure 1**). Forty-five American and Russian vessels will carry the more than 100 pieces (more than 450 000 kg of material) for construction about 450 km above the surface of Earth.

Six astronauts will live on the ISS for three months at a time. Part of their job is to perform experiments in low gravity related to plants, animals, humans, materials, research, crystal growth, chemical reactions, the environment, and other areas. The other part of their job is to find and repair satellites (**Figure 2**).

Figure 1

The International Space Station, ISS

Figure 2

Astronauts working in outer space

Try This Activity Manual Skill

While working outside the ISS, astronauts must wear bulky space suits. At the same time, they must perform delicate procedures with extremely sensitive equipment. To get an idea of how difficult this is, try the following tasks while wearing bulky or heavy gloves:

- tighten a nut on a bolt
- operate a VCR
- use tweezers to pick up a feather

How did it go? Describe any problems you had. Were you able to solve them? How?

Canada's Contribution to the International Space Station

The Canadian Space Agency (CSA) and the National Research Council of Canada (NRC) are involved in Canada's role in space technology. Canada is known for building sophisticated robots, such as the Canadarm (**Figure 3**), as well as visual systems used on the ground and in space. **Figure 4** describes the robotic and visual systems designed, built, and maintained by Canadians.

Figure 3

The Canadarm

Canadian Space Technology

Robotics

- Space System Remote Manipulator System (SSRMS): This "space arm," known as the Canadarm, is used to assemble the ISS. After helping in the construction, the arm will help in moving cargo into and out of the station, docking visiting shuttles, sending satellites out from the station and retrieving them, and helping astronauts working outside the station. This device is controlled by an astronaut inside the station and can move objects weighing up to 100 000 kg!

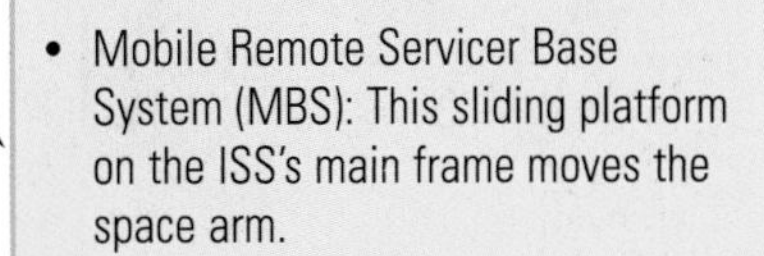

- Mobile Remote Servicer Base System (MBS): This sliding platform on the ISS's main frame moves the space arm.
- Special Purpose Dextrous Manipulators (SPDM): These are two "space hands" used to manipulate tools and other objects in space and conduct repairs.

Vision Systems

- Canadian Space Vision System (SVS): Small cameras are used to locate objects near the ISS and allow astronauts using virtual reality headsets to operate the robotic controls from inside the station.
- Ground-Based Vision Systems: These virtual reality systems are used for testing robotic devices and training astronauts for their mission.

Figure 4

Work the Web

The Canadarm is Canada's major contribution to the ISS. Find out more about the Canadarm and summarize your findings in two or three paragraphs. Start your search by visiting www.science.nelson.com and following the links from Science 9: Concepts and Connections, 4.17.

Understanding Concepts

1. List five uses of a space station.
2. What is Canada's role in the construction of the ISS?

Making Connections

3. How does living on the ISS relate to starting a space colony?

Challenge

1 Do you think the ISS should be a stop on the tour?

3 How would you simplify the information found in your technology information package for students in grades 5 or 6?

Humans in Space

Dr. Valeri Polyakov has spent 22 months in space. He also holds the record for most consecutive time in space: 14 months! Dr. Polyakov was a cosmonaut (a Russian astronaut) aboard the Mir space station. Mir was the first spacecraft designed to keep people in orbit for long periods. The first module was launched in 1986, and several other parts were added later.

Mir spent 15 years in orbit. It was brought down into the South Pacific Ocean on March 23, 2001. It had a very impressive life. During its orbit, 31 spacecraft (with crew) and 64 cargo vessels (without crew) docked at this space station. The American space shuttle docked there nine times. Seventeen space missions went to Mir. Twenty-eight long-term crews served aboard, and 125 cosmonauts and astronauts from 12 different countries visited.

Mir was brought down after the International Space Station was working. The ISS project was launched in 1998.

Figure 1

Canadian astronaut Chris Hadfield

In 1997, a spacecraft attempting to dock crashed into Mir, puncturing part of the station. Fortunately, the crew was able to figure out the damage and stabilize Mir.

Floating in Space

Often, people and objects in orbiting spacecraft appear to be floating (**Figure 1**).

The floating occurs because the spacecraft, and everything in it, is experiencing a special type of **free fall**. Being in free fall means you are falling toward the Earth with only gravity pulling you down. In this case, an object (the ISS, satellite, spacecraft) is falling toward Earth at the same time as it is speeding forward almost fast enough to shoot out into space. The result is that the spacecraft and everything in it remain balanced in orbit. See **Figure 2**.

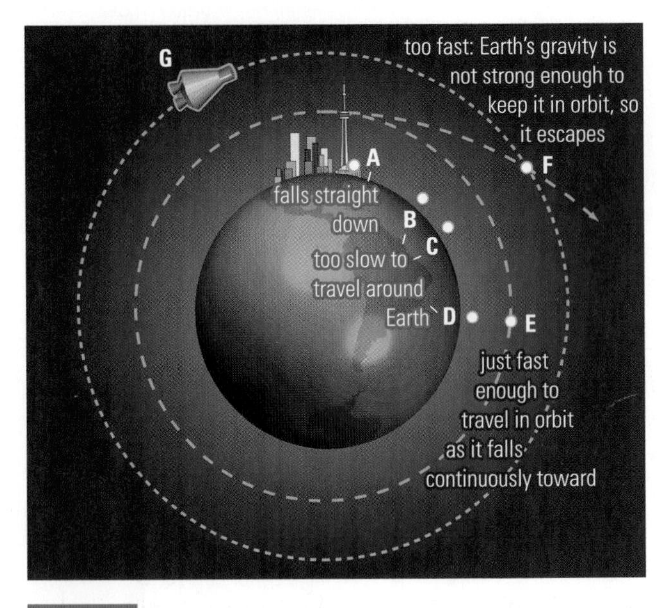

Figure 2

An example of free fall

A problem with free fall, or microgravity, is that the liquid in your body is not pulled down as much as it would on Earth. Two extra litres of blood stay in the top half of your body, swelling the heart and blood vessels. The result is puffy-head bird-legs syndrome. Astronauts have puffy faces and skinny legs while in space. The kidneys are also affected, and the astronauts must urinate frequently. Long-term research on microgravity and its effects on the human body continues.

During long space flights (six to twelve months) astronauts lose an average of 1% of their bone density each month. This makes their bones brittle so they break easily.

Living on the ISS

The cosmonauts and astronauts live on the space station for three months at a time. Most of their meals are made by adding water to a dry, crumbly mixture in a plastic pouch. Right now, the ISS has no ovens or large refrigeration devices. Any fresh vegetables or fruit are grown on the ISS, delivered by the American space shuttle, or delivered by the unpiloted Russian supply vehicle *Progress*.

All the people on the ISS do their share to keep the station clean. If germs or bacteria are allowed to grow, the astronauts could get sick. All parts of the space station are cleaned regularly. Garbage and worn clothing are sealed in airtight bags to return to Earth in the shuttle or to burn up on re-entry in the *Progress*. *Progress* is unpiloted because, for the Russians, it is cheaper to allow *Progress* to be destroyed in space and to build a new one for each mission than it is to staff each flight.

The astronauts must be strapped into a hammock to sleep on the ISS, as shown in **Figure 3**. For entertainment, there are movies, the Internet, e-mail, and scheduled calls home. The astronauts live in a very small space with several other people.

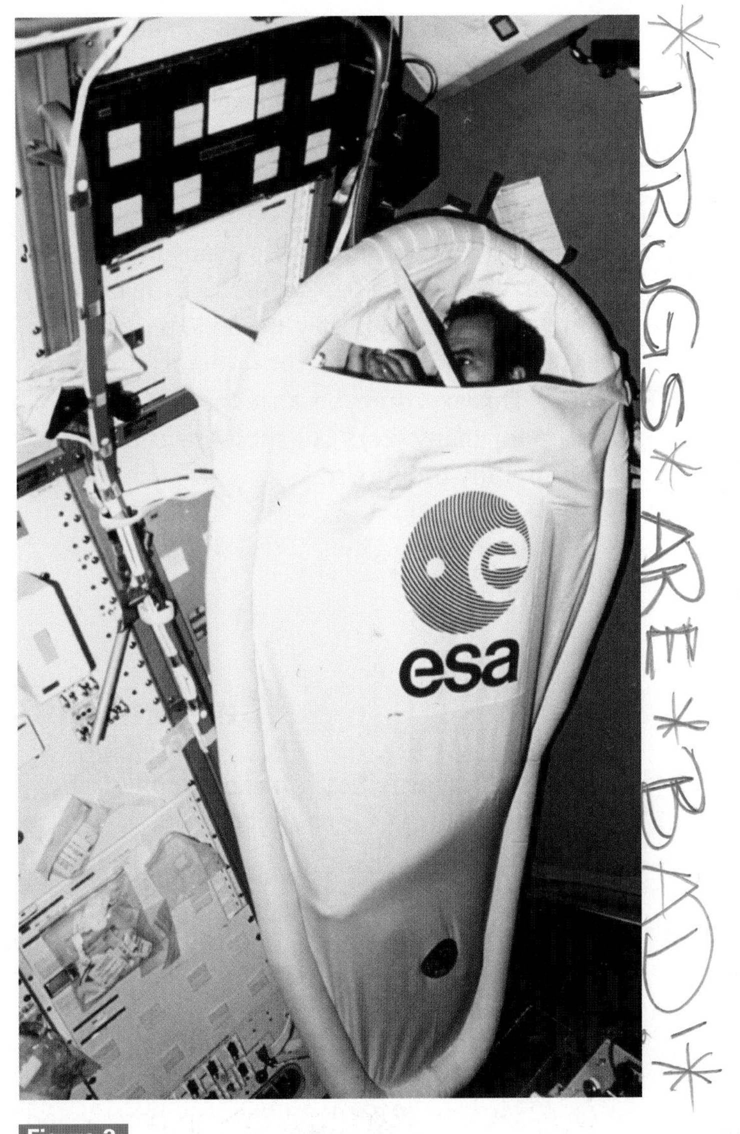

Figure 3

Sleeping in space

The Job

Every person on the ISS has responsibilities. Keeping track of scientific experiments, preparing meals, cleaning the space station, doing space walks, building parts of the station, and repairing the station and satellites are a few of the duties.

The astronauts must follow a special exercise program to keep their blood circulating and their bones and muscles strong. Our muscles are used to working against the full force of gravity. When that force isn't present, our muscles weaken. The men and women of the ISS need to stay healthy to perform their jobs.

Effects of Space Flight

An immediate problem for a human body in orbit is space sickness, a type of motion sickness. During constant free fall, signals from the eyes, skin, joints, muscles, and the balance organs in the inner ear (**Figure 4**) are rearranged. The clues do not match. This leads to dizziness, nausea, and vomiting.

The decreased gravity in space weakens and shrinks your muscles. Regular and energetic workouts are part of the program. The loss of calcium and the resulting increase in the brittleness of bones in space makes daily vitamins a requirement. The workouts also help to increase blood circulation to the lower parts of the body and decrease the puffy-head, bird-legs syndrome.

Another concern in space is exposure to radiation. On a single trip, astronauts are exposed to more radiation than they would experience in several years on Earth. Doctors are not sure of the effects of this exposure, because they may not show up for many years. All space travellers are being carefully monitored.

Figure 4

The drawing shows the structure of the human inner ear. The three fluid-filled tubes are responsible for balance. The fluid moves when the head moves: side to side, spinning around, or up and down.

Since the Space Age began in 1957, 25 976 items have been sent into orbit. About 8733 are still in orbit and 17 243 have fallen down to Earth.

Try This Activity Your Sense of Balance

Here are two activities you can do to test your sense of balance.

1. Stand on one leg with your eyes open. Close both eyes and try to maintain your balance. Describe what happened.
2. Stand on one leg facing a striped sheet or blanket held by two students. Try to maintain your balance as the blanket is moved sideways, as shown in **Figure 5**. Describe what happened.

Have spotters nearby in case you lose your balance.

Figure 5

The moving stripes may fool your brain into thinking that you are moving, so your muscles make up for it, and you fall over!

It is believed that space travel will slow down the aging process. In 1998, 77-year-old astronaut John Glenn went on a space shuttle flight to perform experiments on aging. Some of these experiments were designed by Canadians and included an investigation into osteoporosis, a condition in which the bones become very brittle. This was Glenn's second space flight. His first space flight was in 1962, when he became the first American to orbit Earth (**Figure 6**).

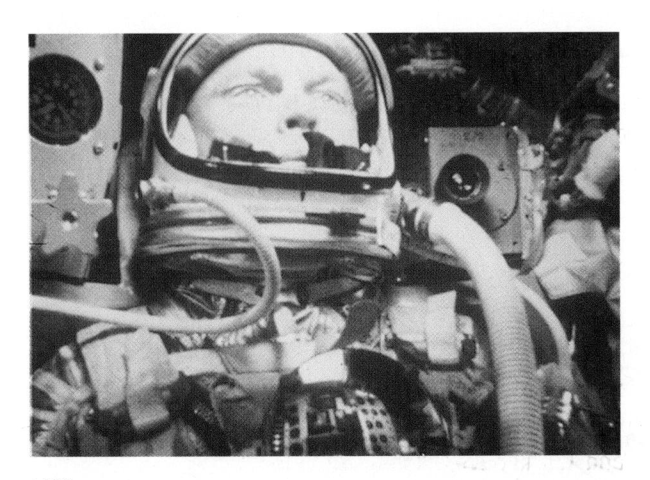

1962

1998

Figure 6
John Glenn

Understanding Concepts

1. In your own words, describe why astronauts appear to be floating in the ISS, even though gravity is pulling on them.
2. Describe two effects of constant free fall on the human body. Why do these effects occur?
3. Astronauts must spend a long time exercising each day. Why?

Making Connections

4. Choose one of the health problems described. How does the research related to it affect the lives of people not involved in space flight? (N) (O)

Challenge

1 Do you think tourists would want to visit some of the places astronauts have been?
2 How does research and knowledge of humans in space help scientists plan for the needs of a space colony?
3 Are there any technologies that were developed specifically for research about humans in space?

Work the Web

What sorts of health problems do the cosmonauts and astronauts aboard the ISS have to watch for? How do they stay healthy? Start your search by visiting www.science.nelson.com and following the links from Science 9: Concepts and Connections, 4.18. Record your answers in point form.

Three of Canada's astronauts, Roberta Bondar, Robert Thirsk, and Dave Williams, are medical doctors working in the field of space medicine. Dr. Williams is director of NASA's Space and Life Sciences program.

Spinoffs of the Space Program

How do the hard plastics used for in-line skates, safety helmets, and many other products relate to space technology? Hard plastics were developed for use in space and then applied to the products we use. This is an example of a **spinoff**—an extra benefit from technology developed for another purpose.

The space industry has produced a huge number of spinoffs (see **Table 1**).

Spinoff of the Hubble Space Telescope

The images that the Hubble Space Telescope (HST) has produced have increased scientists' understanding of space (**Figure 1**).

A spinoff of the HST is the use of its Charge Coupled Device (CCD) chips for the digital imaging breast biopsy system. The results from this type of biopsy are so accurate that both normal and cancerous tumours can be detected without surgery.

More than half a million women in North America need biopsies every year. This procedure saves the patient recovery time and reduces the amounts of pain, scarring, and exposure to radiation. This procedure costs much less than older methods, including surgery.

Figure 1

Spinoff of Plant Research

Hunger is a worldwide problem. The human population is slowly increasing as the areas available for food growth are decreasing.

NASA, with an eye for a possible future not on Earth, developed a method for using plants not only for food, but also for oxygen and water as well. This reduces the need for oxygen and water from other sources. The idea of hydroponics (liquid nutrient solutions) instead of soil to support plant growth was developed to grow food in space. On Earth, hydroponic systems are used by many people to produce quality plants (**Figure 2**).

Figure 2

Procedure

1 Choose a recently developed product and research how its development began with the space program. Use the library or the Internet to help you research.

2 Design a poster that includes the following:

- a title
- an image or drawing of the product or one of its applications
- a clear explanation of what it was developed for in space
- the benefits of the product on Earth
- your opinion as to whether you think this product should continue to be developed (two reasons)

Use colour, sketches, photos, and computer applications as appropriate. Use proper sentences, and check your spelling and grammar.

Understanding Concepts

1. What is meant by the term "spinoff"?
2. Which spinoffs in **Table 1** do you think are likely linked to Canada's contributions to the space program?

Making Connections

3. List four space spinoffs that are now part of your daily life.

Challenge

3 How have technologies developed specifically for space helped people here on Earth? It could be an increase in job availability or the use of a device that makes something easier on Earth.

Work the Web

Find out more about the links between products developed for space and their use on Earth. Add the information to your poster. Start your search by visiting www.science.nelson.com and following the links from Science 9: Concepts and Connections, 4.19.

Table 1 Types and Examples of Spinoffs

Area of research	Examples
microelectronics	digital watches, home computers, pacemakers, handheld calculators
new materials	nylon strips used to fasten clothing and objects, nonstick coating, flame-resistant materials
metal alloys	dental braces
hard plastics	safety helmets, in-line skates (**Figure 3**)
robotics	mining, industry, offshore oil exploration, a situation that is too difficult, too repetitive, or too precise for humans
vehicle controllers	controller for those with disabilities
safety devices	smoke detectors
recycling	water recycling
energy storage	solar cells, chemical batteries
food	freeze-dried convenience foods
pharmaceutics	antinausea medication
pump therapy	method to provide insulin continuously to diabetes patients
scanning	medical scanning using imaging techniques developed for satellites
space vision technology	satellite data applied to improving the efficiency of agricultural spraying
lasers	improved laser surgery (**Figure 4**)

Figure 3

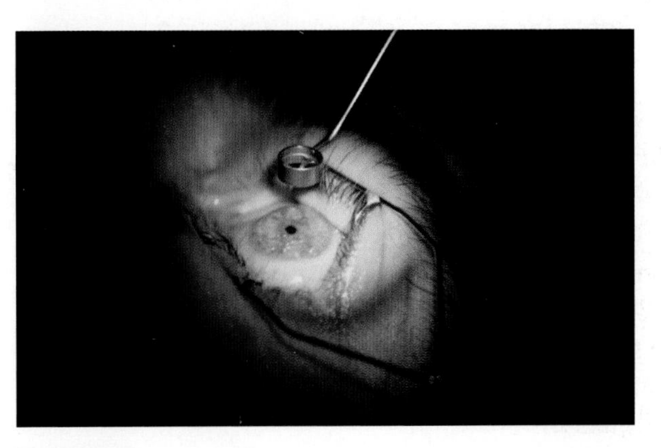

Figure 4

4.20 Explore an Issue

DECISION-MAKING SKILLS
- ● Define the Issue
- ○ Identify Alternatives
- ● Research
- ● Analyze the Issue
- ● Defend a Decision
- ○ Evaluate

Our Future in Space

If space stations are successful, the next stage of space travel will be the exploration of the planets. Maybe the Moon, Mars, or Mars's moons could be mined and the materials used to build structures in space—perhaps even a colony fit for humans. Where would you visit if distance were not an issue? a planet? a star? supernova 1987A (**Figure 1**)? the Cat's Eye Nebula (**Figure 2**)? Use your imagination.

This future is a long way off, and many problems remain. One issue is the way the human body reacts to a free-fall environment. A solution to this is to create an artificial gravity environment by rotating spacecraft, such as in **Figure 3**. This will work only for the voyage. What will happen when humans get to Mars? The force of gravity on Mars is only 40% that of Earth.

The next issues are water, oxygen, and heat. Humans need these resources, and Mars lacks them in the necessary amounts. Maybe small amounts of oxygen and water could be taken from the atmosphere. The frozen ground on Mars and the recycling of wastes could also generate water. Greenhouses could be used to grow food.

Protection from the Sun's radiation is a problem. Mars's atmosphere does not have ozone. Would we live underground? in a bubble?

These are only a few problems that may occur. The first question may be whether we should spend time and money exploring space, or whether we should be looking to improve our lives on Earth.

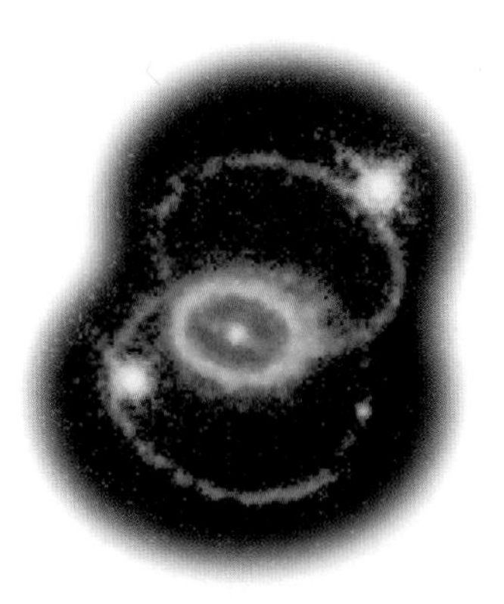

Figure 1

Canadian Ian Shelton observed supernova 1987A, the first recorded supernova in centuries. A supernova is a star that has just run out of fuel and collapsed on itself in a big explosion of light and energy.

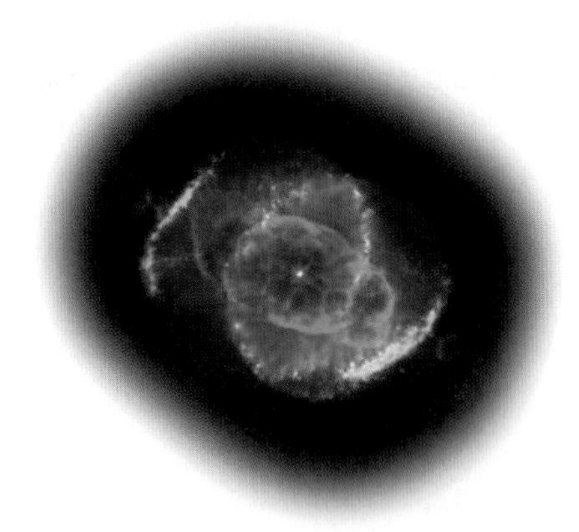

Figure 2

The Cat's Eye Nebula

NASA concept designers say Hollywood-inspired docking ports such as those seen in *Star Trek: Deep Space Nine* could find their way onto real-life space stations. These ports would allow spacecraft to dock faster and more safely.

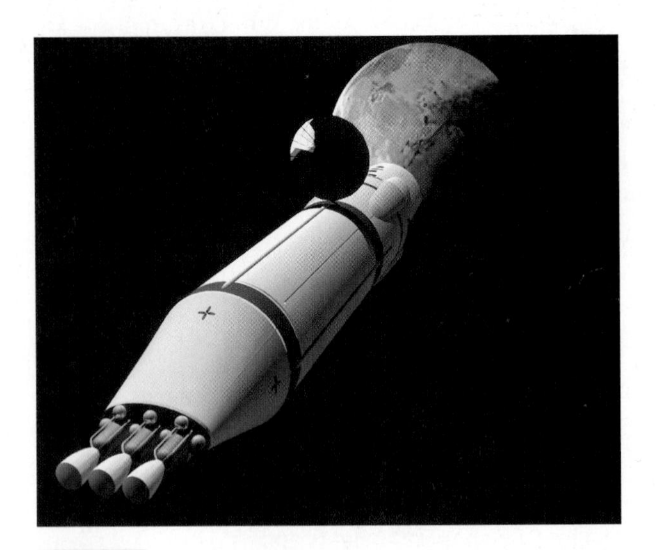

Figure 3

An artist's concept of a spacecraft travelling to Mars

Take a Stand: Space Exploration

People in favour of space exploration can give many reasons why it will benefit the human race. People against space exploration feel strongly that the disadvantages outweigh the benefits.

Working in a group, choose a position on space exploration. Make sure everyone does his or her share of work.

- Brainstorm a list of six reasons that support your point of view. Think about economics, ethics, environmental issues, politics, and available technology as you brainstorm. (I)
- Discuss these reasons as a group. Do more research, if necessary. (I) (J)
- Create a poster that outlines your six reasons. Pictures can be included.
- Share your point of view with the class through a presentation (oral presentation, video tape, audio tape, short skit, and so on). Use your imagination. (S)

Making Connections

1. List five ways that space exploration and research have helped people.
2. Do you think that we should be spending time and money exploring space? Why or why not? Give two reasons.

Work the Web

What are our plans for space? Start your search by visiting www.science.nelson.com and following the links from Science 9: Concepts and Connections, 4.20. Summarize your answer in two or three paragraphs.

Challenge

2 What types of situations should you consider when planning for survival in a space colony? **Figure 4** depicts one artist's concept of a space colony.

Figure 4

Unit 4 Summary

Key Expectations

Throughout this unit, you have had opportunities to do the following:

- recognize and describe parts of the universe (4.1, 4.2, 4.4, 4.5, 4.7, 4.10, 4.11)
- describe the generally accepted theory of the origin and evolution of the universe (4.13, 4.14, 4.15)
- describe, compare, and contrast properties of the parts of the solar system (4.4, 4.5, 4.7, 4.10, 4.11)
- describe the Sun and its effects on Earth and Earth's atmosphere (4.10)
- describe and explain the effects of the space environment on organisms and materials (4.17, 4.18, 4.19)
- identify problems and issues that scientists face when investigating celestial objects, and identify their solutions (4.8, 4.19, 4.20)
- formulate scientific questions about a problem or issue in space exploration (4.16, 4.19, 4.20)
- plan and conduct an inquiry into space exploration (4.4, 4.16)
- select and integrate information (4.4, 4.8, 4.10, 4.12, 4.14, 4.16, 4.19, 4.20)
- organize, record, and analyze information (4.4, 4.6, 4.8, 4.10, 4.12, 4.14, 4.16, 4.19, 4.20)
- communicate results (4.4, 4.6, 4.8, 4.10, 4.12, 4.14, 4.16, 4.19, 4.20)
- conduct investigations on the motion of visible celestial objects (4.4, 4.6, 4.10, 4.16)
- gather, organize, and record data through observations of the night sky or use of software programs to identify the motion of visible celestial objects (4.4, 4.6, 4.10, 4.16)
- identify and assess the impact of developments in space research and technology on other fields (4.18, 4.19)
- relate the beliefs of various cultures concerning celestial objects to aspects of their civilization (4.3)
- provide examples of the contributions of Canadian research and development to space exploration and technology (4.17, 4.18)
- explore careers in science and technology that relate to the exploration of space, and identify their educational requirements (4.9)

Key Terms

asteroid
asteroid belt
astronomer
astronomy
Big Bang Theory
blue shift
celestial
comet
constellations
cosmology
elliptical galaxy
free fall
galaxy
gas giant
inner planet
irregular galaxy
luminous
meteor
meteorite
meteoroid
minor bodies
moon
nebula
nonluminous
North Star
nuclear fusion
orbit
orbital period
outer planet
planet
Polaris
quasar
radar
red shift
reflecting telescope
refracting telescope
retrograde motion
revolution
rotation
satellite
solar flare
solar system
spectroscope
spectroscopy
spinoff
spiral galaxy
star
star cluster
sunspot
telescope
terrestrial planet
universe
visible spectrum
wavelength
zodiac constellation

What HAVE YOU *Learned?*

Review your answers to the What Do You Already Know questions on page 190 in Getting Started.

- Have any of your answers changed?
- What new questions do you have?

Unit Concept Map

Use the concept map to review the major concepts in Unit 4. This map can help you organize the information that you have learned. You can copy the map, and then add more links. You can also add more information in each box.

You can use a concept map to review a large topic on a general level, or you can use it to examine a very specific topic in detail. Select one concept from this unit that you need to study in detail, and make a detailed concept map for it.

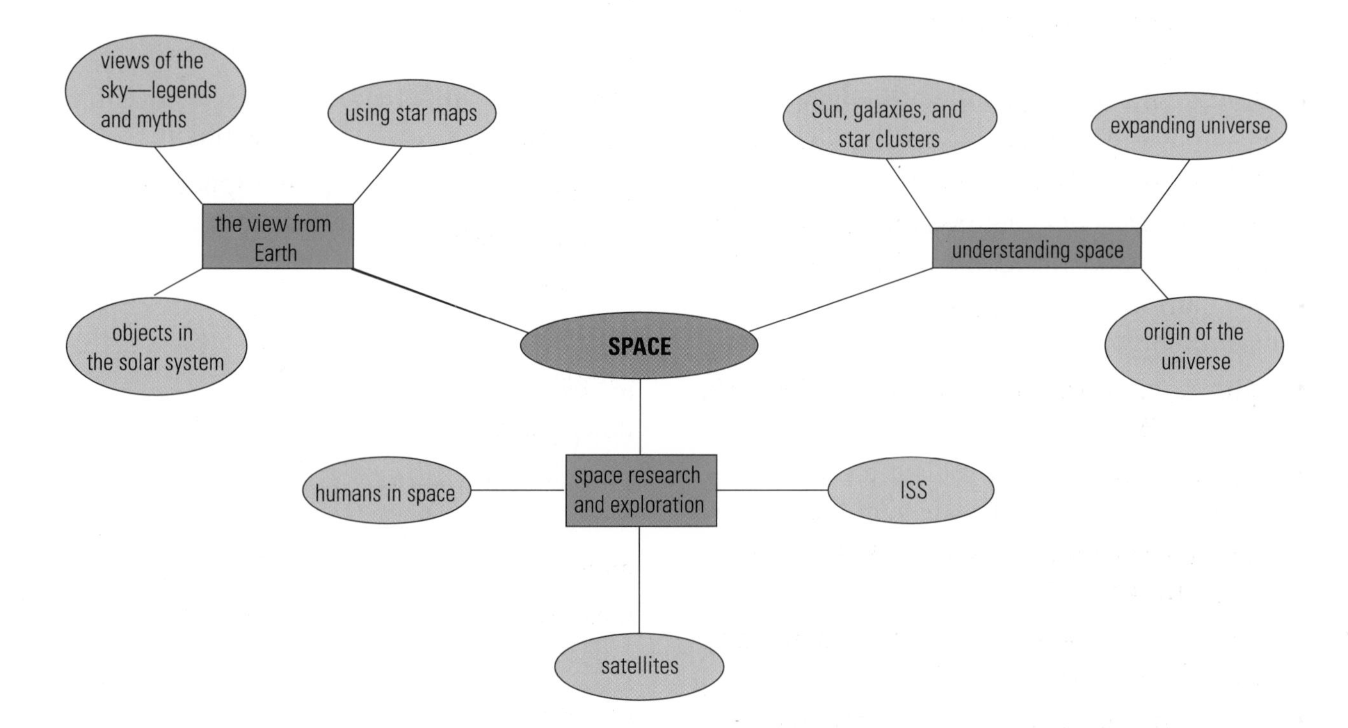

Unit 4 Review

Understanding Concepts

1. Complete the sentences using one of the following words: inner, nuclear fusion, outer, universe, gravity, Pluto, terrestrial, Galileo Galilei, gas giants, spinoff. Use each word only once. Recopy the complete sentence.
   - (a) The ___________ is everything that exists, including all matter and energy everywhere.
   - (b) The four planets closest to the Sun are called the ___________ planets or the ___________ planets.
   - (c) The four planets between the asteroid belt and Pluto are called the ___________ planets or the ___________.
   - (d) ___________ is in a category by itself.
   - (e) ___________ was the first scientist to use a telescope.
   - (f) ___________ is the force of attraction between all objects in the universe.
   - (g) The best explanation of why orbiting astronauts appear to be floating is that they are experiencing ___________.
   - (h) A ___________ is a benefit that comes from space science and technology research.
2. Indicate whether each statement is TRUE or FALSE. If the statement is FALSE, rewrite it so that it is TRUE.
   - (a) The light-year is a unit of time.
   - (b) A comet's tail is visible only when the comet's path is close to the Sun.
   - (c) A meteor is a meteoroid that hits Earth.
   - (d) Auroras are caused by light pollution.
   - (e) Venus is the hottest planet in our solar system.
   - (f) The Sun is in a galaxy called the Milky Way.
   - (g) The Milky Way is an elliptical galaxy.
   - (h) Evidence of an expanding universe comes from violet shift of the spectra of stars and galaxies.
   - (i) Space probes have been sent to explore a variety of objects in outer space.
   - (j) RADARSAT uses light waves.
   - (k) The problems of a puffy head and increased exposure to radiation disappear after astronauts have been in orbit for a week.
3. State one similarity and one difference for each pair of objects below.
   - (a) rotation/revolution
   - (b) asteroid/comet
   - (c) galaxy/star cluster
   - (d) solar flare/solar prominence
   - (e) natural satellite/artificial satellite
   - (f) space shuttle/space station
4. State two differences between inner and outer planets.
5. Each of the following descriptions fits one planet in the solar system. Name the planet described by each sentence.
   - (a) It has more mass than all other planets combined.
   - (b) It is closest to the Sun.
   - (c) Its atmosphere contains oxygen.
   - (d) It is not a gas giant or an inner planet.
   - (e) It has more than 1000 rings around it.
   - (f) It appears reddish in colour.
   - (g) It has the hottest surface temperature because of its carbon dioxide atmosphere.
   - (h) Its rotation is different from the other planets.
6. Choose one planet and name three similarities and three differences between it and Earth.
7. State the three constellations that you can see during each of the four seasons.
8. Why can you see some constellations only during certain seasons?

9. (a) As you observe the night sky in the Northern Hemisphere, which star appears not to move?
   (b) Why does it appear not to move?
   (c) How long do constellations take to travel once around that star?
10. Some cultures in ancient times were able to catalogue stars better than most people do today, although they did not know about most of the stars in the universe. Why were they better at it?
11. Would the asteroid belt be dangerous to travel through in a spacecraft? Why or why not?
12. Why does a meteor appear as a streak of light in the sky?
13. (a) What are sunspots?
   (b) How long is the sunspot cycle?
   (c) When will the sunspot cycle peak again?
14. Describe the effects that the particles from a solar flare have on Earth.
15. What is the Hubble Space Telescope? Why is it important?
16. If you see two stars through a telescope (**Figure 1**) and one looks bluish, the other yellowish, what can you say about the two stars?

Figure 1

17. Describe the Big Bang Theory.
18. List two pieces of evidence that support the Big Bang Theory.
19. What force leads to the formation of planets?
20. Describe the current theory of the formation of the Sun and the planets of our solar system. Use diagrams.
21. (a) What is red shift?
   (b) How does red shift support the Big Bang Theory?
22. As scientists continue to observe the spectra of stars and galaxies, what conclusions should they draw in each of the following cases?
   (a) Red shift continued.
   (b) Red shift was no longer observed.
   (c) Blue shift was observed.
23. What are two purposes of space stations?
24. What is "artificial gravity"? How could it be created on a spacecraft?
25. List three benefits of studying space medicine.
26. State the purpose of RADARSAT.
27. Why are robotics and vision systems important on the International Space Station?
28. What would happen to a spacecraft in orbit around Earth if its speed were too slow?
29. Why are the expressions "zero gravity" or "microgravity" incorrect for an astronaut in Earth orbit? What is the correct term?
30. Identify six ways you have benefited from space exploration in the past few days.
31. If you stood on your head, all your blood would rush to the top part of your body. Relate this to what an astronaut experiences while in Earth orbit.
32. Space probes are sent to explore asteroids and comets. How will this help scientists discover more about the origin of the solar system?
33. Why are we more protected from the Sun's harmful radiation here on Earth than astronauts are in space?

Applying Inquiry Skills

34. In observing the night sky, it is important to tell the difference between stars and planets. Describe how you would do this.

35. If you wanted to see the Big Dipper, how would you find it?

36. Sketch and label a diagram of our solar system. Include the name of each planet, and label the asteroid belt, inner and outer planets, gas giants, and terrestrial planets.

37. Why do all the planets have different surface temperatures?

38. **(a)** A certain comet, visible tonight, has a period of 185 years. When will it be seen again?

(b) Why can we see a comet's tail as it approaches the Sun?

39. The light from two distant galaxies, A and B, is examined using a spectroscope. The two spectra have a similar pattern of lines, but the lines of Galaxy A's spectrum are closer to the red end of the spectrum than are the lines of Galaxy B's spectrum. What can you conclude about the motions of the two galaxies relative to Earth?

40. Describe what conditions would allow you to see a satellite travelling across the night sky.

41. Find the results of a Canadian research project involving students growing seeds that had been in space.

42. Select a career in a space-related field and find out what qualifications and experience you would need to enter this field. If possible, interview someone in this career. Prepare 15 questions you would like answered, and write down the answers.

43. Write a short science-fiction story about space exploration using as many of the key terms from this unit as you can.

44. Make a list of three important questions you would like answered to help you understand more about (a) astronomy and (b) space exploration.

Making Connections

45. What were some skills achieved by astronomers in ancient times?

46. What is the evidence that astronomy is a very old science?

47. Why did some cultures create myths, legends, or even religions to explain the motion of objects in the sky?

48. Find out about one of the many traditional religions or cultures that base their calendar on the movements of the Sun and the Moon. Create a display showing how the dates of various celebrations are fixed.

49. Professional astronomers rely on amateur astronomers to find objects such as comets in the sky. Why?

50. Why is it difficult to observe planets in orbit around distant stars?

51. An observatory (**Figure 2**) at the top of a mountain eliminates light pollution as well as the problem of a thick atmosphere. The thin atmosphere can be a problem for people. They can become ill or light-headed when they first arrive at a high altitude. Research the causes of and solutions for altitude sickness.

Figure 2

52. The table below shows more than 100 known planetary moons in the solar system in 2002. If you checked other reference books published in earlier years, you would find the number of moons as listed below.

Year of publication	Number of known moons
1969	28
1984	44
1987	53
1991	60
1998	63
2002	101

(a) Why has the number of known moons changed so greatly in such a short time?

(b) Do you think the number of known moons will change as much in the next 25 years? Why or why not?

53. Give examples of Canadian participation in discoveries or research in astronomy.

54. In what ways will space probes help increase our understanding of the universe?

55. Before humans are sent to explore other planets, robots are sent to study areas of the planet.

(a) Describe two advantages of using robots rather than humans for this type of exploration.

(b) Describe two advantages of sending humans rather than robots.

56. Do you think computers will become more or less important in the study of astronomy? Why?

57. Many satellites are powered by electricity converted from solar energy by big shiny panels. State two advantages and two disadvantages of this source of energy.

58. What probes are in outer space this year? Research their purpose and find out what data they are sending to Earth.

59. (a) Describe four problems that humans may face as they try to set up a colony on Mars.

(b) Suggest solutions to each of the four problems from (a).

60. Radio telescopes (**Figure 3**) are used to try to detect intelligent life on planets revolving around other stars in the universe. What other methods could you suggest for trying to find evidence of intelligent life beyond Earth (extraterrestrial intelligence)? Explain your answer.

Figure 3

61. Investigate a program called SETI (search for extraterrestrial intelligence). Do you think this is a good program? Why or why not?

62. Defend or oppose one of the following statements. Clearly state three reasons.

(a) Space exploration benefits the human race.

(b) Money spent on space exploration would be better spent cleaning up the global environment.

IIA
Be
Mg
3 4 5 6 7 8 9
IIIB IVB VB VIB VIIB VIII
21 Sc
22 Ti
23 V
24 Cr
25 Mn
26 Fe
27 Co
Ca
40 Zr
41 Nb
42 Mo
43 Tc
44 Ru
45 Rh
Ba
73 Ta
74 W
75 Re
76 Os
105 Unp
106 Unh
107 Uns
La
58 Ce
59 Pr
Th

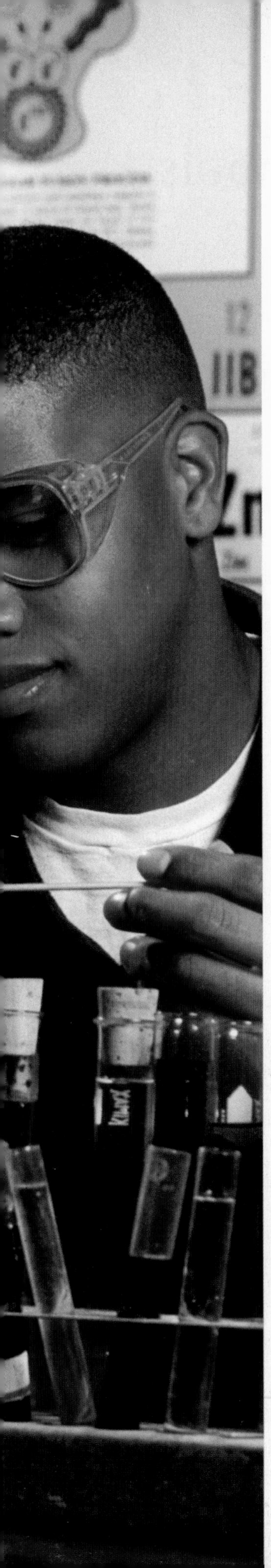

Skills Handbook

(A) Safety Conventions and Symbols

Safety Conventions in *NELSON Science 9: Concepts and Connections*

When you perform the investigations in *Nelson Science 9: Concepts and Connections*, you will find them challenging, interesting, and safe. However, accidents can happen. In this text, chemicals, equipment, and procedures that require extra caution are highlighted in red and are preceded either by the appropriate WHMIS (Workplace Hazardous Materials Information System) symbol (**Table 1**) or by ✋.

Always read cautions carefully and make sure you understand what they mean before you proceed. If you are not sure about anything, ask someone who knows, such as your teacher, a parent, or a classmate.

Table 1 WHMIS Symbols for Hazardous Materials

WHMIS symbol	Risks	Precautions
compressed gas	• could explode due to pressure • could explode if heated or dropped	• ensure container is always secured • do not drop or allow to fall • store in a proper area
flammable and combustible material	• may ignite without warning • may release flammable products when exposed to water	• work in a well-ventilated area • avoid heating • avoid sparks and flames • store in a proper area
oxidizing material	• may cause skin and eye burns • increase fire and explosion hazards • may cause combustible material to explode	• wear body, hand, face, and eye protection • store away from combustible materials • store in a proper container
poisonous and infectious material causing immediate and serious toxic effects	• may be fatal if swallowed or inhaled • may be absorbed through the skin • small amounts are toxic	• avoid breathing dust or vapours • avoid contact with eyes or skin • wear protective clothing, and face and eye protection • work in a well-ventilated area and wear breathing protection
poisonous and infectious material causing other toxic effects	• may cause death or permanent injury • may cause cancer • may cause allergic reactions	• use hand, body, face, and eye protection • avoid direct contact • work in a well-ventilated area • store in appropriate area
biohazardous infectious material	• may cause anaphylactic shock (severe allergic reaction) • includes viruses, yeasts, moulds, bacteria, and parasites that affect humans • includes cellular components (e.g., infected blood plasma)	• avoid breathing vapours • avoid contamination of people and area • work and store in special areas • special training is required to handle materials
corrosive material	• cause eye and skin irritation on contact • severe burns/tissue damage after long period of contact • lung damage if inhaled • may cause blindness if splashed in eyes	• wear body, hand, face, and eye protection • avoid all direct body contact • use breathing apparatus • work in a well-ventilated area • use proper storage containers
dangerously reactive material	• may react with water • may explode if exposed to shock or heat • may release toxic or flammable vapours • may burn unexpectedly	• handle with care, avoiding vibration, shocks, and sudden temperature changes • store in appropriate, sealed containers

Workplace Hazardous Materials Information System (WHMIS) Symbols

The Workplace Hazardous Materials Information System (WHMIS) symbols in **Table 1** were developed to ensure that the labelling of dangerous materials was the same in all workplaces, including schools. Become familiar with these warning symbols and pay attention to them when they appear in your text and on any products or materials that you handle.

Hazardous Household Product Symbols (HHPS)

You are probably familiar with the warning symbols in **Table 2**. They appear on many products that are common around the house. These warning symbols were developed to indicate exactly why and to what degree a product is dangerous. The shape of the symbol (**Figure 1**) indicates the seriousness of the risk. The triangle means that caution is required, the diamond shape identifies a warning, and the octagon (stop sign shape) means danger—the highest level of risk.

Table 2 Hazardous Household Product Symbols

HHP symbols	Risks	Examples of household products
corrosive	• can burn your skin or eyes • will damage your throat and stomach if swallowed	• battery acid (household and automotive) • drain openers • oven cleaners • acids
flammable	• can catch fire quickly (keep this product away from heat, flames, and sparks)	• gasoline • motor oil • barbecue starter fluid • Varsol and other solvents • nail polish
explosive	• container will explode if heated or punctured • metal or plastic can fly out and hurt your eyes or other parts of your body	• aerosol spray cans (e.g., hair spray) • barbecue propane tanks
poison	• may cause serious illness or death if swallowed or licked • fumes (vapours) may be dangerous if inhaled	• paint • pesticides • motor oil • over-the-counter and prescription medicines • household cleaners (e.g., bleach)

Danger

Warning

Caution

Figure 1

For example, of these two symbols, the danger symbol represents the higher risk of igniting.

Warning—flammable

Danger—flammable

(B) Safety in the Laboratory

The Importance of Safety

There are certain safety hazards in every laboratory. You should know about them and about the precautions you must take to reduce the risk of an accident.

Safety in the laboratory requires some common sense and the ability to think about possible accidents that could happen. The activities in this textbook have been tested and are safe, as long as they are done with proper care. Your teacher will review safety rules for your laboratory/classroom and for conducting specific investigations. Take the rules seriously and follow them.

Preventing Accidents

Most accidents in the laboratory are caused by carelessness. Knowing the most common causes of accidents can help you prevent them. The most common causes include

- applying too much pressure to glass equipment (including microscope slides and cover slips)
- handling hot equipment carelessly
- measuring or mixing chemicals incorrectly
- working in a messy or disorganized space
- paying too little attention to instructions
- failing to tie back long hair or loose clothing
- failing to wear protective eyewear and aprons
- unexpected contact with other students

Before You Start

1. Learn the location and proper use of safety equipment, such as safety goggles, protective aprons, heat-resistant gloves, eyewash station, container for broken glass, first-aid kit, fire extinguishers, and fire blankets. Locate the nearest fire alarm.
2. Inform your teacher of any allergies, medical conditions, or other physical limitations you may have. If you wear contact lenses, be extra careful and make sure your teacher is aware of it.
3. Read the procedure of an investigation carefully before you start, and note any safety precautions. Clear your work area of all materials except those you will use in the investigation. If there is anything you do not understand, ask your teacher to explain. If you are designing your own experiment, get your teacher's approval before carrying out the experiment.
4. Wear safety goggles and protective clothing (a lab apron or lab coat), and tie back long hair. Remove loose jewellery. Wear closed shoes, not open sandals.
5. Use stands, clamps, and holders to secure any equipment that may be easily tipped over.
6. Do not taste, touch, or smell any materials unless you are asked to do so by your teacher. Do not chew gum, eat, or drink in the laboratory.
7. Never work alone in the laboratory.

Working with Chemicals

8. Know where the MSDS (Material Safety Data Sheet) manual is kept. Know any important MSDS information for the chemicals you are using.
9. Label all containers. When taking something from a bottle or other container, double-check the label to make sure you are taking exactly what you need.
10. If any part of your body comes in contact with a chemical, wash the area immediately and thoroughly with water. If your eyes are affected, do not touch them but wash them immediately and continuously with cool water for at least 15 minutes, and inform your teacher.
11. Handle all chemicals carefully. If you are asked to smell a chemical, take a few deep breaths before waving the vapour toward your nose (**Figure 2**). Only this method should be used to smell chemicals in the laboratory. Never put your nose close to a chemical.

Figure 2
Safe smelling

12. Put test tubes in a rack before pouring liquids into them. If you must hold a test tube, tilt it away from yourself and others before pouring in the liquid (**Figure 3**).

Figure 3
Safe pouring

13. Clean up any spilled materials immediately, following instructions given by your teacher.
14. Do not return unused chemicals to the containers from which they were taken, and do not pour them down the drain. Dispose of chemicals as instructed by your teacher.

Heating

15. Whenever possible, use an electric hot plate for heating materials. Use a flame only if instructed to do so. If you use a Bunsen burner (**Figure 4**), make sure you follow these procedures:
    - Get instructions from your teacher on the proper way to light and adjust the Bunsen burner.

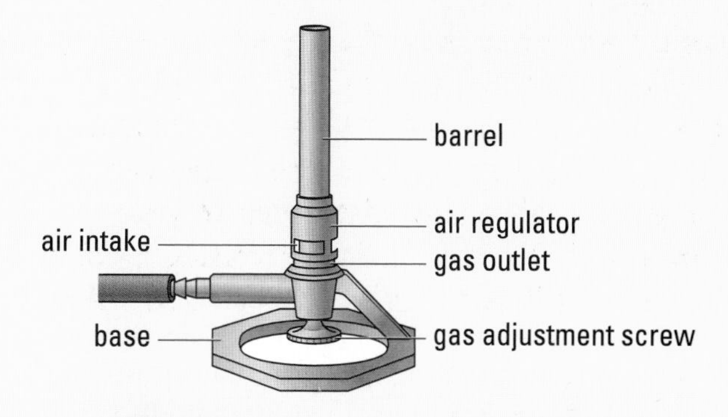

Figure 4
Bunsen burner

    - Do not heat a flammable material (for example, alcohol) over a Bunsen burner. Make sure no flammable materials are nearby.
    - Do not leave a lighted Bunsen burner unattended.
    - Always turn off the gas at the valve, not at the base of the Bunsen burner.
16. When heating liquids in glass containers, make sure you use clean Pyrex or Kimax. Do not use cracked or broken glassware. If the liquid is to be heated to boiling, use boiling chips to prevent "bumping." Always keep the open end pointed away from yourself and others. Never allow a container to boil dry.
17. When heating a test tube over a flame, use a test-tube holder. Hold the test tube at an angle, with the opening facing away from yourself and others. Heat the upper half of the liquid first, then move it gently in the flame to distribute the heat evenly (**Figure 5**).

Figure 5

Safely heating a liquid in a test tube

18. Be careful when handling hot objects and objects that might be hot. Hot plates can take up to 60 minutes to cool completely. Test that they are cool enough to move by touching them first with a damp paper towel. If you hear sizzling or see steam, wait a little longer! If you burn yourself, immediately apply cold water or ice, and inform your teacher.

Other Hazards

19. Keep water and wet hands away from electrical cords, plugs, and sockets. Always unplug electrical cords by pulling on the plug, not the cord. Report any worn cords or damaged outlets to your teacher. Make sure electrical cords are not placed where someone could trip over them.
20. Put broken and waste glass into the specially marked containers. Use a broom and dustpan to sweep up broken glass. Wear heavy gloves if you have to pick up broken glass with your hands.
21. Follow your teacher's instructions when disposing of waste materials.
22. Report to your teacher all accidents (no matter how minor), broken equipment, damaged or faulty facilities, and suspicious-looking chemicals.
23. Wash your hands thoroughly with soap and warm water after working in the laboratory. This is especially important when you handle chemicals, or living or preserved biological specimens.

C Planning an Investigation

Science is more than knowledge and understanding of the world around us. Scientific inquiry (or scientific investigation) is also a way of learning about the world by observing things, asking questions, proposing answers, and testing those answers. Science is also about sharing information that is collected during an investigation.

All scientific investigations use a similar process to find answers to questions. In most cases, this process attempts to find relationships between variables. A **variable** is any factor that can change in an investigation. In some cases the investigator can change or control the variables; in other cases the investigator cannot.

If the purpose of an investigation is to determine the relationship between two variables—for example, between the size of the wire and the current in a circuit—then you can carry out a **controlled experiment**. This means that you can control the variables. If the purpose is to test a suspected relationship between two variables—for example, between smoking and lung cancer—where a controlled experiment is not possible, then you can carry out a **correlational study**. In this case, you cannot ask or expect people to smoke to see if they develop cancer.

Controlled Experiment

A controlled experiment is a test in which one variable is intentionally and steadily changed to find out its effect (if any) on a second variable. All other variables are controlled, or kept the same. The ability to control variables makes a controlled experiment different from a correlational study. For instance, you may observe that more aquatic plants grow in a lake at the end of the summer and ask yourself, "Why?" You may come up with a number of possible answers. Maybe you think aquatic plants are more plentiful in the summer because they grow better in warmer water. This is your hypothesis. A hypothesis is a possible explanation for your observations. From your hypothesis you can make a prediction that you can test in a controlled experiment—for example: "Aquatic plants grow faster in 16°C water than in 6°C water."

In a controlled experiment to test this prediction, you need to control, or keep constant, all other possible causes of the effect. You might conduct an experiment on five identical samples of lake water containing aquatic plants, making sure that all growing conditions—such as volume of water, nutrients, and light—are the same. You would then change one condition—the temperature of the water—and measure the growth of the plants over a period of time.

These different conditions in an experiment are called variables. There are three types of variables:

- **Independent variables** are conditions that you can change. In our example, the water temperature is the independent variable. In a controlled experiment there is usually only one independent variable.
- **Dependent variables** are what you measure to determine a change (or lack of change) as you vary the independent variable. The growth of aquatic plants, possibly measured by height or weight, is our dependent variable.
- **Controlled variables** are all the factors that you keep the same. In a controlled experiment all known possible causes of the result except one are controlled. In our example many factors are kept the same (e.g., volume of water, nutrients, and light) to be certain that the independent variable (temperature) affected the dependent variable (the growth of plants).

The common stages of a controlled experiment are outlined in the flow chart in **Figure 1**. Even though these stages are presented in sequence, usually there are many cycles through the steps during the actual experiment.

1 Once you have decided that a controlled experiment is the best scientific method for your investigation, decide whether you are going to create a concept or whether you are going to test a concept. Indicate your decision in a statement of the **purpose**.

2 Ask a **question** about the relationship between an independent variable and a dependent variable (C1 page 256).

3 Depending on your question, you may create a **hypothesis** and a **prediction**, or just a prediction (C2 page 256).

4 A **design** describes what evidence you will gather, and how (C3 page 257). Your design can be expanded to include a procedure (C5 page 259) and any recording instruments you will need (C6 page 259).

5 Decide on the degree of reliability, precision, and accuracy your measuring instruments will need and list and gather the **materials** you need (C4 page 257).

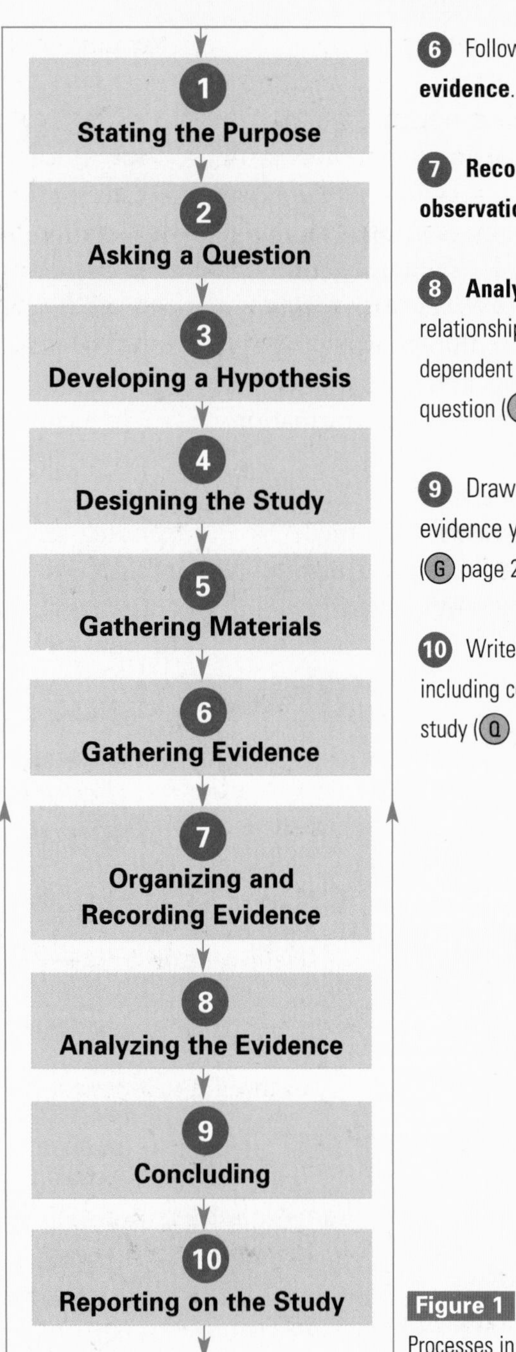

6 Follow your procedure and **gather evidence**.

7 **Record your evidence and observations** (E page 268).

8 **Analyze** your evidence to find the relationship between the independent and dependent variables and your answer to the question (F page 272).

9 Draw a **conclusion** based on the evidence you gathered in the study. (G page 276).

10 Write or present a **report** on the study, including comments on the need for further study (Q page 287).

Figure 1
Processes in controlled experiments

Correlational Study

Is there any truth to the saying "Red sky at night: sailors' delight. Red sky in the morning: sailors take warning"? In trying to answer questions like this, there is no way you can control the variables. You cannot make the sky red! Nor can you control all the other variables concerning the environment and the weather. You would need to collect evidence and carry out a correlational study.

In a correlational study, an investigator tries to determine whether one variable is affecting another variable without intentionally changing or controlling any of the variables. Instead, variables are allowed to change naturally.

In correlational studies it is often difficult to determine whether one factor actually caused a certain result. Any two variables can be compared. It is important for the investigator to determine whether a reasonable link between the two variables is possible. As an extreme example, you could graph the annual iceberg sightings from St. John's, Newfoundland, against the frequency of taxi accidents in Hamilton, Ontario, and discover that the years with the greatest number of icebergs correspond to the years of the greatest number of taxi accidents. However, you could not expect to predict the frequency of accidents in the future by the number of iceberg sightings, because this kind of correlation is likely to be a coincidence.

The flow chart in **Figure 2** outlines the main stages in carrying out a correlational study. By following this format, investigators can do science without doing experiments or fieldwork. You can use this flow chart as a checklist to make sure you use all the steps necessary to complete a proper correlational study.

1 Choose a topic that interests or puzzles you. Determine what kind of study you will carry out, and whether you are going to replicate or revise a previous study or create a new one. Indicate your decisions in a statement of the **purpose**.

2 Ask a **question** about the relationship between two chosen variables.

3 Develop a **hypothesis** describing the relationship between the variables.

4 Write a **design** for the study. Decide how the evidence will be gathered. Will you be using existing databases, or will you be gathering your own evidence through measuring, interviewing, or surveying? You will not be controlling variables, but you should try to list other variables that may be relevant.

5 Write a procedure that you will follow to gather **evidence** and that could be followed by others to replicate your study. Indicate how your sample and sample size are chosen and how your measurements are done.

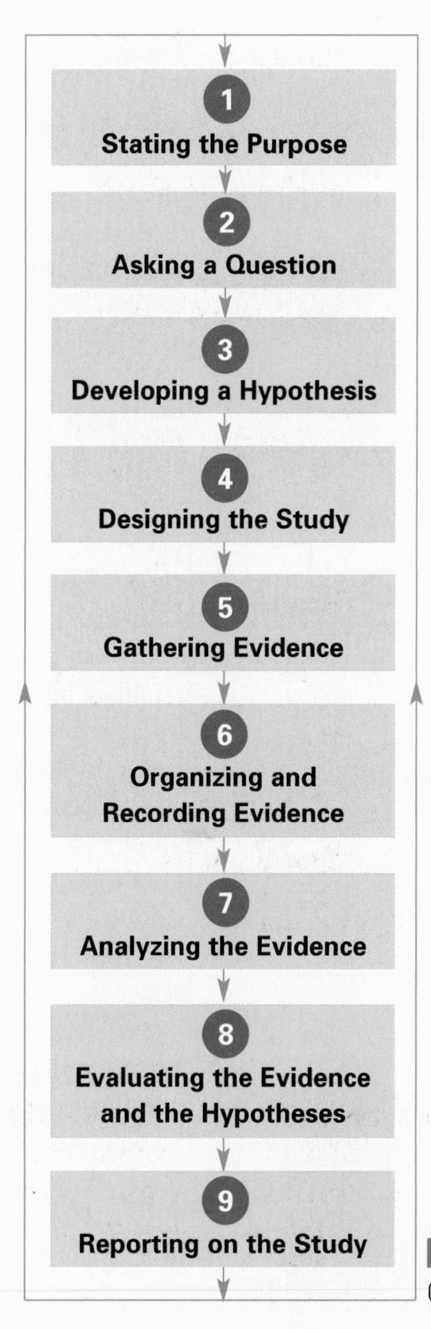

6 Create appropriate tables in which to record and, perhaps, analyze your evidence.

7 Use the evidence to calculate new values or to create graphs showing the relationship (if any) between the variables. Devise and write a hypothesis (tentative concept) by **analyzing** the evidence gathered.

8 **Evaluate** the evidence you gathered in your study, and evaluate the hypotheses you stated in steps 3 and 7. Evaluate the design, materials, and procedure to determine the degree of confidence in the evidence. You can then compare the initial hypothesis generated from the evidence. If the hypotheses are different, decide which hypothesis is best supported. Include your reasons.

9 Write or present your **report** on the study. Suggest the need for replication of this study or the need for different studies.

Figure 2
Conducting a correlational study

C1 Asking the Question

All scientific questions are asked by people who are curious about the world around them. Our experiences and observations of the world may lead us to ask questions that interest us or to express an idea that can be tested. To be considered scientific, you must be able to test the idea and answer the question.

For example, you probably know that balloons stick to walls if you rub them on your head. This observation might have led you to wonder about a number of things: Does a balloon stick better in some situations than others? Does it stick better if you rub it more times? Does it matter how inflated the balloon is or what colour it is?

We can answer each of these questions through a good scientific investigation. However, we cannot answer every question through scientific inquiry—some are too general. Learning to ask questions that can be answered by scientific inquiry is a very important skill in science. A proper question is the basis for every investigation.

Let's look at some of the above questions to see whether they can be answered through scientific inquiry.

- *Why does a balloon stick to a wall after you rub it on your head?* This question is too general to be investigated scientifically.
- *Does a balloon stick better in some situations than others?* The question is narrower; all we need now is a way of determining what we mean by "stick better." We could define "stick better" by the length of time that the balloon sticks to the wall. We also need to decide what situations might make the balloon stick better.
- *Does it stick longer if you rub it more times?* Now we are investigating a variable that can be measured.

A scientific question is often about cause-and-effect relationships. These questions are often stated in this form: "What causes the change in [variable 1]?" and "What will be the effects on [variable 1] if we change [variable 2]?" As you know, a variable is something that can change or be changed in an investigation. In our example, the question asked in proper form could be "How does the number of times you rub a balloon on your hair affect the length of time the balloon will stick to the wall?" or "What will the effect be on the length of time that a balloon sticks to the wall if you change the number of times you rub the balloon on your hair?"

The one variable in the experiment that you intentionally change is called the **independent variable**. For example, the number of times you rub the balloon on your hair is something you can change—you are changing the independent variable. The variable that shows an effect (the length of time the balloon sticks to the wall) is called the **dependent variable** because it "depends" on the independent variable. A scientific question that asks what happens to a dependent variable (e.g., length of time) when you change the dependent variable (e.g., the number of rubs) is a question that can be answered through scientific inquiry.

C2 Predicting and Hypothesizing

A possible explanation of why one variable affects another in a certain way is called a **hypothesis**. If you ask a general question in an experiment, then you can create a hypothesis to answer the question. If you ask a specific question, then a prediction may be made based on the hypothesis. Usually a hypothesis is a general statement that explains how one variable affects another, and a **prediction** is a specific statement that explains a more specific instance of the relationship between the two variables. Continuing with our previous example, a hypothesis might be that rubbing a balloon on your hair produces static electricity; therefore, the more you rub the balloon, the longer it will stick to the wall. A prediction based on this hypothesis could be that as you increase the number of times you rub the balloon on your hair, the length of time the balloon will stick to the wall will also increase. See examples of hypotheses and predictions in **Table 1**.

Table 1 Sample Hypotheses and Predictions

Hypothesis	Prediction
If you increase the number of times you rub a balloon on a person's head,	then the length of time the balloon sticks to a wall increases.
If you increase the amount of salt on a road,	then the amount of rusting of the metal parts of a bicycle increases.
If a plant has bright-coloured flowers,	then it will reproduce more successfully than a plant with dull-coloured flowers.

Predictions and hypotheses go hand in hand. The hypothesis explains the prediction. The prediction is what you test through your experiment. If the evidence gathered confirms the prediction, you can be more confident that the hypothesis is acceptable. If the evidence gathered does not support the prediction, then the hypothesis is likely not an acceptable explanation.

C3 Designing the Investigation

Once you have asked a testable question and developed a hypothesis or a prediction, you can design an investigation that will test the prediction. To design an investigation, you need to think of all the steps in the procedure from beginning to end. The design is a general procedure by which you plan to change one variable (the independent variable), measure the effect on another variable (the dependent variable), and, if possible, keep all other variables constant.

Write your experimental design in a short paragraph (two or three sentences) describing how you intend to answer the question. List all variables, controls, and tests that will be used in the investigation. See the examples in **Table 2**.

Table 2 Questions and Designs

Question	Design
How does the number of times you rub a balloon on your hair affect the length of time the balloon will stick to the wall?	The length of time the balloon sticks to the wall is measured using a stopwatch. The independent variable is the number of times you rub the balloon on your hair. The dependent variable is the time. Some controlled variables are the colour of the balloon, the amount the balloon is inflated, the temperature of the air, and the hair the balloon is rubbed on. Evidence is collected three times.
How does the colour of a plant's flowers affect the success of the plant's reproduction?	The success of the plant's reproduction is measured by the number of seeds the plant produces. The independent variable is the colour of the flowers. The dependent variable is the number of seeds produced. Ten outdoor plants with flowers of different colours will be observed from flowering to seed production.
How is the current in a circuit affected by the diameter of the wire?	The effect of the diameter of the wire on the current in a circuit is determined by setting up a number of circuits with wires of different diameters and measuring the current with an ammeter. The independent variable is the diameter of the wire; the dependent variable is the current in the circuit; and the controlled variables include the length of the wire, the wire material, the voltage in the circuit, and the resistance. Wires of five different diameters are tested.

C4 Choosing Materials

You can do many science investigations using everyday materials and equipment such as plastic cups, straws, and water bottles. You may also need to use more specialized pieces of equipment. Some of these are illustrated in **Figure 3**.

Decide what materials and equipment you will need to perform your experiment. Keep safety in mind: choose the safest materials and equipment possible, and decide whether any particular safety equipment—such as goggles, bike helmets, or rubber gloves—should be worn. For more information on safety, see B Safety in the Laboratory (p. 250).

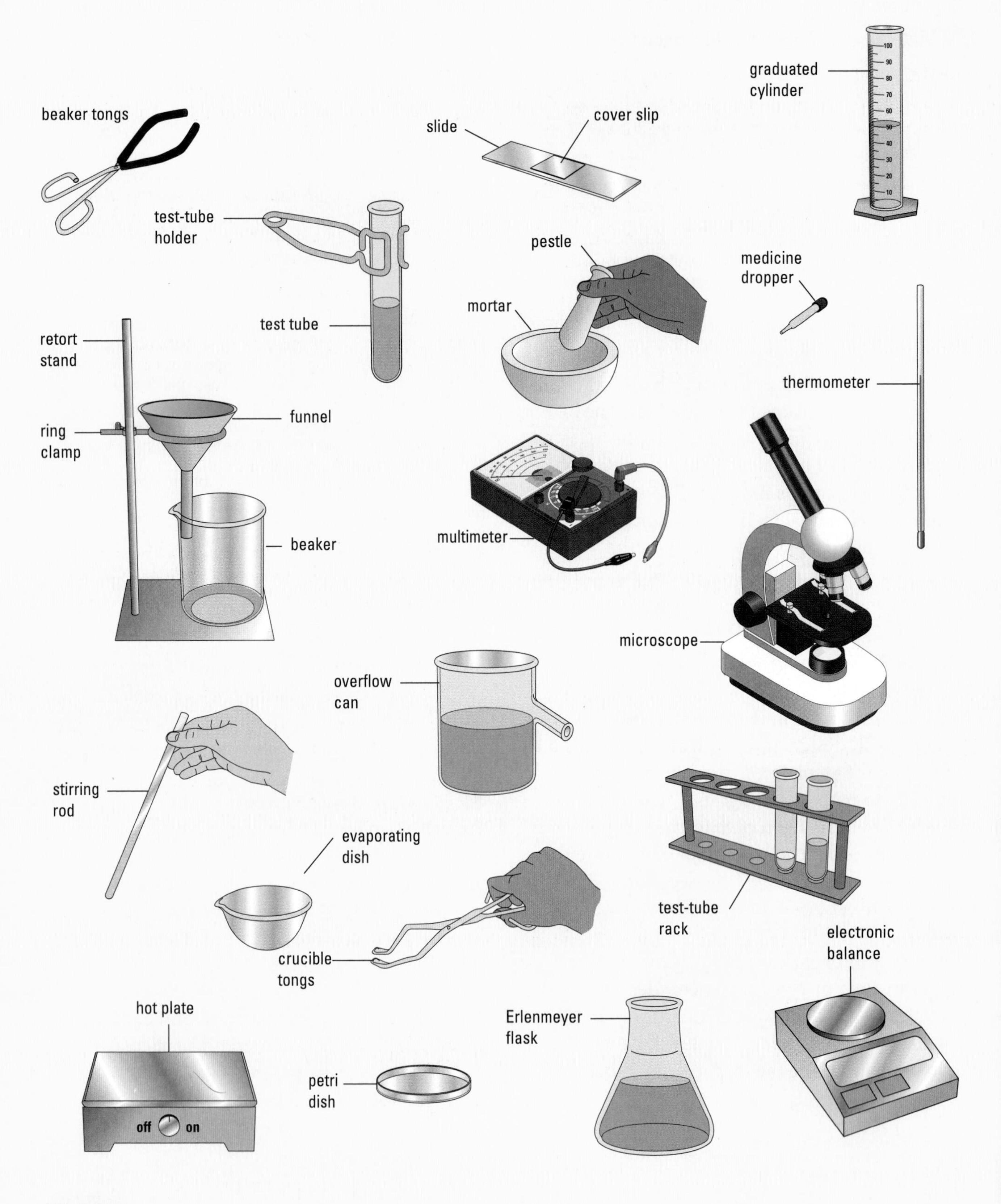

Figure 3

Common laboratory equipment

C5 Writing the Procedure

The procedure is a set of detailed steps that describe how you will conduct your experiment. Anyone who is interested in learning about your experiment should be able to understand how it was performed so that it can be repeated. So it is important to be able to write an experimental procedure clearly, concisely, and accurately. When writing a procedure, use numbered steps with only one instruction in each step. For example, the first three, and the last, steps of a procedure could look like this:

Procedure

1. Wear safety goggles, a lab apron, and rubber gloves.
2. Set up the experiment as shown in the diagram.
3. Measure the temperature of the water to + or –0.2°C every five minutes.

⋮

8. Return materials to the storage cabinet.

Your first step should refer to any safety precautions that must be followed. The last step in your procedure should relate to cleaning up of the area, including any necessary waste disposal or recycling.

C6 Preparing Observation Tables

You need to think carefully about what you are going to measure and how you are going to record these measurements. Where possible, create tables for recording your observations. If you are recording variables, they are generally used as the headings for the columns in the table, as shown in **Table 3** and **4**.

Use the following checklist to help you create an effective table:

- Give each table a short title that describes the information it contains.
- List the dependent variable(s) (the effects) in the last (farthest right) column of the table.
- List the independent variable (the cause) in the next-to-last column of the table.
- List the controlled variables, if necessary, in the first (farthest left) column of the table.
- Be sure to include the units of measurement with each variable in the header at the top of each column, but not inside the table (as in the example in **Table 3**).

Table 3 Current in Wires of Different Diameter

Wire gauge	Wire diameter (mm)	Current (A)
14	1.63	5.33
16	1.29	3.36
18	1.02	2.11
20	0.81	1.32
22	0.64	0.83

Table 4 Preventing Corrosion

	Material					
Day	iron nail	waxed iron nail	aluminum wire	waxed aluminum wire	magnesium strip	waxed magnesium strip
1	dull black colour	white	shiny, silvery grey	white	shiny, silvery grey	white
2						
3						
4						
5	rust coloured	a little rust	slightly dull	no change	slightly dull	no change

D Conducting the Investigation

D1 Using Timing Devices

Several different timing devices are mentioned in this text. Each has unique features and is suited to different tasks.

A **stopwatch** is a small handheld device that measures time. It is turned on and off by an observer as an event occurs. It may be analogue (with hands that point to numbers around the face) or digital (giving its reading as numbers on a screen) (**Figure 4**). Stopwatches are most useful for timing events that take more than a few seconds, and most stopwatches can take only one measurement at a time.

An **ultrasonic probe** is a new technology in schools. It is a small instrument that sends out an ultrasonic (can't be heard by humans) sound wave and detects the reflection of this wave off a nearby object. The time difference between sending and receiving the sound wave is converted to a distance. The sound waves are sent out many times a second so the probe can calculate the movement of the object. These probes are usually attached to desktop computers or graphing calculators, and the object's movement can be shown on a graph.

A **photogate** is a type of switch attached to a timer. In the laboratory, photogates are used to measure times for moving objects such as carts along a track, falling objects, pendulums, and sliders on an air track. Outside the laboratory, it is often used to time sporting events such as ski races. At each end of the race there is a "gate" consisting of a light source on one side and a light detector on the other side. When the athlete crosses the starting gate, he or she interrupts the beam of light, which starts the timer. The timer stops when the athlete crosses the beam of light in the gate at the end of the race. In investigations of motion where photogates are used, the distance is kept constant while the total time to travel the distance is measured.

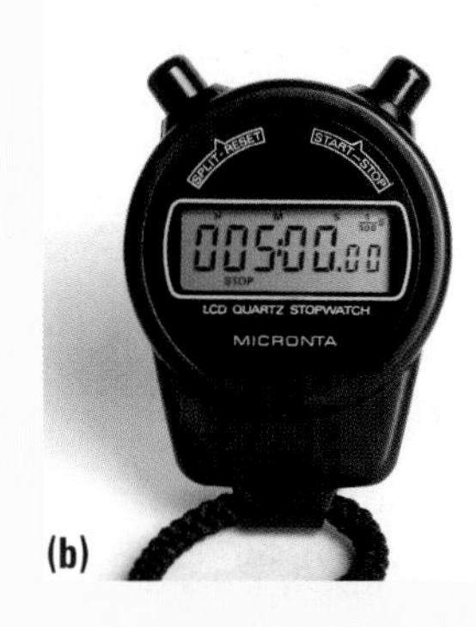

Figure 4

(a) An analogue stopwatch
(b) A digital stopwatch

D2 Using a Mass Balance

To measure mass you must use some type of balance or scale. Several types of balances are available. Which one you choose will depend on what's available, how heavy the object is, and the accuracy you need. The most common type of balance is the triple-beam balance (**Figure 5**). However, if you need to measure with a high degree of accuracy you may use an electronic balance (**Figure 6**).

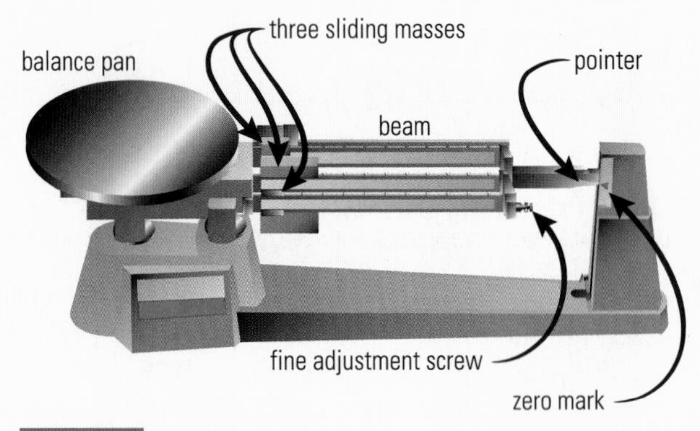

Figure 5
A triple-beam balance

Figure 6
An electronic balance

A Triple-Beam Balance

To find the mass of an object,

1. Make sure the balance is on a level surface and that nothing is on the balance pan.
2. Bring all three sliding masses to the zero point at the left of the beam. Make sure the two larger masses are placed in the notches at the beginning of the beam.
3. If the pointer does not line up with the zero mark, adjust the fine-adjustment screw, which is either at the far left side of the balance or near the left-hand side of the beam.
4. Put the object you want to measure in the centre of the balance pan.
5. Move the largest sliding mass along the beam notch by notch until it is just too heavy (making the balance beam drop below the zero mark at the end). Then put it one notch to the left. Make sure the mass fits properly into the notch.
6. Repeat step 5 with the middle sliding mass.
7. Carefully slide the smallest mass until the pointer is centred on the mark. Use a pencil to make the movements more gradual.
8. The mass of the object is the sum of the readings shown by the positions of the three sliding masses.

Sample Problem

During an investigation you are asked to measure out 250 g of salt from a large container of salt.

1. Place a container (e.g., a watch glass) on the balance and record its mass using the procedure above. Assume the mass of the watch glass is 35 g.
2. Add to this figure the mass of the salt required. Record the total. The total will be 35 g + 250 g, which equals 285 g.
3. With the watch glass still on the balance, move the sliding masses to the total mass calculated in step 2. The balance will now show 285 g, and the pointer at the end of the beam will be at the bottom.
4. Slowly add salt to the watch glass, being careful not to spill any directly onto the pan. Watch the pointer as it moves up, until it reaches the centre zero mark. You now have the correct amount (250 g) of salt.

An Electronic or Digital Balance

To find the mass of an object,

1. Turn on the balance and wait until the display reads zero.
2. Place the object in the centre, and read the number of grams directly.

To find the mass of a substance in a container,

1. Turn on the balance and wait until the display reads zero.
2. Place the empty container at the centre of the electronic balance pan. Press "Tare" or "Rezero." The display will now read "0 g" again.
3. Remove the container, fill it with the substance to be measured, and put it back on the centre of the balance pan. The display will then register the mass of the substance (without the mass of the container.

D3 Drawing and Constructing Circuits

Sources of Electrical Energy

To provide the electrical energy in most of the circuits you use in this course you will be using combinations of dry cells or a special device called a power supply. Power supplies can be set to supply the voltage required.

The source used in the circuits you construct and test yourself will be a "direct current," or DC source of electrical energy. In DC circuits the current only flows in one direction around the circuit. We use DC circuits in this course because the operating voltages are much safer to use, generally below 28 V.

Wall outlets provide a different kind of electrical energy known as an "alternating current" or AC source. In AC circuits the electric current reverses its direction 60 times a second. AC appliances are specially designed for this energy source, and typically operate at 120 V or 240 V.

Drawing Circuit Diagrams

Before building a circuit, it is a good idea to draw a circuit diagram. This will remind you how components should be connected. There are some conventions to follow when drawing circuit diagrams: connecting wires are generally shown as straight lines or 90° angles, and symbols (shown in **Table 5**) are used to represent all components.

Safety Considerations

It is important to observe and use appropriate safety procedures, especially when you see ✋.

- Always ensure that your hands are dry, and that you are standing on a dry surface.
- Do not use faulty dry cells or batteries, do not connect different makes of dry cells in the same battery, and avoid connecting partially discharged dry cells to fully charged cells. Take care not to accidentally short-circuit dry cells or batteries.
- Do not use frayed or damaged connectors.
- Handle breakable components with care.
- Only operate a circuit after it has been approved by your teacher.

Constructing the Circuit

When constructing or modifying a circuit, always follow the instructions. If you are unsure of the procedure, ask for clarification. Check that all the components are in good working order.

- Check the connections carefully when linking connecting dry cells in series or in parallel. Incorrect connections could cause shorted circuits or explosions. Ask your teacher for clarification if you are unsure.
- When attaching connecting wires to meters, connect a red wire to the positive terminal and a black wire to the negative terminal of the meter. This will remind you to consider the polarity of the meter when connecting it in the circuit.
- Sometimes the ends of connecting wires do not have the correct attachments to connect to the device or meter. Use extra, approved attachment devices, such as alligator clips, but be careful to position the connectors so that they cannot touch one another.
- Open the switch before altering a meter connection or adding new wiring or components.
- If the circuit does not operate correctly, open the switch and check the circuit wiring and all connections to the terminals. If you still cannot find the problem, ask your teacher to inspect your circuit again.

Table 5 Circuit Diagram Symbols

	DC CIRCUITS		HOUSEHOLD CIRCUITS (additional symbols)	
Sources/Outlets		cell	120 V	wall outlet
		3-cell battery	R	range outlet
				single outlet
				double outlet (duplex)
			W	weatherproof outlet
				special-purpose outlet
Control Devices		switch		
		fuse		
		circuit breaker		
		switch and fuse		
		distribution panel		
	S	switch		
	S_{WP}	weatherproof switch		
		push button		
Electrical Loads		light bulb		ceiling light
	C	clock		wall light
	M	motor	PS	lampholder with pull switch
	T	thermostat		recessed fixture
		resistor	TV	television outlet
		variable resistor (rheostat)	F	fan
		fluorescent fixture		buzzer
		heating panel		bell
Meters	A	ammeter		
	V	voltmeter		
Connectors		conducting wire		
		wires joined		
		ground connection		

D4 Using the Voltmeter and the Ammeter

As we cannot see electrons flowing in electric circuits, we have to rely on instruments that can detect and measure electricity. There are at least two you are likely to use: the voltmeter (**Figure 7(a)**) and the ammeter (**Figure 7(b)**). These instruments may be digital (providing a digital readout) or analog (indicating voltage or current by the movement of a needle across a scale). You may also use a digital multimeter (**Figure 8**) which can be used to measure voltage, current, or resistance.

(a)

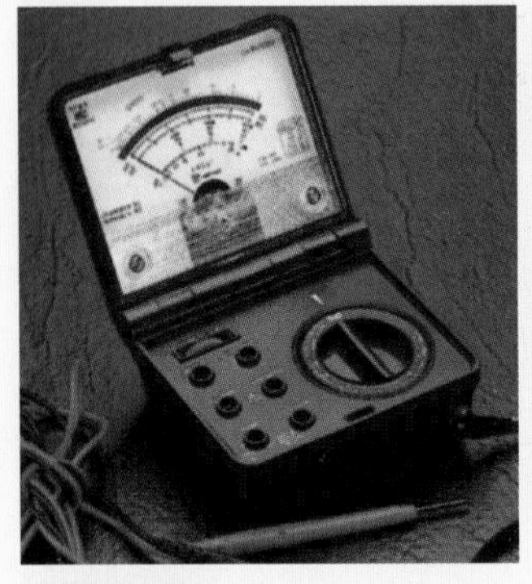

(b)

Figure 7

The Voltmeter

A voltmeter measures the voltage difference between two different points in a circuit. The voltmeter can be connected across the terminals of a cell, to measure the voltage output of the cell, or across another component of a circuit, to measure the voltage drop across that component. In other words, the voltmeter is always connected in parallel with the component you want to investigate.

Reading an Analog Voltmeter

The needle on a voltmeter usually moves from left to right, with the zero voltage being on the left and the maximum voltage on the right of the scale. If the voltmeter scale has only one set of numbers, it is relatively easy to measure the voltage. Be sure that you know the voltage value represented by the smallest division on the scale.

If the voltmeter has several sets of numbers, identify which set of numbers matches the voltage range selected by the switch on the voltmeter.

The two leads that connect a voltmeter to any part of the electric circuit must be attached so that the negative terminal of the voltmeter is connected to the more negative part of the circuit. If the leads were attached incorrectly, the needle would try to move to the left, but would be unable to do so, and would not give a reading.

The Ammeter

An ammeter measures the amount of electric current flowing in a circuit. To measure the electric current, we connect the ammeter directly into the circuit itself. In whatever part of the circuit we wish to measure the current, a wire is disconnected and the ammeter is connected, in series, to complete the circuit. Reading digital and analog ammeters is very similar to reading digital and analog voltmeters. The unit of current is the ampere (A) or milliampere (mA).

The Digital Multimeter

The digital multimeter can measure the voltage, current, or the resistance in a circuit. Using the selector knob, select the electrical value and the range that you wish to measure. Then connect the wire leads properly to the appropriate place in the circuit. Remember that for measuring voltage the meter is connected in parallel; for measuring current it is connected in series. The multimeter will provide a digital readout of the measurement of the electrical value.

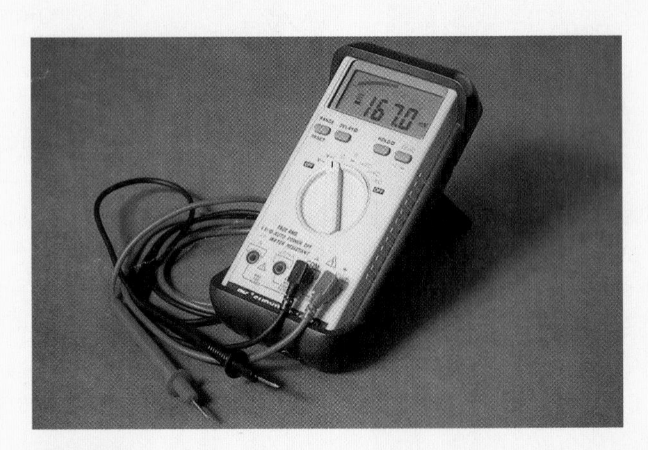

Figure 8

D5 Using the Microscope (Mono or Stereo)

Because cells are small, you must make them appear larger than they really are in order to see and study them. To view cells closely, you will use a compound light microscope (**Figure 9**). It employs two lenses and a light source to make the object appear larger. The object is magnified by a lens near your eye, the ocular lens (sometimes called the eyepiece), and again by a second lens, the objective lens, which is just above the object. The comparison of the actual size of the object with the size of its image is referred to as magnification.

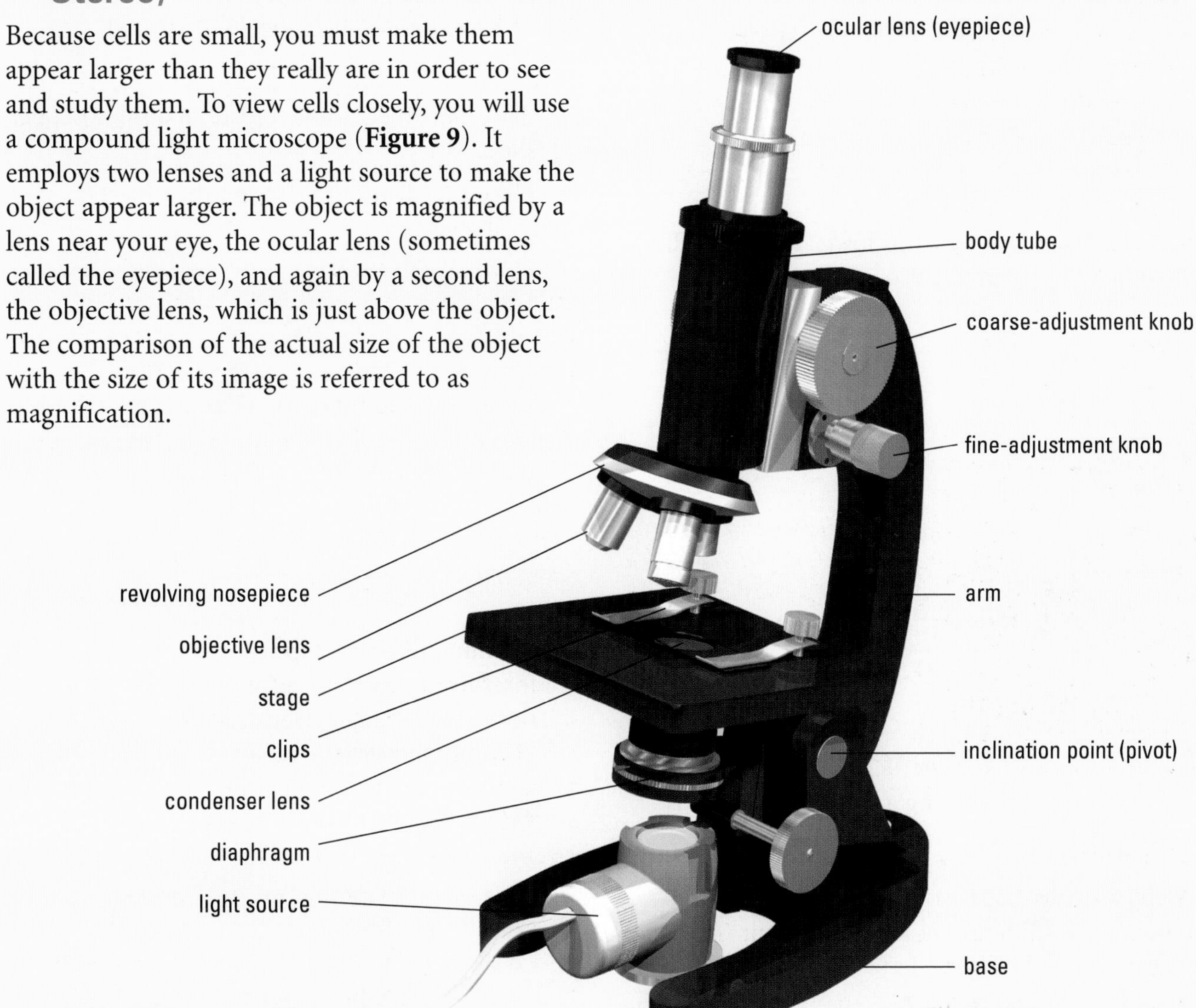

Figure 9 **Parts of the Microscope**

Structure	Function
stage	Supports the microscope slide. A central opening in the stage allows light to pass through the slide.
clips	Found on the stage and used to hold the slide in position.
diaphragm	Regulates the amount of light reaching the object being viewed.
objective lenses	Magnifies the object. Usually three complex lenses are located on the nosepiece immediately above the object or specimen. The smallest of these, the low-power objective lens, has the lowest magnification, usually four times (4X). The medium-power lens magnifies by 10X, and the long, high-power lens by 40X.
revolving nosepiece	Rotates, allowing the objective lens to be changed. Each lens clicks into place.
body tube	Contains ocular lens, supports objective lenses.
ocular lens	Magnifies the object, usually by 10X. Also known as the eyepiece, this is the part you look through to view the object.
coarse-adjustment knob	Moves the body tube up or down so you can get the object or specimen into focus. It is used with the low-power objective only.
fine-adjustment knob	Moves the tube to get the object or specimen into sharp focus. It is used with medium- and high-power magnification. The fine-adjustment knob is used only after the object or specimen has been located and focused under low-power magnification using the coarse adjustment.
condenser lens	Directs light to the object or specimen.

Basic Microscope Skills

The skills outlined below are presented as sets of instructions. This will enable you to practise these skills before you are asked to use them in the investigations in *Nelson Science 9: Concepts and Connections.*

Materials

- newspaper that contains lower-case letter "f," or similar small object
- scissors
- microscope slide
- cover slip
- dropper
- water
- compound microscope
- thread
- compass or petri dish
- pencil
- transparent ruler

Preparing a Dry Mount

This method of preparing a microscope slide is called a dry mount, because no water is used.

1. Find a small, flat object, such as a lower-case letter "f" cut from a newspaper.
2. Place the object in the centre of a microscope slide.
3. Hold a cover slip between your thumb and forefinger. Place the edge of the cover slip to one side of the object. Gently lower the cover slip onto the slide so that it covers the object.

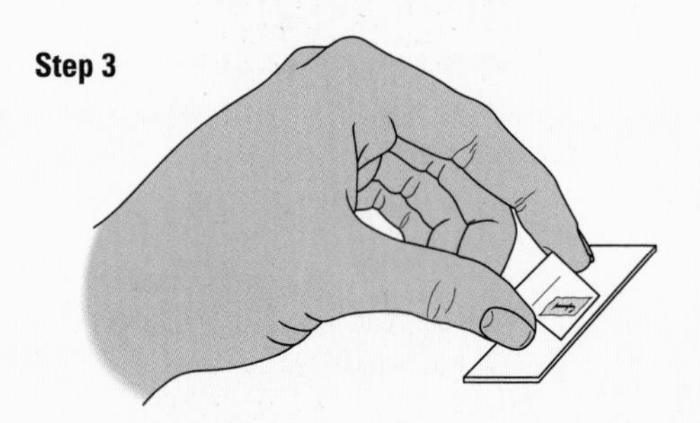
Step 3

Preparing a Wet Mount

This method of preparing a microscope slide is called a wet mount, because water is used.

1. Find a small, flat object.
2. Place the object in the centre of a microscope slide.
3. Place two drops of water on the object.

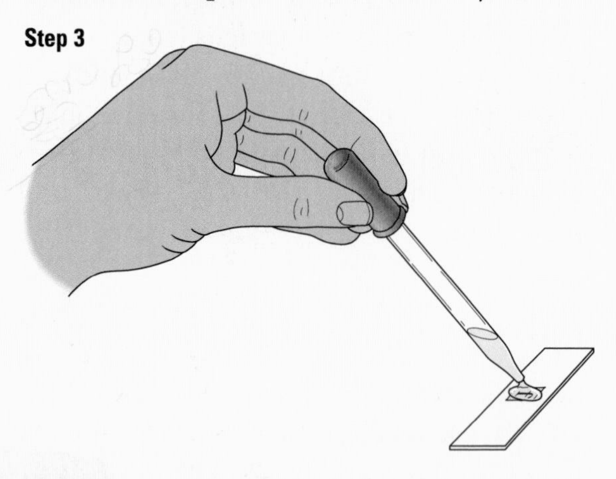
Step 3

4. Holding the cover slip with your thumb and forefinger, touch the edge of the surface of the slide at a 45° angle. Gently lower the cover slip, allowing the air to escape.

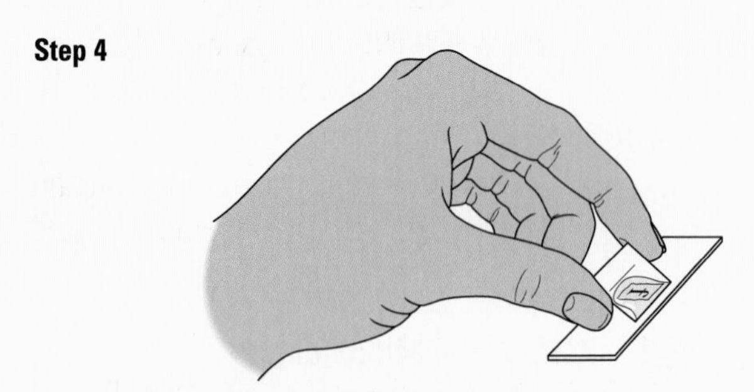
Step 4

Positioning Objects Under the Microscope

1. Make sure the low-power objective lens is in place on your microscope. Then put either the dry or wet mount slide in the centre of the microscope stage. Use the stage clips to hold the slide in position. Turn on the light source.

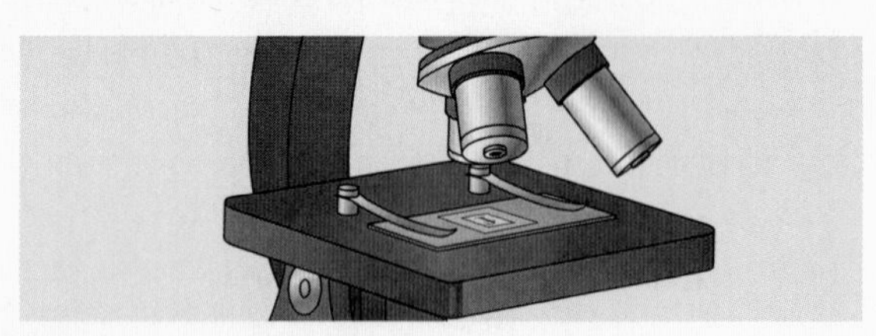
Step 1

2. View the microscope stage from the side. Using the coarse-adjustment knob, bring the low-power objective lens and the object as close as possible to one another. Do not allow the lens to touch the cover slip.

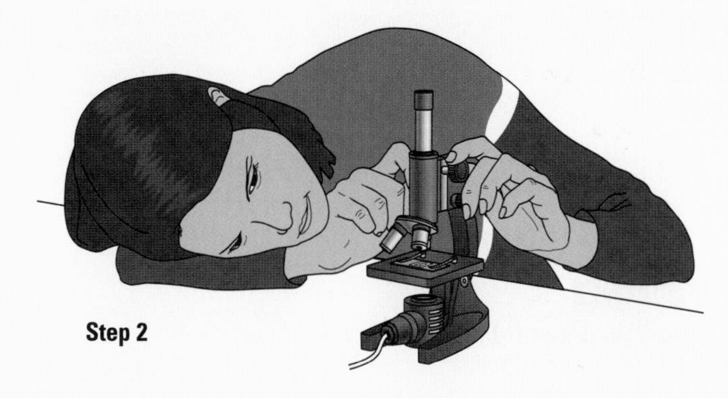
Step 2

3. View the object through the eyepiece. Slowly move the coarse-adjustment knob so the objective lens moves away from the slide, to bring the image into focus. Note that the object is facing the "wrong" way and is upside down.
4. Using a compass or a petri dish, draw a circle in your notebook to represent the area you are looking at through the microscope. This area is called the field of view. Look through the microscope and draw what you see. Make the object fill the same amount of area in your diagram as it does in the microscope.
5. While you are looking through the microscope, slowly move the slide away from your body. Note that the object appears to move toward you. Now move the slide to the left. Note that the object appears to move to the right.
6. Rotate the nosepiece to the medium-power objective lens. Use the fine-adjustment knob to bring the letter into focus. Note that the object becomes larger.

Never use the coarse-adjustment knob with the medium- or high-power objective lenses.

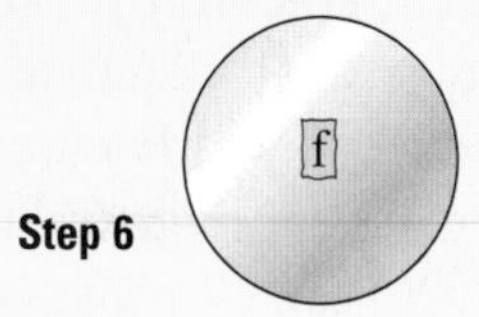

Step 6

7. Adjust the object so that it is directly in the centre of the field of view. Rotate the nosepiece to the high-power objective lens. Use the fine-adjustment knob to focus the image. Note that you see less of the object than you did under medium-power magnification. Also note that the object seems closer to you.

Storage

When you complete an investigation using the micrscope, follow these steps:

1. Rotate the nosepiece to the low-power objective lens.
2. Remove the slide and cover slip (if applicable).
3. Clean the slide and cover slip and return them to their appropriate location.
4. Return the microscope to the storage area.

The Stereo Microscope

The stereo microscope (**Figure 10**), or dissecting microscope, is used for observing any small three-dimensional object. You can use it when it is not appropriate to look at a sample on a slide, for example, to observe live specimens that are too large to fit under a cover slip.

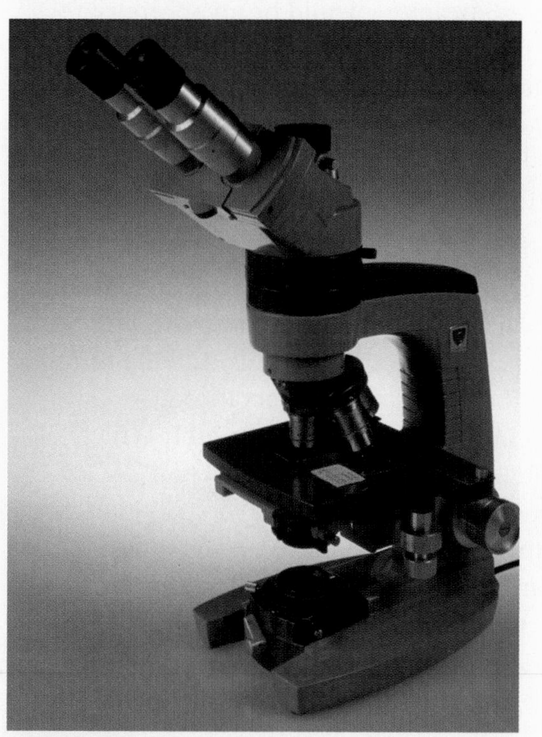
Figure 10

Ⓔ Observing and Recording

To answer the question that forms the basis of a scientific investigation, you must make observations and gather information, or evidence. In many cases you will have to make both qualitative (non-numerical) and quantitative (numerical) observations.

E1 Qualitative Observations

An observation is information that you get through your senses. You observe that a rose is red and has a sweet scent. You may also note that it has sharp thorns on its stem. When people describe the qualities of objects and events, the observations are **qualitative**. Common qualitative observations include the state of matter (solid, liquid, or gas), colour, odour, taste, shininess (lustre), clarity, hardness, and viscosity. These qualities cannot be measured but must be described in words. As you record the results of your investigations, be sure to include your qualitative observations.

Recording Qualitative Evidence

Qualitative evidence can be recorded using words or pictures. A camera would be an appropriate tool for recording this type of evidence. However, sometimes it is more appropriate to draw or sketch your observations.

E2 Quantitative Observations

Observations that are based on measurements or counting provide **quantitative** information, since they deal with quantities of things. The length of a rose's stem, the number of petals, and the number of leaves are quantitative observations.

Look at the two lines in **Figure 11**. Which looks longer? You will find that AB and CD are the same length. Our senses can be fooled—that is one of the reasons quantitative observations are so important in science, and it is also the reason measurements must be made carefully.

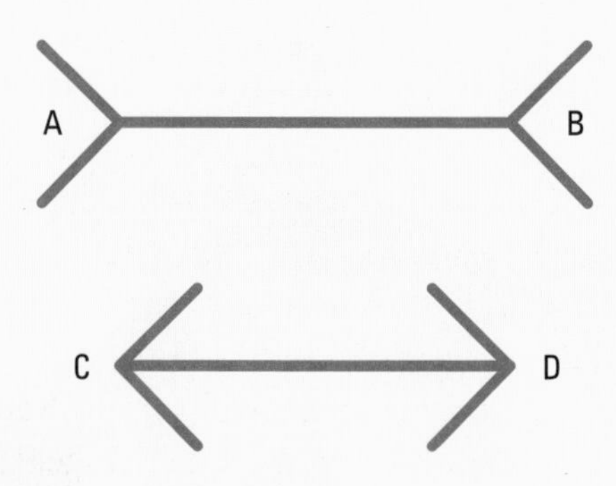

Figure 11
Which line is longer? Use a ruler.

Problems in Measurement

Many people believe that all measurements are reliable (the same over many trials), precise (to as many decimal places as possible), and accurate (representing the actual value). However, many things can go wrong when measuring.

- There may be problems that make the measuring instrument or its use inconsistent (not the same each time).
- The investigator may make a mistake or fail to follow the correct procedure when reading the measurement to the correct number of decimal places.
- The instrument may be faulty. Another similar instrument may give different readings. This indicates a problem with the accuracy of the first instrument.

When measuring the temperature of a liquid, for example, it is important to keep the thermometer at the right depth and the bulb of the thermometer away from the bottom and sides of the container. Otherwise you will be measuring the temperature of the container, not the temperature of the liquid.

To make sure you have measured correctly, repeat your measurement at least three times. If your measurements are similar, calculate the average and use that value. If you repeat the measurements with a different instrument and get the same results, then you can be more certain that the measurements are accurate.

Taking (Estimating) Measurements

All measurements are our best estimates of the actual value. The accuracy of the measuring instrument and the skill of the investigator determine how certain and precise the measurement will be. The usual rule is to estimate a measurement between the smallest divisions on the scale of the instrument. If the smallest divisions on the scale are fairly far apart (greater than 1 mm), then you should estimate to one-tenth (±0.1) of a division (e.g., 34.3 mL, 13.8 mL, and 87.1 mL). If the divisions are closer together (about 1 mm), then you should estimate to two-tenths (±0.2) of a division (e.g., 12.6°C, 11.2°C, and 35.8°C). If the divisions are very close together, then you should estimate to five-tenths (±0.5) of a division (e.g., 13.0 s, 33.5 s, and 42.0 s).

Check your skill in making the best estimate for each of the measurements in **Figure 12**. Estimate to the smallest division.

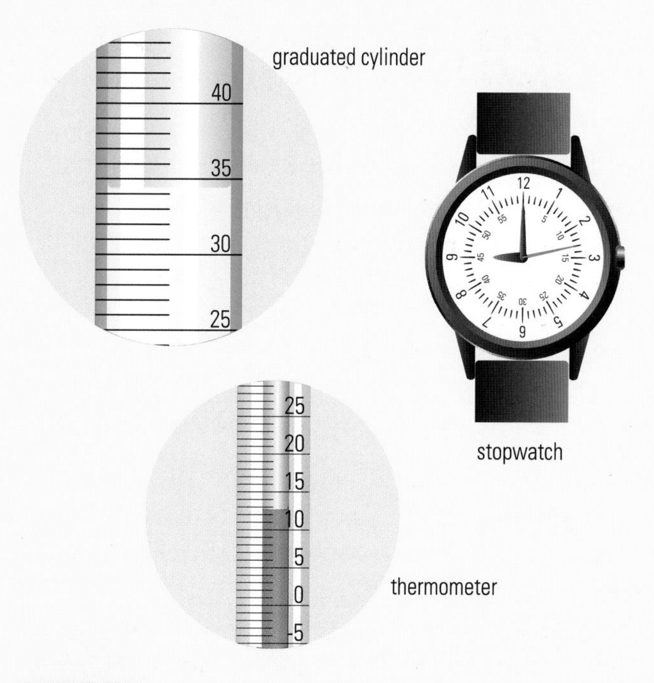

Figure 12

The graduated cylinder contains 34.3 mL of liquid, the thermometer is measuring a temperature of 12.6°C, and the stopwatch shows a time of 13.0 s.

Using a Protractor

A protractor, like other measuring instruments, must be used correctly to get consistent, precise, and accurate measurements. Depending on the distance between the markings on the scale, different protractors give measurements with different precision. For example, a small protractor gives a reading to the nearest 0.5° (**Figure 13(a)**). A larger protractor gives a more precise reading to the nearest 0.2° (**Figure 13(b)**).

Use the following example and **Figures 14, 15,** and **16** to help you use a protractor correctly:

1. Place the protractor with its origin at the apex (point) of an angle (**Figure 15**).

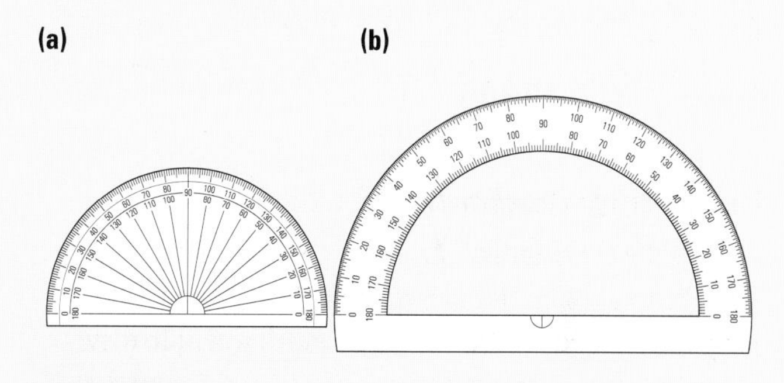

Figure 13

Different protractors can give different precision.

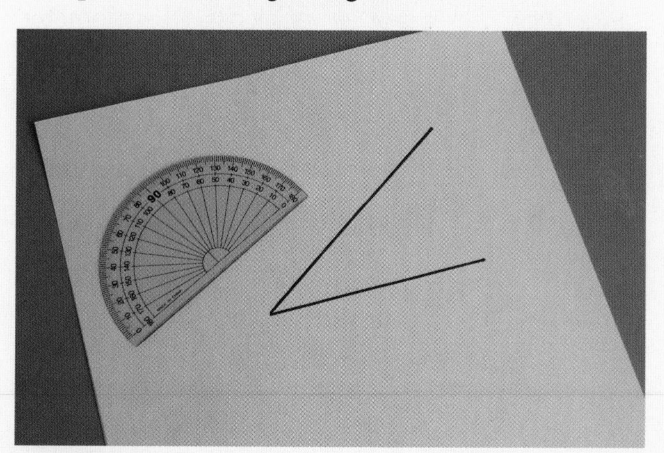

Figure 14

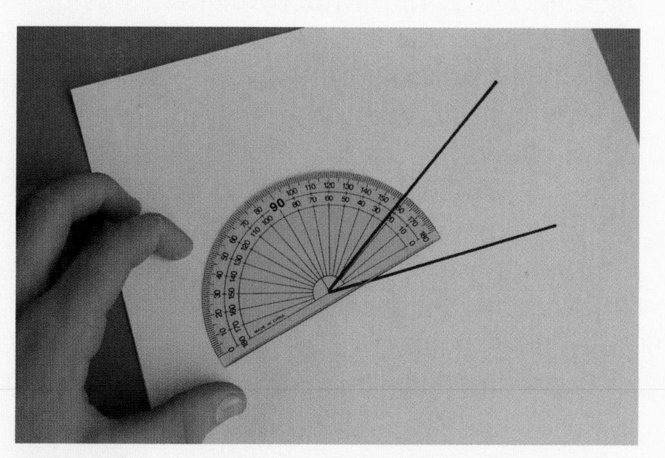

Figure 15

2. Rotate the protractor so its x-axis (0° to 180° line) covers one of the lines of the angle (**Figure 16**).
3. Read the angle to the nearest 0.5 of a degree.
4. Record the angle to a decimal fraction of a degree. In this case, the angle is 35.5°.

Figure 16

E3 Scientific Drawing

Scientific drawings are done to record observations as accurately as possible. They are also used to communicate, which means they must be clear, well labelled, and easy to understand. Following are some tips that will help you produce useful scientific drawings.

Before You Begin

- Obtain some blank paper. Lines might obscure your drawing or make your labels confusing.
- Find a sharp, hard pencil (e.g., H or 2H). Avoid using pen, thick markers, or coloured pencils. Ink can't be erased—even the most accomplished artists change their drawings—and coloured pencils are soft, making lines too thick.
- Plan to draw large. Ensure that your drawing will be large enough that people can see details. For example, a third of a page might be appropriate for a diagram of a single cell or a unicellular organism. If you are drawing the entire field of view of a microscope, draw a circle with a reasonable diameter (e.g., 10 cm) to represent the field of view.
- Leave space for labels, preferably on the right side of the drawing.
- Observe and study your specimen carefully, noting details and proportions.

Drawing

- Simple, two-dimensional drawings are effective.
- Draw only what you see. Your textbook may act as a guide, but it may show structures that you cannot see in your specimen.
- Do not sketch. Draw firm, clear lines, including only relevant details that you can see clearly.
- Do not use shading or colouring in scientific drawings. A stipple (series of dots), as shown in **Figure 17**, may be used to indicate a darker area. Use double lines to indicate thick structures.

Figure 17

Label Your Drawing

- All drawings must be labelled fully in neat printing. Avoid printing labels directly on the drawing.
- Use a ruler. Label lines must be horizontal and ruled firmly from the structures being identified to the label (**Figure 18**).
- Label lines should never cross.
- If possible, list your labels in an even column down the right side.
- Title the drawing, using the name of the specimen and (if possible) the part of the specimen you have drawn. Underline the title.

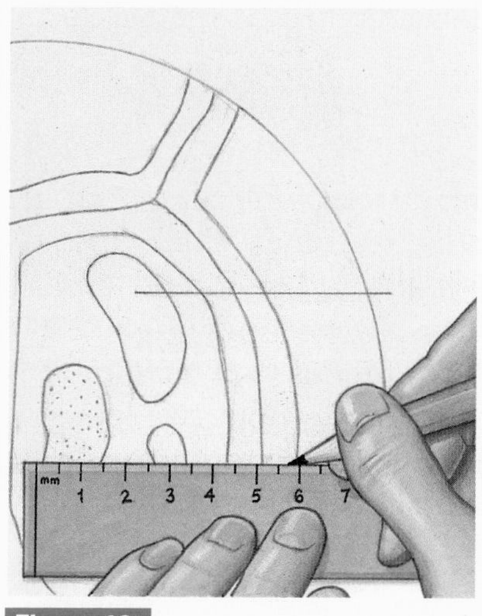

Figure 18

Scale Ratio

- To show the relation of the actual size to your drawing size, print the scale ratio of your drawing beside the title.

$$\text{scale ratio} = \frac{\text{size of drawing}}{\text{actual size of the specimen}}$$

For example, if you have drawn a nail (**Figure 19**) that is 5 cm long and the drawing is 15 cm long, then the scale ratio, which in this case is a magnification, is

$$\frac{15\text{ cm}}{5\text{ cm}} = 3\text{X}$$

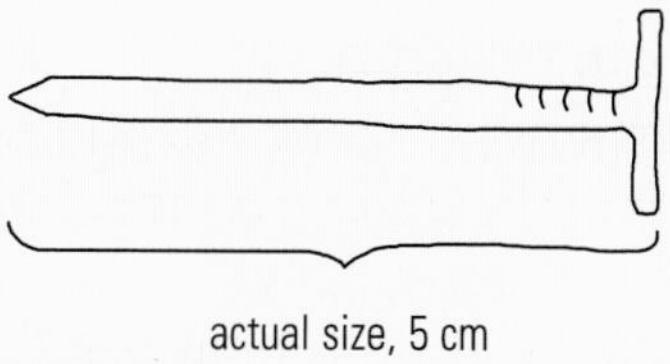

actual size, 5 cm

Figure 19

The magnification is always written with an "X" after it. In a fully labelled drawing, the total magnification of the drawing should be placed at the bottom right side of the diagram. If the ocular lens magnified a specimen 10X, the low-power objective (4X) was used, and the scale ratio was 3X, the total magnification of the diagram would be as follows:

$$\begin{aligned}\text{Total Magnification} &= \text{Ocular Lens} \times \text{Low-Power Objective Lens} \times \text{Scale Ratio}\\ &= 10 \times 4 \times 3\\ &= 120\text{X}\end{aligned}$$

The total magnification should be written on the bottom right-hand side of the diagram, as shown in **Figure 20**.

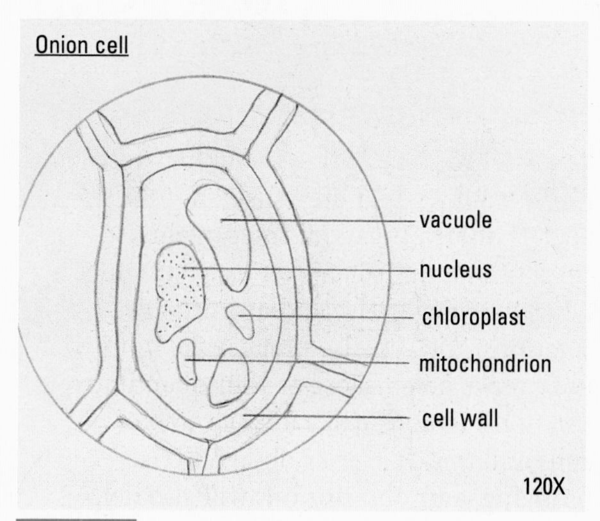

Figure 20

Try This Draw It

Obtain a specimen from your classroom. It could be a piece of your hair, chalk dust, or something else you would be interested in looking at under a microscope. Prepare a dry mount slide and focus your specimen under the medium-power objective lens. Complete a scientific drawing of your specimen. Use the checklist to ensure that your diagram is accurate and complete. When your drawing is complete, exchange it with your friend's drawing. Note the strengths and weaknesses of the drawing, keeping in mind all the features of a good scientific drawing. Evaluate your friend's drawing using the checklist.

Checklist for Good Scientific Drawing

✔ Use blank paper and a sharp, hard pencil.
✔ Draw as large as necessary to show details clearly.
✔ Do not shade or colour.
✔ Draw label lines that are straight and parallel and run outside the drawing. Use a ruler for this!
✔ Include labels, a title, and the total magnification.

F Analyzing Results

F1 Interpreting Observation Tables

While observation tables are useful for organizing your observations, they are usually not the final product of an investigation. You can almost always get more information from your tables by studying your qualitative and quantitative observations. You can also plot graphs from your tables to make better sense of your quantitative observations. Graphs make it easier to see patterns and trends in your observations.

Studying your observations will also help you to identify any errors in your measurements. Any measurement that is clearly very different from the others should be carefully checked and possibly removed from your analysis. For example, in an investigation involving the growth rate of five identical plants, if one of the plants dies (and therefore does not grow at all) you obviously would not consider that plant in your analysis.

F2 Interpreting Graphs

A graph is an easy way to see whether a relationship or pattern exists between two or more variables. It also allows you to see what the relationship is so it can be accurately described in words and by mathematics. Graphs will also help you determine whether the observations support your hypothesis or prediction.

Figure 21 is a point-and-line graph that shows the data from **Table 6**. The graph shows the relationship between the two variables as a fairly straight line. The line of best fit has been drawn on the graph. See W2 Constructing a Line Graph (p. 296). The graph could be described by saying that as the current through a circuit steadily increases, the voltage in the circuit also steadily increases. This is a simple direct relationship between current and voltage and is easier to see in the graph than it is in the table.

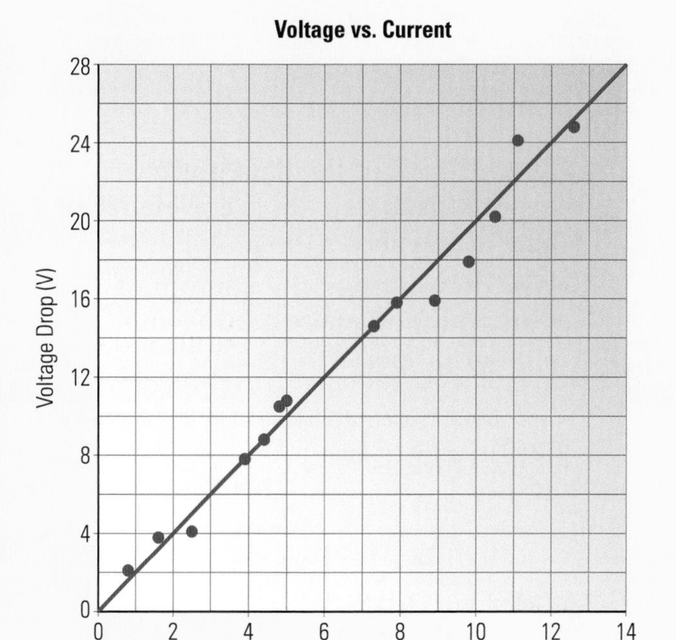

Figure 21

Table 6 **Electric Current and Voltage in a Simple (one-resistor) Electric Circuit**

Power supply setting	Current (mA)	Voltage drop (V)
1	0.8	2.1
2	1.6	3.8
3	2.5	4.1
4	3.9	7.8
5	4.4	8.8
6	4.8	10.5
7	5.0	10.8
8	7.3	14.6
9	7.9	15.8
10	8.9	15.9
11	9.8	17.9
12	10.5	20.2
13	11.1	24.1
14	12.6	24.8

Reading a Graph

When you are interpreting a graph, try to answer the following questions:

- What variables are represented?
- What is the dependent variable? What is the independent variable?
- What are the units of measurement?
- What is the range of values (the difference between the highest and lowest values)?
- What do the highest and lowest values represent on the graph?
- What patterns or trends exist?
- If the graph is a straight line, what might the slope of the line tell me?
- Is this the best type of graph for the data?

Using Graphs for Predicting

If a graph shows a regular pattern (for example, a straight line), you can use it to make predictions. For example, you could use the graph in **Figure 21** to predict what the current in the circuit will be if the voltage is increased to 40 volts. To do this, you would extend the graph beyond the measured points, assuming that the observed trend would continue.

Try This Read a Graph

Until the 1970s, small amounts of lead were added to gasoline to improve the performance of car engines. Eventually, when people became aware of the hazards of lead, leaded gasoline was phased out. As you are probably aware, leaded gasoline is no longer used in Canada.

Lead pollutes the air. In one experiment in Ontario, the amount of airborne lead was measured near expressways and at several other locations. **Figure 22** shows the results of this study. Take a close look at the results and answer the following questions:

1. In what year was the level of airborne lead highest?
2. Was the amount of lead in the air increasing or decreasing during the time shown in the graph?
3. In any year, is the amount of airborne lead near expressways more than or less than the average of all sites?
4. (a) On the basis of the data, form a hypothesis to explain the presence of lead in the air.
   (b) Why do you think the amount of lead pollution changed over the measuring period?

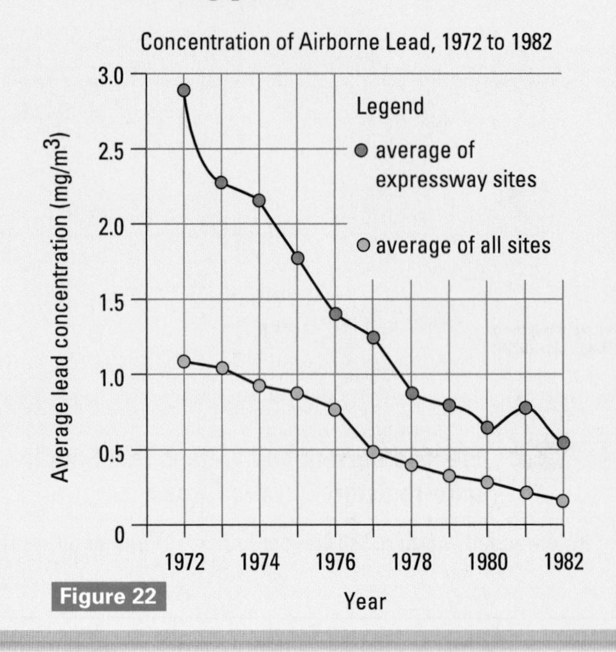

Figure 22

F3 Using Star Maps

What Is a Star Map?

A star map shows the most easily seen stars in the sky, with many of the stars joined by lines into constellations. Each star map is designed for a range of latitudes, such as locations about 45° north of the equator. Thus, a star map designed for southern Canada cannot be used in Australia.

You can use a star map to help you recognize what you can see in the sky, and to observe the motions of objects as Earth goes through its cycles of rotation and revolution.

Maps for All Seasons

Different parts of the sky are visible during different times of the year. To show the different stars and constellations visible, different star maps have been designed for each of the seasons or even months.

Another way to take into consideration the changing skies is to use a seasonal star map (**Figure 23**) in which a "window" can be rotated to expose different parts of the map. Each visible section represents the portions of the sky that are visible at different times of the year. Section 4.4 gives instructions on making this type of map.

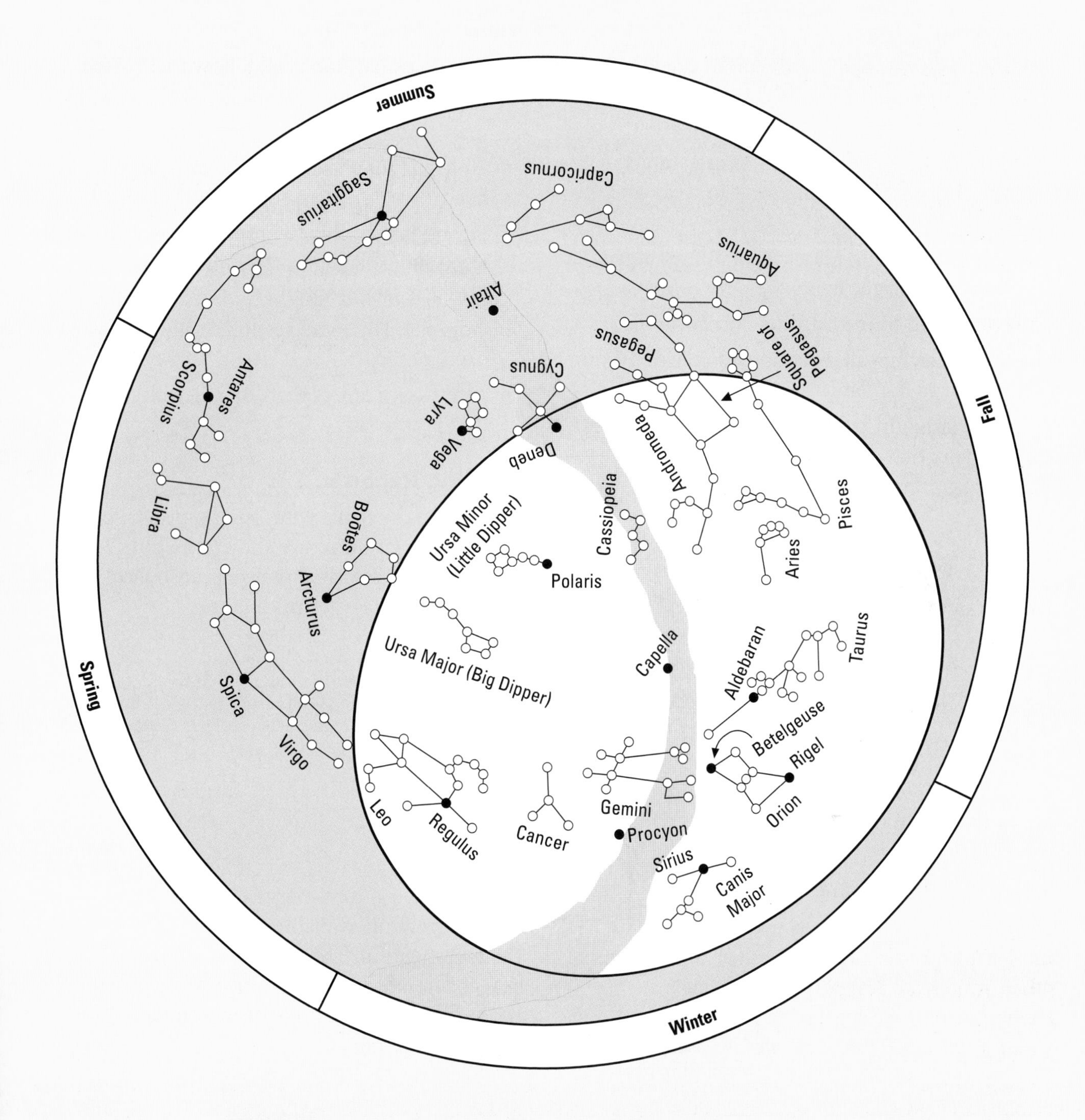

Figure 23

Only the unshaded part of the seasonal star map is visible during the winter months.

Stargazing Trips

When you want to observe the night sky, consider these tips:

- Plan the trip in advance, taking into consideration the weather forecast, safety, transportation, location, what to wear, and what to bring.
- Choose a location far away, or at least screened away, from bright lights.
- Be prepared to record your observations.
- Before viewing, allow your eyes at least 10 min to adapt to the dark.
- Use a flashlight covered with red cellophane to view your star map.

Using a Star Map

To use a seasonal star map, follow these steps:

- Rotate the window to expose the part of the map closest to the current season or month.
- Hold the map, facing downward, above your head.
- Rotate the map so the top part (away from the window) is facing north. This means that Orion is facing south.
- Compare what you see in the sky with what is on the map.
- Also notice any planets or other objects besides stars.

Keeping Records

Use a table to record your observations. Possible titles for the columns are shown in **Table 7**. Be careful when recording dates because Dec. 15 becomes Dec. 16 after midnight.

Table 7

Date	Object seen	Description (including a diagram)	Location (including angles)	Questions I want answered
?	?	?	?	?
?	?	?	?	?

F4 Answering the Question

The goal of carrying out an investigation is to answer the question that you ask at the beginning. When you create a hypothesis or make a prediction, you are suggesting an answer to the question. This suggested answer is based only on the information available to you before starting the investigation. The analysis of the information you collect during the investigation may provide the evidence you need to answer the question with more certainty.

G Concluding

Scientific investigations mean something only if the scientist states a conclusion about the results. A conclusion is a statement that explains the results of an investigation. This statement should refer back to the original hypothesis or prediction. It should reveal whether the results support, partially support, or do not support the prediction or hypothesis. Don't worry if your hypothesis is not supported—scientists usually need to revise and repeat experiments many times to obtain the solution they are looking for. Remember, science is a repetitive process. For example, experiments have been repeated many times and many new hypotheses have been created in order to learn what we now know about cancer treatments.

Reflecting on Your Work

It is always important to reflect on events in order to learn from them. This is one aspect of science that is sometimes neglected. Once you have been through the process of scientific inquiry, it is always important to step back and think about what you did and how you did it. What went well? What were the challenges? What would you do differently if you were to go through the process again?

When reflecting on the process of scientific inquiry, we must judge how good the evidence is in order to decide whether it supports the hypothesis or prediction. We need to identify the types and sources of errors that may have been made while the information was being collected. Any sources of error should be identified in the conclusion or in the discussion of your results and should be used as the basis for improving the process the next time.

For example, let's say you completed an experiment and were able to conclude that plants do indeed grow toward sunlight. What sources of experimental error could have occurred while you carried out this experiment? Possible errors could include variations in air temperature, differences in exposure to light, or differences in soil composition. It is important to be as specific as possible when identifying and describing possible errors. It is not good enough to simply say, "It was due to human error." If there appear to be no sources of error, you could say something like "The procedure was acceptable because it produced valuable evidence."

During reflection you should also think about other aspects of the process:

- Were there any flaws in the design? To answer this question, look at the plan, the variables, and the controls. The plan should lead to an answer to the original scientific question.
- Were the materials of suitable quality to do the experiment? Check the list of materials used. Did any of the materials give incomplete or inaccurate information?
- Did the steps of the procedure produce enough evidence? Check to see whether all required evidence was obtained.
- Did you or your partners have the skills needed to use the equipment? Poor-quality information may suggest that the investigator needs to improve his or her measurement skills.

Once you have identified possible errors in each part of the experiment listed above, make an overall judgement about the quality of the evidence (such as low, medium, or high quality). The evidence must be of reasonable quality in order to decide whether it supports the hypothesis or prediction.

Suggesting Further Experiments

Work in science does not end with a single experiment. Sometimes other investigators repeat the experiment to see if the evidence is the same. More often, the analysis and evaluation of one experiment can be used to create new, related experiments. The following guidelines may help you suggest further experiments:

- Review the evaluation of the evidence. You may suggest repeating the experiment using
  - a new or improved design that fixes any flaws that were noted
  - better equipment to make more precise measurements
  - a revised procedure to collect additional evidence
  - better skills in using the equipment
- Review the design of the experiment and focus on the variables. Can a controlled variable become the independent variable in another experiment? Can the dependent variable be changed or measured in a different way?
- Review the hypothesis or prediction. Look for new ways to test it. If the evidence did not support the hypothesis, suggest a new hypothesis that would lead to new experiments.

Decision-Making Skills

In science, there are often differences of opinion about certain matters. The process outlined in **Figure 1** will help you to examine supporting evidence and arguments related to the disagreement in order to reach a decision.

1 Identify both the issue and the question to investigate.

2 Suggest the possible points of view that different stakeholders might have. This could be based on prior knowledge and experience with the issue. If the issue is new to you, make an "educated guess" about the possible points of view.

3 Plan, identify potential sources of information, and gather, sort, select, and organize relevant information.

4 Evaluate the information and the sources of information. Describe the points of view, and distinguish fact from opinion. Identify and evaluate the possible solutions to the issue, and complete a risk–benefit analysis on each solution.

5 Make a decision or take a position. Defend the decision or position. Communicate your decision or position, and act on the decision if possible.

6 Reflect on and evaluate the decision itself and the process you followed in arriving at it.

Figure 1
The decision-making process

H Defining the Issue

The first step in the process of making a decision is to define the issue. An issue is a problem that affects different people in different ways and to which there is usually more than one possible solution. For most issues there is no right or wrong answer. What may be right for one person is not acceptable to another.

For example, an issue in Atlantic Canada is whether the cod fishery should be reopened. For some people, the answer is an obvious "yes"; for other people, the answer is "no"; and for yet others, the answer is "maybe." The first step in understanding such an issue is to explain why it is an issue and to identify the different groups who may be affected by it.

- Describe what problems arise out of the issue for individuals and groups, and why they are problems.
- Identify the stakeholders—individuals or groups who may be positively or negatively affected by the issue.
- Suggest questions that will help you research the issue. Begin with the six questions used in journalism: Who? What? Where? When? Why? How?
- Develop background information on the issue by determining the facts and identifying the important characteristics of the problem.

(I) Identifying Alternatives/Positions

Think of as many possible solutions to the problem as you can. At this point it doesn't matter if they seem unrealistic. A possible solution for the cod fishery issue is to open the fishery to everyone with no restrictions. That may not be the best solution, but it is one that should be examined.

To begin the process of analyzing these solutions, you must examine the issue from different points of view. Stakeholders may bring different viewpoints to an issue, and these will determine their position on the issue. **Table 1** lists some points of view from which issues may be analyzed.

To identify possible points of view on an issue, put yourself in the place of one of the stakeholders. How do you think he or she would feel about the solution? In our cod fishery example, what would be the point of view of some of the possible stakeholders: a marine biologist? a fisher? a politician? an environmentalist?

Remember that one person could have more than one point of view. It is even possible for two people looking at an issue from the same point of view to disagree about the issue. For example, two politicians might disagree about the solution to the cod fishery issue.

Table 1 **Points of View on an Issue**

Point of view	Explanation
aesthetic	having to do with art and beauty
cultural	involving habits and practices of a particular group
ecological	involving an interaction among organisms and their habitat
economic	focusing on the production, distribution, and consumption of goods and money
educational	dealing with learning
emotional	having to do with feelings
environmental	affecting physical surroundings
moral/ethical	holding beliefs about what is good/bad, right/wrong
legal	involving laws describing the rights and responsibilities of individuals and groups in society
spiritual	having personal belief systems or religions
political	dealing with the aims of some group or party
scientific	having a basis in research
social	affecting human relationships, the community, or society
technological	involving machines or industrial processes

(J) Researching the Issue

Begin your research by deciding on a question that helps to narrow or define the issue. Then develop a plan to find and select reliable sources of information related to the question. This information must then be evaluated and organized. Gather information from several different sources, including books, newspapers, magazines, and the Internet. For more information on how to conduct your research, see (N) General Research (p. 283) and (O) Internet Research (p. 284).

Ⓚ Analyzing the Issue

This step in the process will help you explain where you stand on the issue. Five steps must be completed to analyze an issue.

(1) Setting Criteria for Evaluating Information

In evaluating the information, you determine how important it is to understanding the issue. Is the information directly related to the issue? If this information were not available, would you fully understand the issue? You must also decide which points of view are most important. For example, is an ecological point of view more important than a political point of view?

(2) Evaluating the Sources of Information

It is important to separate fact from opinion when evaluating information. Evaluating the source of the information will help you decide whether it is valuable. Is the source reliable and honest? Do the facts agree with those presented by other sources? What point of view does the author share? What biases is the author likely to have? The PERCS model (see Ⓝ General Research, p. 283) is useful for determining whether information is trustworthy.

(3) Identifying and Challenging Assumptions

In presenting their positions on an issue, people sometimes make assumptions that may or may not be supported by facts. For example, an environmentalist who is writing about the cod fishery might assume that the cod population is threatened, while a fisher might assume that the cod population is stable. Neither of these positions may be supported by scientific data. If there is not enough evidence to support the position, then the assumptions must be questioned.

(4) Determining Relationships

Often an issue is not simply right or wrong. There may be consequences that are not directly related to the original issue. For example, one of the problems associated with closing the cod fishery is that the fishers have reduced incomes. That may cause families to move to nearby towns or other provinces in search of work, which could put pressure on social services in those towns or provinces.

(5) Evaluating Alternatives

The last part of analyzing the issue is to evaluate the alternative solutions. You may decide to carry out a risk–benefit analysis, a decision-making tool that helps you look at each possible solution and decide whether to proceed.

In carrying out a risk–benefit analysis (see the sample risk–benefit analysis in **Table 2**),

1. Predict the results of a proposed solution and determine whether each result is a risk or a benefit.
2. On a scale of 1 to 5, rank the size or importance of each result (risk or benefit).
3. Estimate the probability (or likelihood) of each risk or benefit happening.
4. Multiply the rank of each result by the probability of it happening. This will give you the probability value of each result.
5. Add up all the probability values of all the risks and all the benefits.
6. Compare the sums to help you decide which of the proposed solutions is most appropriate.

Table 2 **Risk–Benefit Analysis of Using Fireworks**

Risks				Benefits			
Possible result	**Rank of that result (scale of 1 to 5)**	**Probability of that result happening**	**Probability value (rank × probability**	**Possible result**	**Rank of that result (scale of 1 to 5)**	**Probability of that result happening**	**Probability value (rank × probability**
Many injuries result from unsafe use of fireworks	very serious 5	likely 60%	300	Traditional way to celebrate in some cultures	high 5	very likely 80%	400
Fireworks displays pollute the environment	a minor issue 2	somewhat 20%	40	People will use them anyway—outside the city where they are a greater hazard	slight 2	likely 50%	100
Fireworks are expensive and last for only a short time—waste of money	serious 4	likely 60%	240	Fireworks displays promote tourism and produce economic benefits	high 5	very likely 70%	350
Total probability value (risks)			**580**	**Total probability value (benefits)**			**850**

L Defending a Decision

Once you have completed your analysis of the information, you can answer the original question and take an informed position on the issue. You should be able to defend your position to people who have different opinions and different points of view. Ask yourself the following questions:

- Do I have supporting evidence from a number of sources?
- Can I state my position clearly?
- Can I show why this issue is important to society?
- Do I have solid arguments to support my position?
- Have I considered the arguments against my position, and can I point out their faults?

When you are able to successfully defend your position, you can communicate it in an appropriate way. For example, you can

- Write a brief essay.
- Participate in a debate.
- Make a presentation.
- Participate in a role-play activity, taking the role of a stakeholder.
- Write a letter to the editor of a newspaper or magazine.

(M) Evaluating

The last phase of decision making involves thinking about the decision itself and the process you used to reach the decision. The following questions may help you think about your decision making:

- How did I determine what the issue was?
- What was my first reaction to the issue?
- How has my point of view changed since my first reaction?
- What facts did I consider most important in making my decision?
- Whose points of view did I consider?
- What solutions did I consider? Did I consider how each solution would affect the stakeholders?
- How did I make my decision? What steps did I follow?
- Does my decision solve the problem?
- How might my decision affect the stakeholders?
- Am I satisfied with my decision?
- What reasons would I give to explain my decision?
- If I had to make this decision again, what would I do differently?

 General Research

An incredible amount of information is available to you from many different sources. If this information is to be useful, you must first be able to find it and then to determine whether it is valuable and of good quality. **Table 1** lists information resources that may be available to you.

Collecting Information

The following tips will help you find the best information in the least amount of time:

1. Clearly define the topic or the question you will be researching.
2. List the most important words associated with your topic so you can search for related information.
3. Search out and collect information from a variety of resources.
4. Ask yourself, "Do I understand what this resource is telling me?"
5. Check when the resource was published. Is it up-to-date?
6. Keep organized notes or files while doing your research.
7. Keep a complete list of the resources you used so you can quickly find a source again if you need to, and so you can make a bibliography for your report.
8. Review your notes. Do they cover the topic you selected? Do they answer your question? After your research, you may want to change your original position or research in a slightly different direction.

Table 1 **Information Resources**

Information consultants (people who can help you locate and interpret information)	
teachers	business people
nurses	scientists
public servants	librarians
volunteers	politicians
lawyers	farmers
parents	doctors
members of the media	senior citizens
veterinarians	

Reference materials (sources of packaged information)	
encyclopedias	bibliographies
magazines/journals	newspapers
videotapes	slides
databases	almanacs
yearbooks	dictionaries
textbooks	maps
filmstrips	charts
biographies	films
pamphlets	records
television	radio

Places (sources beyond the walls of your school)		
public libraries	shopping malls	art galleries
parks	colleges	museums
universities	government offices	hospitals
historic sites	zoos	volunteer agencies
research laboratories		
businesses		
farms		

Electronic sources
World Wide Web (WWW)
CD-ROMs
online search engines
online periodicals
computer programs

Assessing the Information Sources

From all of the information available to us, how do we know what to believe and what not to believe? Is all of the information correct? Are all of the sources reliable?

A useful framework for evaluating information sources is PERCS (**Table 2**). This framework uses a series of questions that can help you evaluate the information you have collected.

Table 2 **The PERCS Checklist**

Perspective	From whose viewpoint are we seeing, reading, or hearing?
Evidence	How do we know what we know? What's the evidence and how reliable is it?
Relevance	So what? What does it matter? What does it all mean? Who cares?
Connections	How are things, events, or people connected to one another? What is the cause, and what is the effect? How do they "fit" together?
Supposition	What if...? Could things be different? What are or were the alternatives? Suppose things were different.

O Internet Research

The Internet is a vast network of information that is continually growing. There are several ways of finding this information. **Table 3** shows four main ways of searching the Internet for documents and Web pages.

Search Engines

Search engines are programs that create an index of Web pages in the Internet. When you are using a search engine, you are not actually searching the Internet, you are searching through an index.

Table 3 **Ways of Searching the Internet**

Search engine	Meta search engine	Subject gateway (or directory)	E-mail, discussion lists, databases
Searches using keywords that describe the subject you are looking for	Enables you to search across many search engines at once	Provides an organized list of Web pages, divided into subject areas. Some gateways are general and cover material on many subjects.	Puts you in touch with individuals who are interested in your research topic
AltaVista Canada www.altavista.ca Lycos Canada www.lycos.ca Excite Canada www.excite.ca Google www.google.net Go Network http://infoseek.go.com/ HotBot www.hotbot.com Webcrawler www.webcrawler.com	MetaCrawler www.go2net.com/index.html CNET Search www.search.com	About.com www.about.com Looksmart www.looksmart.com Yahoo www.yahoo.com Librarians' Index http://lii.org Infomine http://infomine.ucr.edu WWW VirtualLibrary http://vlib.org/overview.html SciCentral www.scicentral.com/index.html	

When you type some words into a search engine, it searches through the index to find the Web pages that contain the key words. It then lists the pages where these words are found.

Search Results

Once you have done a search you will be provided with a list of Web pages, and a number of "matches" for your search. If your key words are general, you are likely to get a high number of matches and you may need to refine your search. Most of the search engines provide online help and search tips. Always look at these to find tips for better searching.

Every Web page has a URL (Universal Resource Locator). The URL can sometimes give you a clue to the usefulness of the site. The URL may tell you the name of the organization, or it can give you a clue that you are looking at a personal page (often indicated by a ~ symbol in the URL). The Web address includes a domain name, which also contains clues to the organization hosting the Web page (**Table 4**). For example, the URL http://weatheroffice.ec.gc.ca/jet_stream/index_e.html is a page showing a weather map of Canada; "ec.gc.ca" is the domain name for Environment Canada—probably a fairly reliable source.

Table 4 **Some Organization Codes**

ca	Canada
com or co	commercial
edu or ac	educational
org	nonprofit
net	networking providers
mil	military
gov	government
int	international organizations

Evaluating the Sources

Since it is so easy to put information on the network, anybody can post just about anything on the Web without any proof that the information is accurate. There are almost no controls on what people choose to write and publish on the Web. Because of this, it is very important that you evaluate the information you find on the Web.

Use the following questions to determine the quality of a Web resource. The greater the number of questions answered "yes," the more likely the source is of high quality.

- Is it clear who is sponsoring the page? Does the site seem to be permanent or part of a permanent organization?
- Is information available on the sponsoring organization? For example, is a phone number or address available to contact someone for more information?
- Is it clear who developed and wrote the information? Are the author's qualifications provided?
- Are the sources of factual information given so that they can be checked?
- If information is presented in graphs or charts, are they labelled clearly?
- Is the page presented as a public service? Does it present a balance of points of view?
- If there is advertising on the page, is it clearly separated from the content?
- Are there dates to indicate when the page was written, put online, or last revised? Are there any other indications that the material is updated periodically?
- Is there an indication that the page is complete and not still "under construction"?
- Is it clear whether the entire work or only part of it is available on the Web?

In your future career, whether or not you work in the science field, you will probably be required to write a report of some kind. These reports may be of different forms and intended for different audiences. Of course, there are other methods of communication besides written reports—for example, you may give an oral or an electronic presentation.

No matter how and why you communicate, it is important both in science and at work to report accurately and clearly.

(P) SI Units

The international scientific community of many countries, including Canada, has agreed on a system of measurement called **SI** (Système international d'unités). In this system, all physical quantities can be expressed as a combination of seven SI units, called base units (for example, length, mass, time). The seven SI base units are listed in **Table 1**.

Table 1 The Seven SI Base Units

Quantity name	Unit name	Unit symbol
length	metre	m
mass	kilogram	kg*
time	second	s
electric current	ampere	A
temperature	kelvin	K**
amount of substance	mole	mol
light intensity	candela	cd

* The kilogram is the only base unit that contains a prefix.

** Although the base unit for temperature (T) is kelvin (K), the common temperature (t) unit is degree Celsius (°C).

For example, the speed of an object is relative to the distance travelled during a specified time period. The unit for speed is metres (distance) per second (time). Other common quantities and their units are listed in **Table 2**.

Table 2 Common Quantities and Units

Quantity name	Quantity symbol	Unit name	Unit symbol
distance	d	metre	m
area	A	square metre	m^2
volume	V	cubic metre	m^3
		litre	L
speed	v	metre per second	m/s
acceleration	a	metre per second per second	m/s^2
concentration	c	gram per litre	g/L
temperature	t	degree Celsius	°C
pressure	p	pascal	Pa
heat	q	joule	J
energy	E	joule	J
power	P	watt	W

An important feature of SI is the use of a common set of prefixes to express small or large quantities. These prefixes (**Table 3**) act as factors to increase or reduce the size of a quantity in multiples of 10.

Table 3 Common SI Prefixes

Prefix	Symbol	Factor by which the unit is multiplied	Example
giga	G	10^9 = 1 000 000 000	10^9 m = 1 Gm
mega	M	10^6 = 1 000 000	10^6 m = 1 Mm
kilo	k	10^3 = 1 000	10^3 m = 1 km
		10^0 = 1	
centi	c	10^{-2} = 0.01	10^{-2} m = 1 cm
milli	m	10^{-3} = 0.001	10^{-3} m = 1 mm
micro	μ	10^{-6} = 0.000 001	10^{-6} m = 1 μm
nano	n	10^{-9} = 0.000 000 001	10^{-9} m = 1 nm

Converting Units

Sometimes it is necessary to convert, or change, to larger or smaller values of a unit. For example, if we need to change kilometres to metres we can multiply by a conversion factor. In the example below, we know that 1 km = 1000 m. Therefore, if we want to convert 1256 m to kilometres we only have to multiply 1256 by the conversion factor to get the answer.

$$1256\ \cancel{m} \times \frac{1\ km}{1000\ \cancel{m}} = 1.256\ km$$

Notice how the initial units "m" cancel, leaving "km" as the new unit.

Conversion factors can be used for any unit, such as 1 h = 60 min and 1 min = 60 s. Sometimes we have to use two or more conversion factors. In the example below, we are converting a speed from five metres per second to kilometres per hour.

$$5\ \frac{\cancel{m}}{\cancel{s}} \times \frac{1\ km}{1000\ \cancel{m}} \times \frac{60\ \cancel{s}}{1\ \cancel{min}} \times \frac{60\ \cancel{min}}{1\ h} = 18\ \frac{km}{h}$$

Notice that the units "m," "s," and "min" cancel to leave "km" and "h."

Q Writing a Lab Report

A lab report is written to communicate how an investigation was carried out, including the results that were obtained, an analysis of these results, and any conclusions that can be made. All investigators use a similar format. This has been modified to be more suitable for high-school learners.

Your lab report should follow the process that was used in the investigation and should follow the headings in the sample lab report (**Figure 1**). Try to write your report so that another investigator can follow it and repeat the investigation.

Title: At the beginning of your report, write the number and title of the investigation. Include the date and the names of your lab partner(s).

Purpose: State the purpose of the investigation. Why are you doing it?

Question: This is the specific question that you are attempting to answer in the investigation.

Prediction/Hypothesis: The prediction is the answer to the "question" that you asked before the investigation, based on common sense or on a concept that has already been covered. The hypothesis, if required, explains the prediction.

Design: Provide a brief description of the investigation. List the dependent, independent, and controlled variables. Identify any control that was used in the investigation.

Materials: Make a detailed list of all materials used, including sizes and quantities where appropriate.

Procedure: List the steps you took in carrying out the investigation.

Observations: This includes both qualitative and quantitative observations that you made. If you have only a few observations, this could be a list; for controlled experiments or many observations, present them in a table.

Analysis and Conclusion: Interpret the evidence using tables, graphs, or illustrations. Include any calculations. Describe any patterns or trends that you observed. The final part of this section should answer the experimental question. Were you able to obtain suitable, quality evidence to allow you to do this? At the same time, you should be able to state whether the prediction and hypothesis were supported or rejected by the results you obtained.

1.11 Classifying Elements
October 15, 2002

By: Ted G. and Barry L.

Purpose

The purpose of this investigation is to determine whether elements can be grouped together by the properties that they share.

Question

How can a selection of elements be classified?

Prediction

The elements that are shiny solids can be grouped together, with the remainder, which are dull solids, grouped together.

Materials

safety goggles
apron
small pieces of paper or paper baking cups
magnet
electrical conductivity apparatus
elements: aluminum, iron, nickel, tin, lead, copper, magnesium, silicon, zinc, carbon

Procedure

1. Put on the apron and safety goggles.
2. Obtain a sample of each element and place it on a labelled piece of paper or into a paper baking cup.
3. Examine each element and note the colour and lustre.
4. Try to bend or break each element.
5. Pick up each element to determine how heavy or light they are relative to their sizes.
6. Use a magnet to determine which elements are magnetic.
7. Test each element to determine whether it conducts electricity.

Observations

Element	Colour	Lustre	Bend/break	Heavy/light
A	black	√	X	heavy
B	grey	√	bends	light
C	black	X	break	light
D	yellow	X	break	light
E	red/brown	√	bend	heavy

Element	Magnetic	Conducts electricity
A	√	√
B	X	√
C	X	X
D	X	X
E	X	√

Analysis and Conclusion

Category 1: elements A, B, E. These three elements are shiny and conduct electricity. Elements B and E are also malleable—they can be bent.

Element A could not be bent or broken. Perhaps this was due to the thickness of the sample. Also, because element A was the only magnetic element, element A may require its own category.

Category 2: Elements C and D. These two elements were both dull, brittle, and did not conduct electricity. They were also much lighter than the other elements.

From our observations we can conclude that you can classify elements according to their physical properties.

Figure 1

A sample lab report

Ⓡ Writing for Specific Audiences

In the working world, both individuals and companies often need detailed information on a particular topic to help them make informed decisions. In preparing to write your findings in a report, consider the purpose of your report: are you presenting facts, presenting choices to your readers, or trying to change readers to your way of thinking? You should know your audience.

Research Reports

The purpose of a research report is to present factual information in an unbiased way. Your readers must be able to understand, without being talked down to. Be sure to indicate your sources with a complete list of references.

Issue Reports

Decisions are rarely based only on facts. When preparing an issue report, consider the issue from many points of view. Issues are controversial because there are various points of view. Try to consider and address as many of them as possible.

Position Papers

When planning a position paper, you may

- Start with a position on the issue, and then conduct research to support your position.
- Start by researching the issue, and then decide on your position based on your research.

Once you have taken a position, support your arguments with evidence, reasoning, and logic.

Letters to the Editor

A letter to the editor of a newspaper or magazine is typically a shorter version of a position paper. Space is limited, so express your thoughts as briefly as possible. You do not have to support every point in your argument with scientific facts, but indicate that such support is available.

Magazine Articles

Magazine articles are usually written in response to a request from the editor. The editor will tell you how long the article should be, so there is no point in writing more than you are asked for. As you write the article, try to stick to factual information. You must also be prepared to support your statements if you are challenged, maybe by letters to the editor.

Environmental Impact Reports

Environmental impact reports are a relatively new type of report. Until recently, the environment was not a consideration when developments or actions were planned. Today we must consider, study, and explain the possible results of any intended action.

Carrying out an environmental impact assessment begins with a listing of existing species, both plant and animal. If a similar development or action has been carried out elsewhere, you may want to observe what is happening there. You may also want to find out what experts think. All this information must then be brought together and presented in a report.

There is no single format for an environmental impact report. However, the following elements should be included:

- an introduction stating what the report will contain
- a description of how the data were collected
- an analysis of the data
- conclusions
- possible long-term results
- suggestions for further study to be conducted before a decision is made

(S) Oral Presentations

You may be asked to make oral presentations to present the results of an investigation, take part in a debate on an issue, or role-play a situation. In all of these situations, you are communicating with others using mainly your voice.

While some people find this method of communication stressful, the following tips will help to reduce or eliminate the stress, improve your presentation, and help you effectively deliver your message to the audience:

- Plan your oral presentation well in advance. Waiting until the last minute increases stress levels dramatically.
- Find out how much time you are being given for your presentation.
- Write the key points of your presentation on cue cards or into a computer slide show.
- Practise your presentation to make sure that it fits into the allowed time.
- When you are giving the presentation, speak clearly and make eye contact with your audience.

Presenting Results

This is a very factual type of oral presentation. It must be clear and to the point. Your presentation should include answers to the following questions:

- What was the reason for doing the investigation?
- What question were you trying to answer?
- What was your prediction or hypothesis?
- How did you carry out your investigation?
- What were your results?
- Were there any problems, or sources of error, that might cause you to question your results?
- What conclusion did you reach?
- On the basis of your results, is there another investigation that could be done?

Debating

A debate is basically an organized argument. A group takes a position on an issue, while another group takes a different position. Each group tries to win the audience over to its way of thinking. The following suggestions may be helpful if you are taking part in a debate:

- Research the issue thoroughly. Make notes as you go.
- Pick four or five major points, with examples, to support your position. Write them out in point form.
- Present your argument logically and clearly, and within the allowed time period.
- Listen closely to the arguments presented by those who have taken the opposite position. Make notes on any points that you feel you can argue against in your rebuttal.
- At all times, show respect for the opposition. While you can question their evidence, never resort to name-calling or rudeness.

Role-Playing

Role-playing is simply an extension of the debate, where you are expected to take on a "character." You then present an opinion, position, or decision from the point of view of that person. You are asked to deliver a speech to provide information on the issue and to convince the audience that your position or recommendation should be followed. Follow these steps as you prepare for and present your point of view:

- Research both your topic and your assigned character. Your arguments must appear to be those of your character.
- Include personal examples from your character's life to support the position you are taking.
- In making your presentation, stay "in character." Use "I" and "my" to convince your audience that you are indeed the character you are portraying.

Electronic Communication

Electronic communication is being used more often to communicate effectively with a wide audience. To create an effective electronic presentation,

- Be sure to have a definite purpose that you can state in your own words.
- Know who your audience will be.
- Begin by identifying your main goals, and include them in both your introduction and your conclusion.
- Create a flow chart for your presentation (**Figure 2**). Remember to present only content that your audience will see as useful. Be sure to make the links between what you are presenting and how the audience is able to use it.
- Draw storyboards for your presentation (**Figure 3**). Be sure to consider all media—what is the audience going to see and hear at any given moment?
- Choose special effects that make your presentation more memorable and clarify the information that you are presenting.

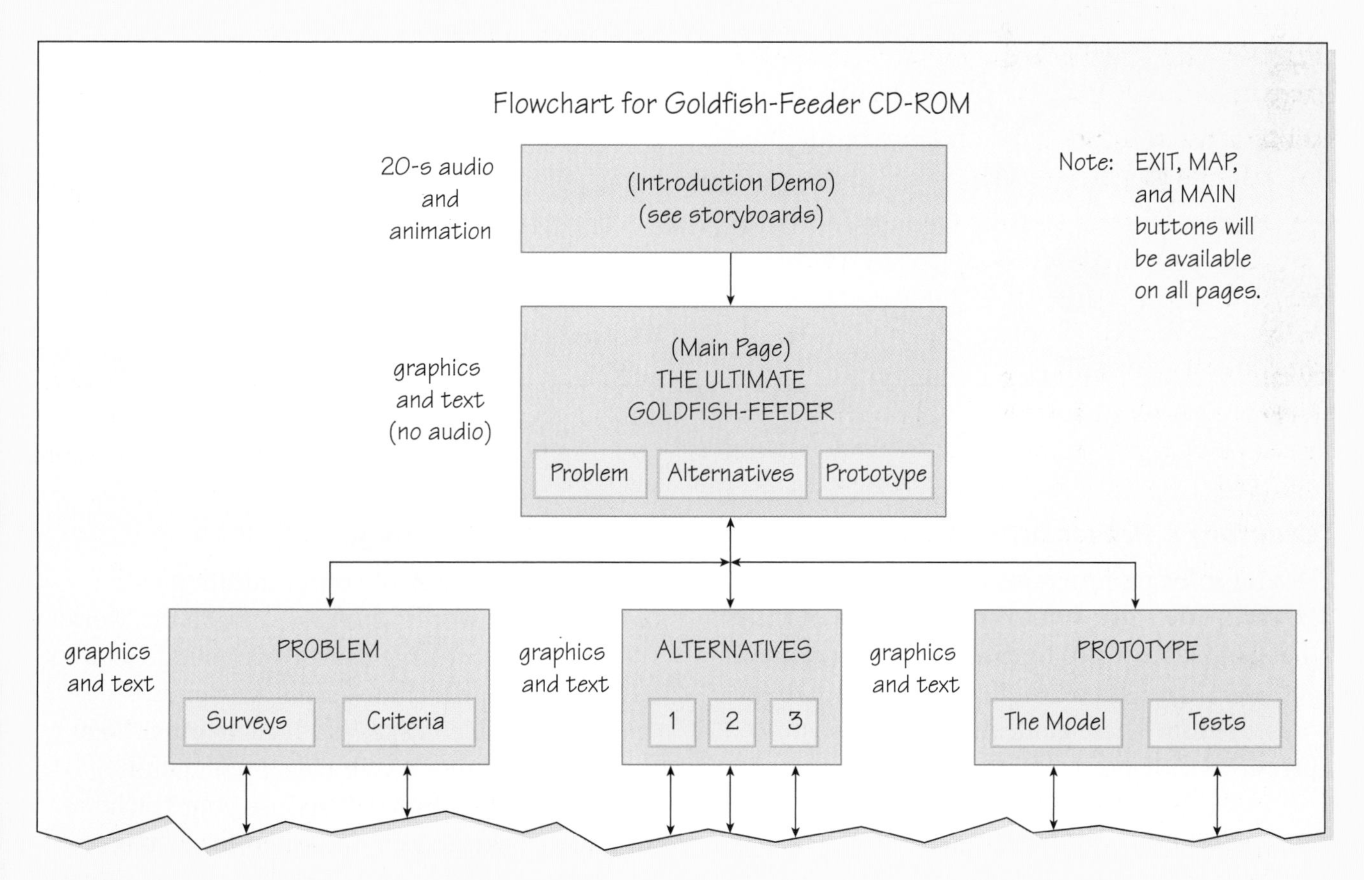

Figure 2

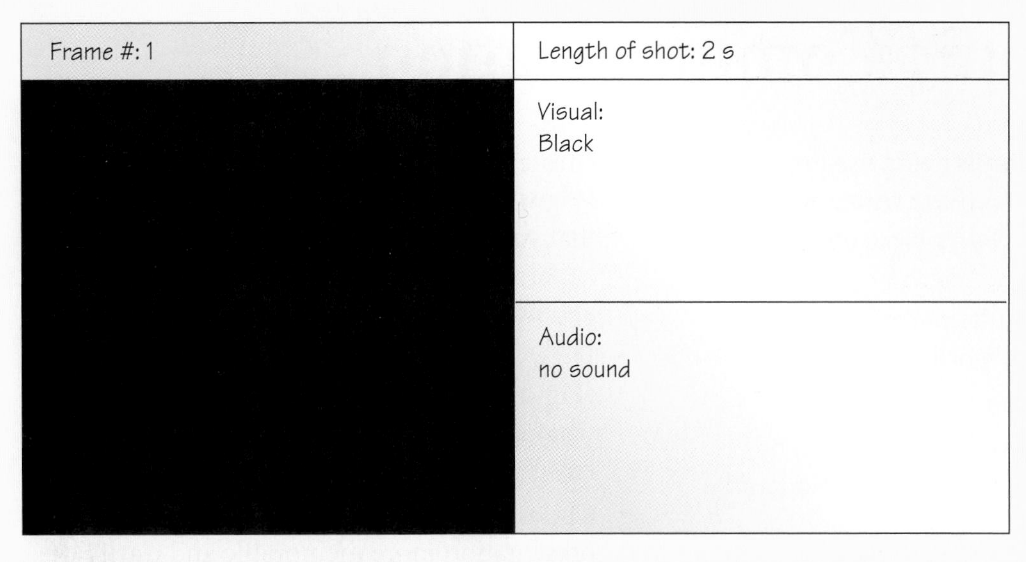

Frame #: 2	Length of shot: 3 s
	Visual: Fade-in on front-view of goldfish. Bubbles coming from mouth. Mouth opening and closing. Background becoming lighter—turning a clear blue.
	Audio: Bubbling sound of fish tank.

Figure 3

Creating a Presentation Using Overheads

Several software programs (such as Microsoft PowerPoint) allow you to create a series of slides that can be projected onto a screen. You can even add sound, music, or your recorded voice to the presentation. These guidelines may help you prepare a slide presentation:

- Use a storyboard to write your slides as point-form notes.
- Limit yourself to no more than 10 lines of text on each slide.
- Use a font that is large enough to be read by the audience.

Creating a Web Page

A Web page is a way of communicating with anyone in the world who has access to the World Wide Web. Every page has its own unique address (URL) and may include graphics, sounds, animations, video clips, and links to other Web pages. If you think a Web page presentation would be of benefit to others, ask your teacher or go to www.science.nelson.com for help in creating one.

Ⓤ Science Projects

Presenting Your Project

After you have completed your project, you may want to display your results in a science fair or present them to your class (**Figure 4**).

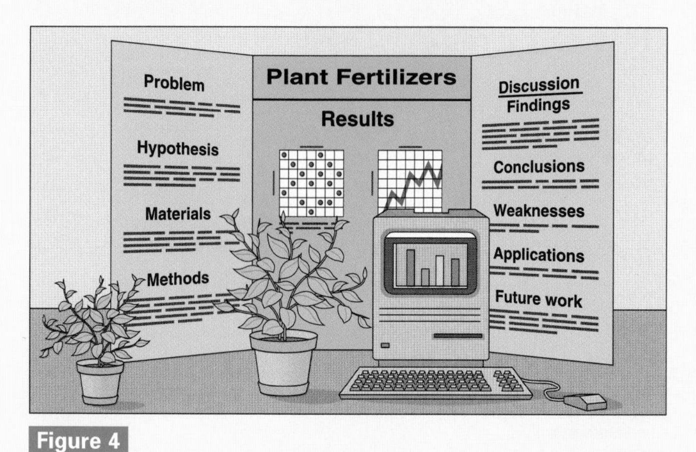

Figure 4
Your display

When constructing a display board, there are some things to keep in mind.

- Use a sketch to plan the layout.
- Collect a variety of display items (photos, sketches, charts, graphs, text).
- Use different sizes of letters for the text on your board. The most important ideas should have the largest letters.
- Place the title in a central location.
- Make all lettering neat and easy to read.
- Simple is best. Use the same kind of lettering throughout. If you are using colours or shapes to highlight important features, use only a few.
- Make sure your diagrams, graphs, and charts are neat.
- Place your results in a prominent position on the board.
- Try to place all your materials on the board before you start gluing or stapling. Have a classmate or a parent "critique" your display.
- Place only the most important information on the display board. If there is lots of space around wording and diagrams, the display will be more attractive.
- Make sure that nothing on your board is blocked by objects that will be in front of the board.
- Make sure things flow in a logical, easy-to-follow sequence.

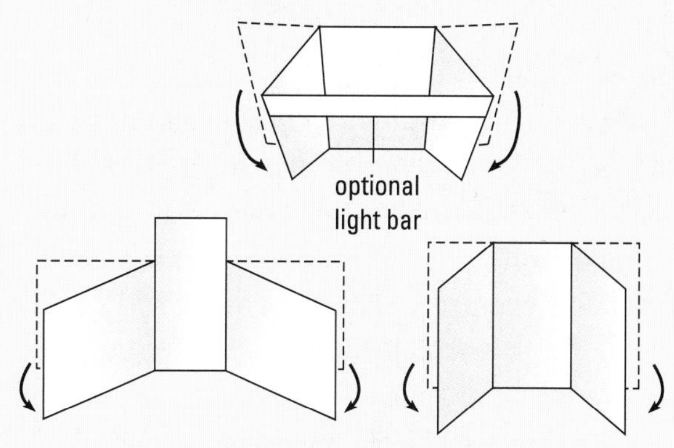

Three different display boards

You may also want to display your project on the Internet or put together a computer-based presentation. Have a look at on-line science fairs on the Internet for ideas on how to put together your electronic display.

Your Oral Presentation

Plan your oral presentation well in advance. Prepare a point-form "speech" on cue cards or a computer-based presentation and be sure to consider the following questions:

- What was the goal of your project?
- Why was the topic interesting?
- How did you carry out your project?
- What were the major results of your project?
- What conclusion did you reach?
- How might your findings help others?
- On the basis of your results, is there another project that could be done?

It is best not to read your speech when presenting. But don't memorize every word. Just practise it several times. If you get stuck, use your display board as a guide. Another trick is to memorize the seven questions above, or write them on cards, and answer them as you speak.

 Using the Calculator

A calculator (**Figure 1**) is an electronic instrument, or tool, to help you do mathematical calculations easier, faster, and with more accuracy. The following instructions will help you use this tool efficiently.

Figure 1
Scientific and graphing calculators

General Points

- Most calculators follow the usual mathematical rules for order of operations—multiplication and division before addition and subtraction.
- Calculators do not keep track of significant digits. For example, 12.0 is the same as 12 to a calculator. You will need to round your answer appropriately.
- All scientific calculators have at least one memory location where you can store a number (M+ and STO are the common keys). If you want to recall the number later, use the MR or RCL keys.

Multiplication and Division

- Any division problem can be expressed as a fraction. For example, 16 divided by 4 can be written as $\frac{16}{4}$. Division requires that you enter the numerator of the fraction first. Then press the "divide by" key (÷), followed by the denominator of the fraction.
- Multiplication and division may include parentheses: ( ). Any operation in parentheses must be done before the multiplication or division. Using parentheses forces the calculator to perform the calculation inside the parentheses first, before continuing with the rest of the calculation.

Scientific Notation

You may remember that scientific notation can be used to represent both very large and very small numbers. Each number is written as a number between 1 and 10 multiplied by a power of 10.

On many calculators, scientific notation is entered using a special key labelled EXP or EE. This key includes "× 10" from the scientific notation, so you need only enter the exponent. For example, to enter

7.5×10^4 press [7] [.] [5] [EXP] [4]

3.6×10^{-3} press [3] [.] [6] [EXP] [+/−] [3]

All mathematical operations can be carried out with numbers in scientific notation just like ordinary numbers.

Ⓦ Graphing

W1 Types of Graphs

You can use many types of graphs to organize and present your data. Three of the most useful kinds are **bar graphs**, **circle (pie) graphs**, and **point-and-line graphs**.

Bar Graphs

When at least one of the variables is divided into categories, use a bar graph. For example, if you surveyed your class to determine the students' favourite type of music, you could make a bar graph (**Figure 2**) to display the results.

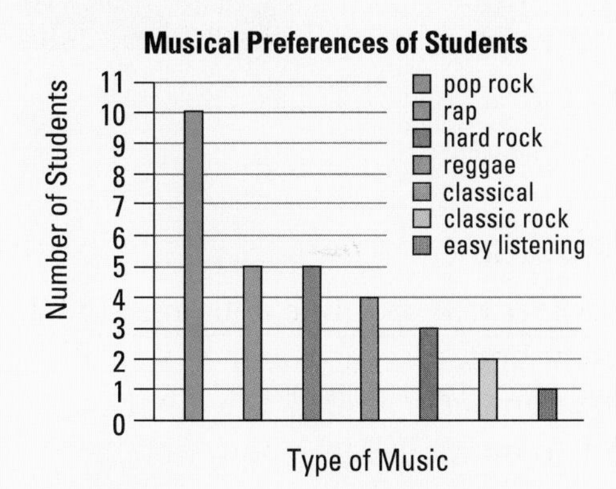

Figure 2
A bar graph

In this kind of graph, each bar stands for a different category—in this example, a type of music.

Circle Graphs

Circle graphs and bar graphs are both used for similar types of data. If your quantitative variable (the number of students in **Figure 2**) can be changed to a percentage of a total quantity, then a circle graph is a useful way to display the data.

Each bar (number of students favouring a particular kind of music) in **Figure 2** can be converted into a percentage of the whole class (30 students) by completing the mathematical calculation described in **Table 1**. First convert the values of your quantitative variable into percentages, and then into decimal form. In the sample, each number of students who prefer a type of music was turned into a percentage of the total number of students.

Table 1

Type of music	Number of students who prefer that type	Percentage of total (% and decimal)	Angle of piece of pie (degrees)
rap	5	17% = 0.17	61.2
pop rock	10	33% = 0.33	118.8
classical	3	10% = 0.10	36.0
reggae	4	13% = 0.13	46.8
easy listening	1	3% = 0.03	10.8
classic rock	2	7% = 0.07	25.2
hard rock	5	17% = 0.17	61.2
TOTAL	30	100% = 1.00	360.0

$$\text{percentage} = \frac{\text{number}}{\text{total}} \times 100\%$$

$$\text{percentage for rap} = \frac{5}{30} \times 100\% = 17\%$$
(decimal version = 0.17)

Then, multiply the decimal version of each percentage by 360 (there are 360° in a circle) to get the angle of each "piece of the pie" within the circle.

$$\text{angle of piece of pie for rap} = 0.17 \times 360° = 61.2°$$

The resulting circle graph in **Figure 3** displays each kind of music as a piece of the whole circle. The size of one piece represents the percentage of the whole class who prefers that particular kind of music.

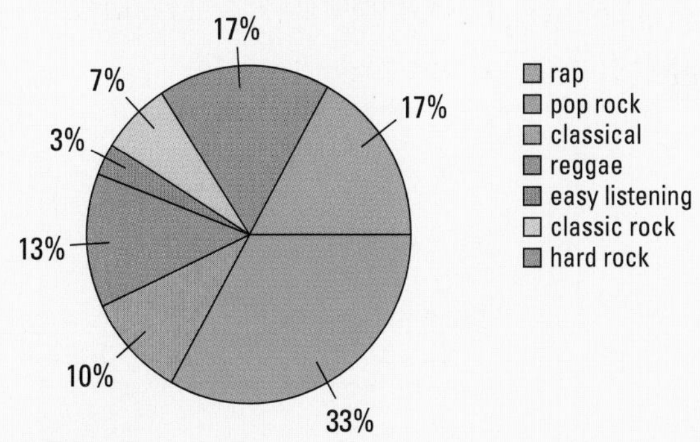

Figure 3
A circle graph

Point-and-Line Graphs

When both variables are quantitative, use a point-and-line graph. **Table 2** clearly shows that "days after germination" and "seedling height" are both quantitative variables.

Table 2 **Height of Seedling vs. Time**

Days after germination	Seedling height (cm)
1	1.0
2	2.5
3	2.9
4	3.2
5	3.5
6	4.0
7	4.2
8	5.0
9	6.0
10	6.6
11	7.3

The resulting point-and-line graph (**Figure 4**) shows the growth of the seedlings once the tiny plants break the soil.

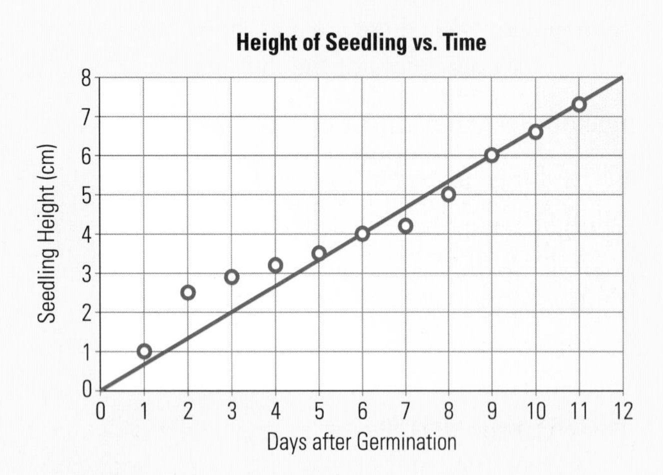

Figure 4

(W2) Constructing a Line Graph

Let's see how the data in **Table 2** was used to produce the graph in **Figure 4**.

1. Use graph paper and construct your graph on a grid. The horizontal edge on the bottom of this grid is called the *x*-axis. The vertical edge on the left is called the *y*-axis.

2. Decide which variable goes on which axis. The independent variable (time) is usually plotted along the *x*-axis. The dependent variable (seedling height) is usually plotted along the *y*-axis. Label each axis, including the units of measurement (**Figure 5**).

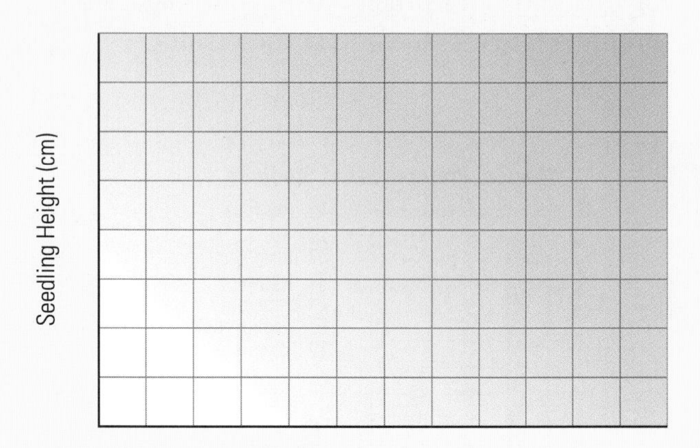

Figure 5

3. Determine the range of values for each variable. The range is the difference between the largest and smallest value of each variable. For time in **Table 2,** the range is 11 days. For seedling height, the range is 8 cm.

4. Choose a scale for each axis. The scale will depend on how much space you have and the range of values for each axis. In **Figure 6,** we have chosen to use one line for each day on the *x*-axis, and one line for each 1 cm on the *y*-axis.

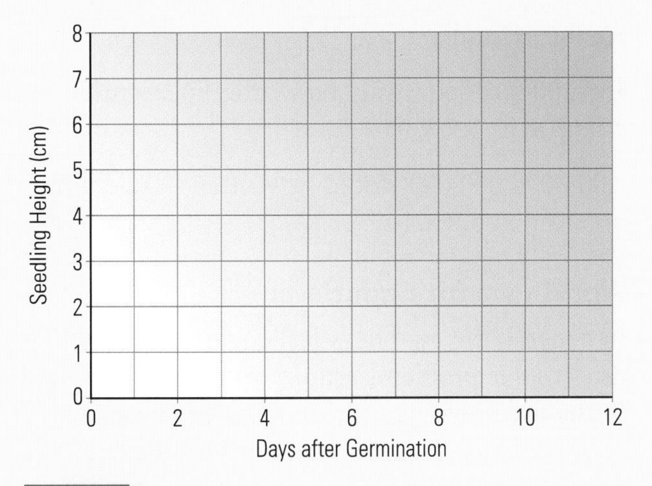

Figure 6

5. Plot the pairs of values from **Table 2**. The first pair of values from the table is 1 day and 1 cm. This point is located above 1 day on the *x*-axis, and opposite 1 cm on the y-axis. The second point is located above 2 days on the *x*-axis, and half way between 2 cm and 3 cm on the *y*-axis. You can plot the remaining points the same way.

6. After you plot all the pairs of values as points, draw a line that touches, or comes close to as many of the points as possible. This is called the line of best fit. This line shows the relationship between the two variables.

7. Finally, provide a short, meaningful title for the graph. For example, you might call the graph in **Figure 7** "Seedling Growth."

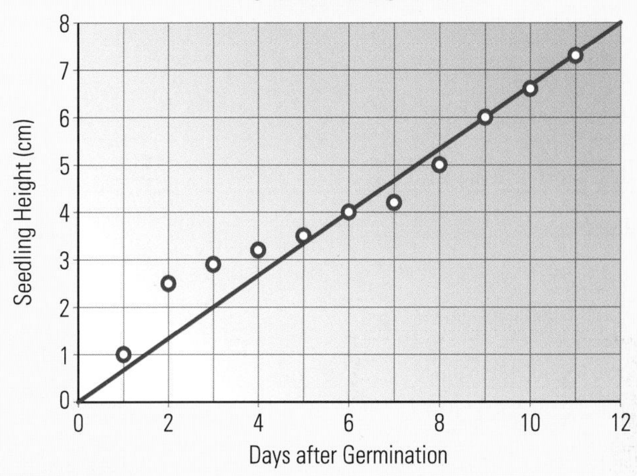

Figure 7

Calculating

Calculation Rules

Two types of quantities are used in science: exact values and measurements. Exact values include defined quantities, such as the number of metres (1000) in a kilometre. Counted values, such as when you count out five tablets, are also exact values. Exact values are said to be error-free.

However, no measurement is exact or error-free. The amount of possible error in any measurement depends on both the instrument being used and on the skill of the person making the measurement.

When making a measurement, look at eye level directly over or in front of the scale on the instrument you are using. Record all the digits available on the scale. Estimate the final digit of the measurement, which should fall between the smallest division lines on the scale (see E2 Quantitative Observations, p. 268).

After measurements are made, they are often used in mathematical calculations. After you have finished the mathematical calculation, you must round off the answer to the number of digits allowed by rule. The rules to cover rounding off are as follows:

- Determine the number of significant digits in each quantity to be used in the calculation by counting all digits except leading zeroes in decimal fractions. See **Table 3**.

Table 3 Significant Digits

Example	Number of significant digits
120.50 mm	5
0.9 mL	1
2502 g	4

Table 4 Significant Digits in a Calculated Answer

Sample calculation	Maximum significant digits allowed in answer	Answer	Rounded answer
$V = I \times R$ $I = 0.833$ A (four significant digits) $R = 144$ ohms (three significant digits)	3	119.952 V	120 V
$I = \frac{V}{R}$ $V = 120$ V (three significant digits) R = 14 ohms (two significant digits)	2	8.571 A	8.6 A
$R = \frac{V}{I}$ $V = 120$ V (three significant digits) $I = 1.5$ A (two significant digits)	2	80.0 Ω	80 Ω

- An answer obtained by multiplying or dividing measurements is rounded to the same number of significant digits as the measurement with the fewest number of significant digits (**Table 4**).
- If the digit after the last one to be kept is less than 5, leave the last digit as it is. If the digit after the last one to be kept is greater than 5, then round up (**Table 4**).

Scientific Notation

When working with measured values that are either very large or very small, it is easier to use a mathematical abbreviation known as scientific notation. To convert a number to scientific notation,

- Count the number of places you have to move the decimal point to get a value between 1 and 10. This counted number is the exponent.
- The exponent will be positive if the original number is greater than 10, and negative if the original number is less than 10.

So, for example,

150 000 000 000 would be written in scientific notation as 1.5×10^{11}.

0.000 000 000 050 would be written in scientific notation as 5.0×10^{-11}.

Working with Equations

In general, the mathematical equations you will use to solve problems require you to solve for one of the unknown quantities in the equation. To solve for this unknown, you need to isolate it on one side of the equal sign. To do this you need to follow only two logical rules:

- The same quantity can be added or subtracted from both sides of the equation without changing the equality.
- The same quantity can be multiplied or divided on both sides of the equation without changing the equality.

For example, you can calculate the voltage drop across a load in an electrical circuit if you know the resistance of the load:

$$V = I \times R$$

If you were given the voltage drop and the resistance of a particular load, such as an electric light bulb, and were asked to calculate the current in the circuit, you would isolate the equation for the unknown variable *I*. To do this, divide both sides of the equation by *R*.

$$\frac{V}{R} = \frac{I \times \cancel{R}}{\cancel{R}}$$

Cancelling the *R*'s on the right side, the resulting equation is

$$\frac{V}{R} = I$$

more commonly written as

$$I = \frac{V}{R}$$

Working Together

Teamwork is just as important in science as it is on the playing field or in the gym. Scientific investigations are almost always carried out by teams of people working together. Ideas are shared, experiments are designed, data are analyzed, and results are evaluated and shared with other investigators. Group work is necessary and is usually more productive than working alone.

Several times throughout the year you may be asked to work with one or more of your classmates. Whatever the task your group is assigned, a few rules need to be followed to ensure a productive and successful experience.

General Rules for Effective Teamwork

- Keep an open mind. Everyone's ideas deserve consideration.
- Divide the group task among all group members. Choose a role on the team that is best suited to your particular strengths.
- Work together. Take turns; encourage, listen, clarify, help, and trust one another.

Investigations and Activities

This kind of work is most effective when done by small groups. Here are some more suggestions for effective team performance during these activities:

- Make sure each group member understands and agrees to the task assigned him or her.
- Take turns doing various tasks during similar and repeated activities.
- Safety must come first. Be aware of where other group members are and what they are doing.
- Take responsibility for your own learning. Make your own observations and compare them with those of other group members.

Explore an Issue

When there is research to be done,

- Divide the topic into several areas, and assign one to each person.
- Keep records of the sources used by each person.
- Decide on a format for exchanging information (e.g., photocopies of notes, oral discussion, etc.)
- When the time comes to make a decision and take a position on an issue, allow for the contributions of each member of the group. Make decisions by compromise and consensus.
- Communicating your position should also be a group effort.

Evaluating Teamwork

After you've completed a task with your team, evaluate the team's effectiveness—the strengths and weaknesses, opportunities, and challenges. Answer the following questions:

- What were the strengths of your teamwork?
- What were the weaknesses of your teamwork?
- What opportunities were provided by working with your team?
- What challenges did you see with respect to your teamwork?

Try This Your Team Must Survive

Your team is stranded in the middle of nowhere, the result of a breakdown in the public transportation you were using in a wilderness area of the province. It will be five days before you are missed and searchers come looking for you.

You have gathered your combined resources from your luggage. As a team, review the following list of items and choose the five things that you feel are essential for survival over the five days. The final decision on the items should be reached by consensus.

Share your decision with the rest of the class.

flashlight + 4 batteries
4 hats
box of matches
6 cans of beans
map of area
fruit (4 days' supply)
water
chewing gum
Monopoly game
can opener
laptop computer
12 chocolate bars
groundsheet
pen, pencil, paper
knife
rope
calculator
books
compass
first-aid kit
comic books

Z Setting Goals and Monitoring Progress

Think back to the time you spent in school last year. What classes did you really do well in? Why do you think you were successful? What classes did you have some difficulty with? Why do you think you had those troubles? What could you do differently this year that would help you do better than last year?

By answering these questions, you are reflecting on your past behaviours in an attempt to make positive changes. The things that you want to accomplish today, this week, and this year are all called **goals**. Learning to set goals and to make a plan to achieve them takes skill and practice.

Setting Goals

Assess Your Strengths and Weaknesses

The process of setting goals starts with honest reflection. Maybe you have noticed that you do better on projects than you do on tests or exams. You may perform better when you are not pressured by time. Inattention in class and poor study habits may be weaknesses that result in poor performance.

Set Realistic Goals That You Can Measure

Do not set yourself up to fail by setting goals that you cannot possibly achieve. "I will have the best mark in the class at the end of the semester" may not be realistic. However, setting a goal "to increase my test marks by 10% this semester" may be achievable. You will find it easier to reach your goals if you can tell whether you are getting closer to them. A goal "to increase my test marks by 10% this semester" is a goal that is easy to measure.

Share Your Goals

People who know you can often help you set and clarify goals. Someone who knows your strengths and weaknesses may be able to think of possibilities you haven't considered. Sharing your goals with a trusted friend or adult will often provide needed support in reaching your goals.

Planning to Meet Your Goals

Once you have made a list of realistic goals, create a game plan to achieve them. A successful game plan usually consists of two parts: an action plan and target dates.

The Action Plan

First, make a list of the actions, or behaviours, that might help you reach your goals (**Figure 1**).

> Goal: To increase my test results by 10% by the end of the semester
>
> Possible actions:
>
> - Arrange to work with a partner, or other students in the class, to list things you must know for the test.
> - Ask the teacher to explain anything on the list that you cannot explain yourself or do not understand.
> - Choose a study area at home or in a public place. Use it.
> - Use an organizer to make a weekly schedule and to keep a record of all evaluations.

Figure 1

If you have made an honest assessment of your strengths and weaknesses, then you know what you have to do to improve. If poor test marks are to be improved, try to arrange to work with others to prepare. This might include using a weekly planner and organizing your study area at home. Identify what is preventing you from achieving your goal. Plan ways to overcome these obstacles.

Setting Target Dates

If you want to improve test results by 10% by the end of the semester, how much time do you have? How many tests are scheduled between now and then?

Work back from your target date at the end of the semester. Determine the date of each test between now and the end of the semester. These dates will give you short-term targets that, if you hit them, will make it easier to hit your overall target. A working schedule appears in **Figure 2**.

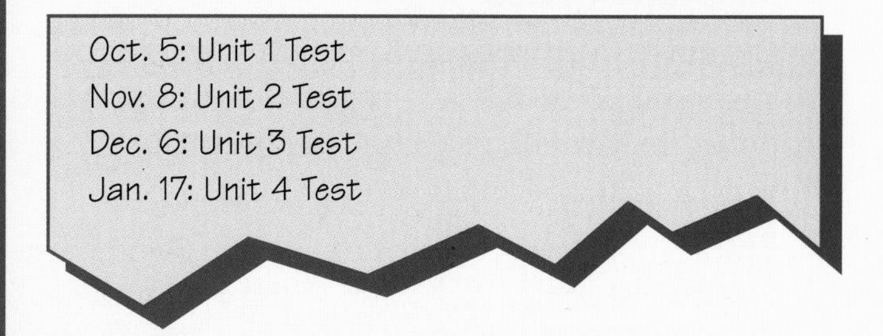

Figure 2

Once you have built your schedule in your planner, transfer it to a calendar in your study area. Make it a habit to refer to either your planner or your calendar every day.

Monitoring Progress

Do not wait until your final target date to determine whether you have achieved your goal. It is important to measure your progress along the way. You might decide, for example, to check your progress after the first test. Did your test result show a 10% increase?

If the 10% increase was not achieved, you may need to change your plan. For example, maybe the group of friends you're "studying" with is spending more time talking than identifying what is to be learned. It is always possible to change your plan or even to adjust your goal. The most important thing is to keep moving forward and to remain committed to improvement.

AA Good Study Habits

Study takes many forms. Developing good study skills can help you to learn.

Your Study Space

- **Organize your work area.** The place where you study should be tidy and organized. Papers, books, magazines, or pictures all over your work area may distract you from the work you have to do.
- **Maintain a quiet work area.** Where possible, make sure your work area is free of distractions—a phone, music, TV, other family members. If your study space is not at home, you can usually find an appropriate space at the school or public library. Any quiet space, free from interruptions, can be a productive workspace.
- **Make sure you are comfortable in your work area.** If possible, personalize your work area. You will be most productive when you work in an area where you feel at ease.
- **Be prepared—bring everything you need.** It is important to have all the necessary materials and equipment that you will need when you begin to study. If you find yourself continually getting up to find something you need, your productivity will decrease.

Study Habits

- **Take notes.** Take notes during class. Outside class, review the appropriate section of the textbook. Read or view additional material on the topic from other sources, such as newspapers, magazines, the Internet, and television.
- **Use graphic organizers.** You can use a variety of graphic organizers to help summarize a concept or unit. (See BB, Graphic Organizers, p. 303.)
- **Schedule your study time.** Use a daily planner and take it with you to class. Write any homework assignments, tests, or projects in it. Use it to create a daily "to do" list. This will help you to prepare material for handing in and to avoid last-minute panic, the most ineffective way to study.
- **Take study breaks.** It is important to schedule breaks into your study time. For example, you could decide to take a study break after completing one or two items on your "to do" list.

Study Checklist

- Choose a place to study. Is it quiet? comfortable? organized?
- Take notes in class.
- Summarize. Use graphic organizers to help you connect concepts.
- Schedule study time that includes study breaks.
- Know your strengths and weaknesses. Find ways to compensate for your weaknesses. You may have to ask for help.

Your Study Space

Draw a picture of your study space (at home, at school, or at the public library). List five good points about your study space. Now list five things that you could change to improve your study space and make it a more effective place to learn.

BB Graphic Organizers

Sometimes, it may be helpful to write down your ideas so that you can see them. When you are trying to organize your thoughts, or when you are trying to describe objects and events, you may want to use a **graphic organizer** instead of using complete sentences. Graphic organizers are visual representations of what you think. They represent, at a glance, your understanding of an event, idea, or concept.

Several types of graphic organizer are described below.

A PMI Chart

Purpose: To examine an issue.

Plus	Minus	Interesting

Directions:

1. Record your ideas on positive aspects of the topic or issue in the **P** (plus) column.
2. Record your ideas on negative aspects of the topic or issue in the **M** (minus) column.
3. Record your notes on interesting or controversial issues in the **I** (interesting) column.

A KWL Chart

Purpose: To help identify what you already know from previous experience and to help you record what you have already learned.

Know	Want to Know	Learned

Directions:

1. Before you begin a new topic, list what you already know about it in the **Know** column.
2. List what you would like to know in the **Want to Know** column.
3. After completing the new topic, list what you have learned in the **Learned** column.

The Venn Diagram

Purpose: To show similarities and differences between two or more concepts.

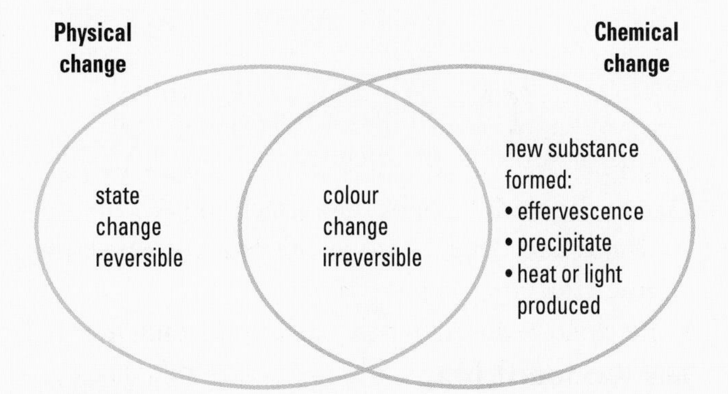

Figure 3
A Venn diagram

Directions:

1. List the key points of one concept in the left circle.
2. List the key points of the second concept in the right circle.
3. Write all similarities or common ideas in the overlapping section of the circles.

The Fish Bone Diagram

Purpose: To identify causes and effects.

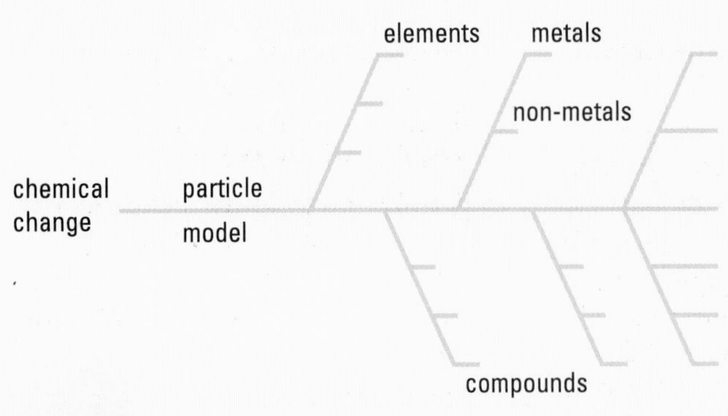

Figure 4

A fish bone diagram

Directions:

1. At the head of the fish, identify the **topic** or **effect**.
2. At the end of each major bone, identify a major subtopic or category.
3. On the minor bones that attach to each major bone, add details such as the possible causes of the effect or result.

The Concept Map

Purpose: To show connections among ideas and concepts. These ideas and concepts can be represented by words or by pictures.

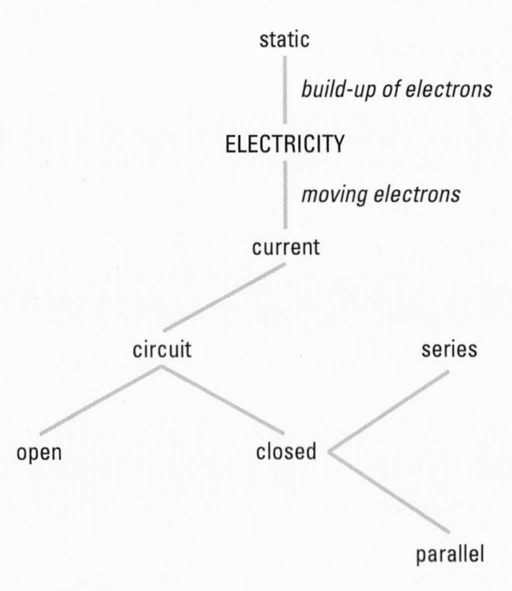

Figure 5

A concept map using words

Directions:

1. Put the central idea or word in the middle of a piece of paper.
2. Put the idea or words that connect to the central idea in the space around the centre of the page.
3. Draw arrows or lines to connect the ideas or words that are related.
4. On each arrow or line, write a brief description of how the ideas or words are related to each other.
5. Connect additional related words and ideas to those already listed.

Try This: *Science 9: Concepts and Connections* at a Glance

Develop a graphic organizer for each unit in *Science 9: Concepts and Connections.* Identify the key concept of each unit. Decide which graphic organizer is appropriate. Use the graphic organizer you have chosen to connect the key concept to other concepts and connections described in each section. When you are finished, share your graphic organizer with a partner. Other ideas and connections may come to you during this sharing session. Use a different colour of pen or pencil to modify your organizer as needed.

Glossary

A

alloy a metal made by combining two or more different metals or metals and nonmetals

ammeter a device that measures the amount of electric current flowing past a point in a circuit

amnion a fluid-filled sac that surrounds the embryo, protecting it from infection, dehydration, impact, and changes in temperature

anaphase the phase of mitosis during which each chromosome splits and the two halves move to opposite ends of the cell

asexual reproduction a form of reproduction that results in offspring with identical genetic information to the parent

asteroid a small rocky object

asteroid belt the ring of asteroids around the Sun

astronomer a person who sky watches; often the term is used to refer to a professional astronomer

astronomy the study of what is in space beyond Earth

atom a particle found in elements and compounds

atomic mass the average mass of one atom of an element

atomic number the number of protons in an atom

atomic radius the average distance from the nucleus to the outer edge of the atom

B

benign a term used to describe a cancerous tumour that cannot spread to other parts of the body, and is generally harmless

Big Bang theory the event when, 10 billion to 15 billion years ago, a small dense hot mass began to move apart to form the beginning of the universe that we know

binary fission the form of asexual reproduction in which the organism splits directly into two equal-sized offspring, each with the parent's genetic material

blue shift a movement toward the blue end of the spectrum signalling the galaxy or star is moving toward Earth

Bohr-Rutherford diagram a diagram that summarizes the numbers and positions of all three subatomic particles in an atom, that comprises electrons in a series of concentric circles (orbits) drawn around a central nucleus, containing the number of protons and neutrons in the element

boiling point the temperature at which a liquid begins to boil

bond the connection, representing electrons, that holds atoms together in models of molecules

budding the form of asexual reproduction in which the offspring begins as a small outgrowth of the parent and eventually breaks off, becoming an organism on its own

C

cancer the term for a group of diseases associated with uncontrolled, unregulated cell division

capacitor a device that can store small amounts of electrical energy

carcinogen any substance or energy that can cause cancer

celestial a synonym for space, for example celestial bodies (e.g., planet) and celestial events (e.g., lunar eclipse)

cell cycle the sequence in events in a cell from carrying out their normal functions to the process of cell division

cell membrane a covering around a cell that controls the movement of materials into and out of the cell

cervix the muscular opening between the vagina and the uterus

charging by contact the process of electrons moving from one place to another

chemical change the alteration of a substance into one or more different substances with different properties

chemical formula a formula that states which elements are present and the number of atoms of each element found in that substance

chemical property the behaviour of a substance as it becomes a new substance

chemical symbol an abbreviation for the name of an element

chromosome a threadlike structure that contains genetic information

cilia tiny hair-like organs that line the oviduct and move the zygote towards the uterus

circuit breaker a device that automatically shuts off when there is a current overload

circuit diagram a diagram of a circuit that uses symbols

cloning the process of forming identical offspring from a single cell or tissue

combining capacity the number of connections an element is able to make when combining with another element

combustibility a chemical property that describes the ability of a substance to burn

combustion the chemical reaction that occurs when a substance reacts rapidly with oxygen and releases energy

comet an object that travels around the Sun, the head of which is composed of a mass of frozen gases, ice, and dust; comets have fixed and very long orbits

composite a material formed by combining to other materials

compound a pure substance that consists of only one kind of molecule containing more than one kind of atom

conception the process of fertilization in which the head of the sperm cell breaks through the cell membrane of the egg

conductor a substance that carries electrical energy without much resistance so electrons lose little energy

conjugation the simplest form of sexual reproduction in which two cells come together and exchange genetic information

constellation a group of stars that seem to form a shape or pattern

corrosion a slow chemical change that occurs when a metal reacts with oxygen to form oxide

cosmology the study of the origin and changes of the universe

crossbreeding the process in which pollen is taken from one plant and used to fertilize the eggs of another plant

current electricity the movement of an electric charge from a source of electrical energy along a controlled path

cytokinesis the process in cell division in which the cytoplasm and its contents separate into equal parts

cytoplasm the area of a cell where nutrients are absorbed, transported, and processed

D

density the ratio of an object's mass to its volume

deoxyribonucleic acid (DNA) the genetic chemical found in all living things

diatomic molecule a molecule that forms when two atoms of the same element join together

distribution panel the location where all circuit breakers connect to each circuit

DNA fingerprinting the process of using DNA material from a few cells to identify an individual

E

efficiency the degree to which electrical energy is changed to useful energy by the electrical device

egg the female sex cell

electric circuit a basic part that is made up of an energy source, connecting wires, an electrical device (load), and a switch

electric current the product of electric charges as they move from one place to another

electric meter a device that measures the total amount of electrical energy used in a building

electrical load the device that converts electrical energy into a needed form of energy, such as heat or light

electron a negatively charged particle that moves rapidly in the space around the nucleus, or core, of the atom

element a pure substance that consists of only one kind of atom

elliptical galaxy a galaxy that has a football shape

embryo the stage in development when a fertilized egg has divided to form a hollow ball of over 100 cells

endometrium the thick lining of the uterus that provides nourishment for the zygote and embryo

energy input the energy that goes into an electrical circuit

epididymis a coiled structure near the testes where sperm development is completed

external fertilization a process in which the male sperm cells are united with the female egg cells outside the female's body

F

family a group of elements in the periodic table, arranged in columns, that share similar properties and characteristics

fetus the term used to describe a developing unborn baby from the eighth week after conception until the moment of birth

flagellum a whiplike tail that propels the sperm cell

flammable the chemical property of a substance that allows it to burn when exposed to flame and oxygen

follicle a group of cells, including a reproductive cell, inside the ovary of the female

fragmentation the form of asexual reproduction in which a new organism is formed from a part that breaks off from the parent

free fall a state where an object or person is falling toward Earth with only gravity pulling the object or person down

fuel cell a cell that produces electricity from hydrogen and oxygen; cars that use fuel cells would not create harmful byproducts

fuse an object that will melt when the temperature in a circuit becomes too high

G

galaxy a huge collection of gas, dust, and hundreds of billions of stars and planets

gametes specialized sex cells that combine to form a zygote

gas giants those planets that do not seem to have a solid surface, but rather low-density gases—Jupiter, Saturn, Uranus, Neptune

gene a small unit of DNA that contains the information necessary for all cell functions and that determines the characteristics of an individual

genetic engineering the transfer of genes from the DNA of one organism to another

grafting the process of attaching branches from one type of tree on to another type of tree

ground fault circuit interrupter (GFCI) an outlet that includes a circuit breaker that responds to very small changes in current and interrupts the flow of electricity; GFCI outlets should be used in all bathrooms, kitchens, laundry rooms and any other area where water is found

ground terminal the third round hole in an outlet

ground wire a wire that helps to protect against electric shock from a short circuit

grounding pin the third prong on a plug that allows electrical current to be grounded to the metal pin rather than passing through a person

H

hardness a measure of the resistance of a solid to being scratched or dented

heavy metal a dense metal that is required in small amounts in healthy plants and animals but can cause damage in large amounts

hermaphrodite an organism that creates both male and female sex cells within the same body

heterogeneous mixture a mixture in which two or more substances can be seen or felt

homogeneous mixture a mixture in which two or more substances mix together so completely that the mixture looks and feels like it is made of only one substance

hormone a chemical that acts as a messenger between cells

hot or **live wire** the black wire in an electrical circuit that carries voltage into a house

I

inner planets those planets closest to the Sun (Mercury, Venus, Earth, Mars)

internal fertilization a process in which the male sperm is united with the female egg cell inside the female's body

interphase the period between cell divisions during which the cell grows and duplicates the chromosomes in the nucleus

irregular galaxy a galaxy with no familiar shape

joule the SI unit for measuring energy; its symbol is **J**

L

law of electric charges the law that states that like charges repel each other, and unlike charges attract each other

light sensitivity the chemical property of a substance that allows it to become a new substance when it interacts with light

luminous the state of an object, such as the Sun, that can be seen because it gives off light

M

main breaker switch a switch that, when shut off, cuts all electricity to the building; the main breaker switch is controlled by a circuit breaker

malignant the term used to describe a cancerous tumour that spawns cells that can spread to other parts of the body

mass number the sum of the protons and neutrons in the nucleus of an atom

melting point the temperature at which a solid melts

menstruation the process in nonpregnant women during which the endometrium is shed from the uterus through the vagina

metalloid an element that has the properties of both metals and nonmetals

metallurgy the technology of separating a mineral from its ore

metaphase the phase of mitosis during which double-stranded chromosomes line up in the middle of the cell

meteor a bright streak in the sky caused by the burning meteoroid; most meteors completely burn up before hitting Earth

meteorite a meteor that hits Earth before it completely burns up

meteroid a lump of rock or metal that is pulled through Earth's atmosphere

mineral a naturally occurring compound, sometimes containing metal combined with oxygen, sulfur, or other elements

minor bodies asteroids and comets

miscarriage a spontaneous abortion

mitosis the process by which nuclear material is divided during cell division

mixture a substance that consists of two or more pure substances

modern periodic law if the elements are arranged according to their atomic number, a pattern can be seen in which similar properties occur regularly

molecule a combination of two or more atoms

monomer one of the individual molecules that make up polymers

moon a natural satellite; several planets have more than one moon

mutation a change in the genetic code

N

nebula a spinning mass of gas and dust; stars are born from nebulas

negative charge an excess of electrons

neutral wire the white wire in an electrical circuit that allows the current to leave the house after it has been used

neutron a neutral particle located in the nucleus, or core, of the atom

nonluminous the state of an object, such as the planets, that makes them visible only when light is reflected off of them

North Star (see Polaris)

nuclear fusion the process during which substances fuse to form new substances that results in heat and light energy given by the Sun and all stars

nucleus the main organelle of the cell, which directs the cell's activities of atom

O

orbit the act of an object revolving around a much larger object

orbital period the time it takes a planet to complete its revolution around the Sun

ore a rock formation containing a valuable mineral

organelle a specialized structure inside plant and animal cells

outer planets those planets further from the Sun (Jupiter, Saturn, Uranus, Neptune, Pluto)

ovary the primary reproductive organ of the female

oviduct the part of the female reproductive system where fertilization of the egg cell takes place

ovulation the process during which the ovary wall bursts and the egg cell is released into the oviduct

P

parallel circuit a method of wiring electrical circuits in which current follows more than one path, electrical energy is not shared, and each electrical device can be on or off within the circuit

period a horizontal row of elements in the periodic table that do not share similar properties

periodic anything that repeats according to the same pattern

periodic table an organized arrangement of elements that explains and predicts physical and chemical properties

physical change a change in the state or form of a substance that does not change the original substance

physical property a characteristic or description of a substance that may help to identify it

placenta an organ formed in the womb that allows nourishment of the embryo from blood vessels

planet a sphere-shaped object that follows a certain path around a star

Polaris also known as the North Star, a star in the Little Dipper, which can be seen all year in the Northern Hemisphere

polarized plug a plug that fits into the outlet only one way to ensure that the neutral wire is not connected to the electrical device switch

pollen the male sex cells of a flower

pollination the process where pollen is moved from the anther to the female egg cells to fertilize those cells

polymer a material made up of long chains of molecules

polymerization the process of chemically bonding monomers to form polymers

positive charge a deficiency of electrons

potential difference the voltage drop between two points in an electric circuit

prophase the phase of mitosis during which the individual chromosomes shorten and thicken, and the membrane around the nucleus starts to dissolve

proton a positively charged particle located in the nucleus, or core, of the atom

pure substance a substance that consists of only one kind of atom or molecule

Q

quasar a distant object that looks like a faint star but gives off up to 100 times more energy than the Milky Way galaxy

R

radar acronym for radio detection and ranging; a device that emits bursts of radio waves and picks up their reflections to detect the location of objects and determines how far away they are

red shift the movement in the visible spectrum toward the red end signaling that the wavelength is getting larger (the galaxy is moving away from Earth)

reflecting telescope a telescope that gathers light; Sir Issac Newton built the first reflecting telescope in 1668

refracting telescope a telescope in which light rays bend as they pass through a light-gathering lens

renewable energy source a source that cannot be used up; hydro, solar, wind, and tidal are renewable energy sources

resistance the ability to slow down the flow of electrons in conductors

resistor an electrical device that slows the flow of current into an electrical object

retrograde motion the apparent backward loop made by Mercury, Venus, Mars, Jupiter, and Saturn that occurs once per year

revolution the movement of one object around another

rotation the spinning of an object around its axis

rusting an example of corrosion involving iron's reaction with oxygen from air, water, and other chemical substances dissolved in water

S

satellite an object that revolves around a planet or another object; satellites can be natural (e.g., a moon) or artificial (e.g., a weather satellite)

selective breeding a method of reproduction that results in several generations of offspring all having the same desired characteristics

seminiferous tubules tiny twisting tubes inside the testis that produce sperm cells

series circuit a method of wiring electric circuits in which current follows one path, electrical energy is shared, and all electrical devices must be on or off at the same time

sexual reproduction reproduction in which genetic information from two cells is combined to produce a new organism

solar system a system that comprises the Sun and all the objects that travel around it

solubility the ability of a substance to dissolve in a solvent such as water

spectroscope a device used to look at patterns of light emitted and identify which elements are present in a particular star

spectroscopy the spectral analysis of light

sperm the male sex cell

spinoff an extra benefit from technology developed for another purpose

spiral galaxy a galaxy that has a spiral shape

spore formation the form of asexual reproduction in which the organism undergoes cell division to produce smaller, identical cells, called spores, that are usually housed within the parent cell

standard atomic notation an internationally recognized system used to identify chemical substances

star a large body of matter that emits huge amounts of energy

star clusters groups of stars that are close together and that travel together

static electricity a charge that does not move

structural diagram a drawing to explain molecules in which atoms are represented by chemical symbols and bonds are shown as straight lines connecting the symbols

subscript the small number written after an element, below the line, in a chemical formula to indicate how many atoms of the element are present

sunspot a dark region on the Sun's surface that represents a storm

surrogate mother a woman who carries a baby to term and returns it to the genetic parents

synthetic a nonnatural substance

T

telescope a device that gathers light and allows users to see and record images not visible to the naked eye

telophase the phase of mitosis during which the chromosomes reach the opposite poles of the cell and a new nuclear membrane starts to form around each set

terrestrial planets the small rocky planets (Mercury, Venus, Earth, Mars) closest to the Sun

testis the reproductive organ of the male mammal

trait a characteristic of a plant, including its size, hardiness, yield, or flavour

trimester one of three stages in human pregnancy

tumour a mass or lump formed by abnormal rapid cell division

U

umbilical cord the cordlike structure that connects the embryo to the placenta

universe the term used to describe all of space

useful energy output the energy that is produced by the electrical device as light, sound, heat, or mechanical energy

uterus the organ in the female reproductive system in which the fertilized egg becomes embedded and develops into a fetus

U

vagina birth canal leading from the opening of the vulva to the cervix

vegetative reproduction the form of asexual reproduction in which a plant produces slender stems that take root and develop into new plants

viscosity a measure of how easily a liquid flows

visible spectrum a spectrum in which the colours are always in the same order: red, orange, yellow, green, blue, and violet with red having the longest wavelength and violet the shortest wavelength

voltage the force that moves electric charges in a circuit

voltmeter a device that measures the voltage drop in a circuit

W

wall outlet a plastic outlet located in a wall so that electricity can be safely delivered to a room

Z

zodiac constellation a constellation named after an animal

zygote a fertilized egg cell; the product of sexual reproduction

Index

G

H

I

J

L

M

N

O

P

Q

R

S

T

U

V

W

Z

Credits

Cover Image: © NASA/Science Photo Library

Table of Contents: p. iv left Corbis/Digital Art, right Lennart Nilsson/Albert Bonniers Forlag AB, *A Child Is Born*; p. v left Photodisc/Getty Images, right ©NASA STSCI.

Unit 1 Exploring Matter

Unit Opener: Ron Avery/Superstock; p. 4 Corel; p. 6 left © Lester Leftkowitz/ Getty Images, right American Aluminum Company; p. 7 Visuals Unlimited; p. 8 left © Paul A. Souders/Corbis, right Photodisc/Getty Images; p.12 Dick Hemingway Photographs; p.13 left Robert Brimson/Getty Images, right Photodisc/Getty Images; p.14 top left © Wolfgang Kahler/Corbis, top right © Maxx Images, bottom left © Henry Diltz/Corbis, bottom right Foodpix/Getty Images; p.15 top Unicorn Images, bottom Tek Image/Science Photo Library; p.16 © Michael Yamashita/Corbis; p. 18 Dick Hemingway Photographs; p. 19 Dick Hemingway Photographs; p. 20 top Peggy Rhodes, middle © Gianni Dagli/Corbis, bottom © Galen Rowell/Corbis; p. 23 Eyewire/Getty Images; p. 25 Meredith White-McMahon; p. 27 Photodisc/Getty Images; p. 28 top © Paul Souders/Corbis, bottom Dick Hemingway Photographs; p. 30 Jeremy Jones, bottom Photodisc/Getty Images; p.32 top to bottom John Cancalosi/ VALAN PHOTOS, Bob Newman/Visuals Unlimited, A.J.Copely/Corbis, Jose Manuel Sanchis Calvete/Corbis, Ken Lucas/Visuals Unlimited, James L. Amos/Corbis, Dane S. Johnson/Visuals Unlimited; p. 33 Paul A. Souders/ Corbis; p. 34 Dick Hemingway Photographs; p. 35 Dick Hemingway Photographs; p. 36 top Photodisc/Getty Images, bottom Superstock; p. 37 Federal Programs Division, Environmental Protection Branch; p. 40 Dave Starrett; p. 44 top Corbis/Digital Art, bottom Dave Starrett; p. 46 Dick Hemingway Photographs; p. 48 top Rich Treptow/ Visuals Unlimited, bottom Burndy Library; p. 52 © V. Whelan/J.A. Wilkinson/VALAN PHOTOS; p. 53 Corel; p. 54 © David Parker/IMI/University of Birmingham High Technology Consortium/Science Photo Library.

Unit 2 Reproduction: Processes and Applications

Unit Opener: Dr. Gopal Murti/Science Photo Library; p. 64 left © Lester V. Bergman/Corbis, right Jerome Wexler/Visuals Unlimited; p. 65 © A. Barter/ Custom Medical Library; p. 66 left © Paul Campbell/Getty Images, right Bill Bomaszewski/Visuals Unlimited; p. 67 © Dr. Howard Smedley/Science Photo Library; p. 68 top © CB066090 Magma, bottom left VALAN PHOTOS, bottom right © Kurt Kreiger/Corbis; p. 69 Dr. Yorgos Nikas/Science Photo Library; p. 70 © Guy Stubbs/Corbis; p. 75 Andrew Bajer/University of Oregon; p. 79 left SIU/Peter Arnold, right © A.F. Michler/Peter Arnold; p. 82 Al Harvey/Slide Farm; p. 85 top left A.B. Dowsett/Science Photo Library, top middle John Cooke/ Oxford Science Films, top right © Jeffery L. Rotman/Corbis, bottom left CNRI/Science Photo Library, bottom right Harry Taylor/Oxford Science Films; p. 86 © Sacha Ajbeszyc/Getty Images; p. 88 Runk/Schoenberger from Grant Heilman/ Miller/Comstock; p. 96 top © Owen Frank/Corbis, bottom G.A. Maclean/Oxford Science Films; p. 97 First Light.ca, p. 99 Meredith White-McMahon; p. 100 top J.A.L. Cooke/ Oxford Science Films, bottom N. Pecnick/Visuals Unlimited; p. 101 top Harold Taylor/Oxford Science Films, bottom John D. Cunningham/ Visuals Unlimited; p. 104 © Lennart Nilsson/Albert Bonniers Forlag AB, *A Child Is Born*; p. 105 Andy Walker, Midland Fertility Services/Science Photo Library; p. 108 Oxford Science Films; p. 109 © Lennart Nilsson/Albert Bonniers Forlag AB, *A Child Is Born*; p. 110 © A. Tsiaras/Photo Researchers Inc.; p. 112 © Lennart Nilsson/Albert Bonniers Forlag AB, *A Child Is Born*.

Unit 3 Electrical Applications

Unit Opener: © Christopher J. Morris/CORBIS/MAGMA; p.122 left Fundemental Photographers, right Dave Starrett; p. 123 Dick Hemingway Photographs; p. 124 Jeff J. Daly/Visuals Unlimited; p. 125 Photodisc/Getty Images; p. 126 top ©Roy McMohon/Corbis, middle © Doug Wilson/Corbis; bottom courtesy of Natural Resources Canada; p. 128 Dick Hemingway Photographs; p. 129 top "Physics: /Algebra/Trig" by Eugene Hecht, bottom Barry Cohen; p. 130 © top William James Warren/Corbis, bottom © William James/Corbis; p. 131 Ferrell McCollough/Visuals Unlimited; p. 132 courtesy of Boreal; p. 134 courtesy of Creative Publishing International; p. 135 J.A. Wilkinson/VALAN PHOTOS; p. 137 Dick Hemingway Photographs; p. 138 Larry Stepanowicz/Visuals Unlimited; p. 139 J.A. Wilkinson/VALAN PHOTOS; p. 140 top J.A. Wilkinson/VALAN PHOTOS, bottom ZAP® zapworld.com; p. 141 Dan McCoy/Rainbow; p. 142 Dave Starrett; p. 144 top Jeff Greenberg/Visuals Unlimited, bottom courtesy of Boreal; p. top146 Martyn F. Chillmaid/Science Photo Library, bottom Dave Starrett; p. 149 © Carl and Ann Purcell/ Corbis; p. 150 top Larry Stepanowicz/Visuals Unlimited, bottom J.A. Wilkinson/ VALAN PHOTOS, Jeff Greenberg/Visuals Unlimited, Photodisc/Getty Images, © Ed Eckstein/Corbis; p. 151 Hulton Archive/ Getty Images; p. 152 Steve Callahan/Visuals Unlimited; p. 153 Loren Winters/Visuals Unlimited; p. 154 Dick Hemingway Photographs; p. 156 top Robert Williams, left Dick Hemingway Photographs, right © Kelly-Mooney Photography/Corbis; p. 157 John Patterson; p. 158 top Eastcott/E. Momatiuk/VALAN PHOTOS, bottom Dick Hemingway Photographs; p. 160 J.A. Wilkinson/VALAN PHOTOS; p. 161 Jeremy Jones; p. 162 top Dave Starrett, bottom John Fowler/VALAN PHOTOS; p. 164 top Image Network, bottom Maxx Images; p. 165 Alan Sirulnikoff/Firstlight.ca; p. 166 top courtesy of Innovative Technologies Inc., bottom Corel; p. 167 courtesy of Natural Resources Canada p.168 Motorcites.com, middle Motorcites.com, bottom courtesy of U.S. Department of Energy; p. 169 Ballard Power Systems; p. 170 top to bottom James R. Page/VALAN PHOTOS, top right Jeannie R. Kemp/VALAN PHOTOS, Val Wilkinson/VALAN PHOTOS, R. Moller/VALAN PHOTOS; p. 173 © Nik Wheeler/Corbis; p. 174 top Werner Otto/Firstlight.ca, bottom © Ontario Power Generation; p. 175 Al Hirsch; p. 176 Photodisc/Getty Images; p. 177 John Patterson

Unit 4 Space Exploration

Unit Opener: ©NASA/Visuals Unlimited; p. 186 left David Aguilar, Harvard-Smithsonian Center for Astrophysics, right © Mike Brinson/Getty Images; p. 187 © NASA/Visuals Unlimited; p. 188 left Space Telescope/Science Photo Library, right Victor Habbick Visions/Science Photo Library; p. 189 NOAO/AURA/NSF; p. 190 © NASA; p. 191 ©NASA/Science Visuals Unlimited; p. 192 Ian Davis-Young/VALAN PHOTOS; p. 193 top to bottom John Sanford/Science Photo Library, John Sanford/Science Photo Library, Eckhard Slawik/Science Photo Library, John Sanford/Science Photo Library, Pekka Parviainen/Science Photo Library, John Sanford/ Science Photo Library; p. 195 ©Roger Ressmeyer/Corbis; p. 196 John Cancalosi/VALAN PHOTOS; p. 197 Cheryl Hogue/Visuals Unlimited; p.200 top NASA/JPL/Cal Tech, bottom NASA-JDC/Visuals Unlimited; p. 201 top NASA-JDC/Visuals Unlimited, middle Earth Imaging/Getty Images, bottom © NASA; p. 202 top © NASA/Science Photo Library bottom © NASA/Science Photo Library; p. 203 top left © NASA/Science Photo Library, top right © NASA-LDC/Visuals Unlimited, bottom © NASA; p. 204 © Richard H. Johnston/Getty Images; p. 206 top © NASA/Oxford Science Films, bottom left © NASA, middle Visuals Unlimited, bottom right © NASA; p. 207 © NASA; p. 208 © NASA; p. 210 NOAO/AURA/NSF; p. 211 top David Nunuk/Firstlight.ca, bottom Space Telescope Sky Institute; p. 212 © NASA-JDC/Visuals Unlimited; p. 213 courtesy of Ron Thorpe; p. 214 © John Wilkinson/Corbis; p. 215 Pekka Parviainen/Science Photo Library; p. 216 Science/Visuals Unlimited; p. 218 top © NASA, middle Jeanne Lorre/Science Photo Library, bottom Pekka Parviainen/Science Photo Library; p. 219 top left © NASA-ESO-JDC/Visuals Unlimited, top right © NASA, bottom Science VU/Visuals Unlimited; p. 220 Imperial College Dept. of Physics/ Science Photo Library; p. 221 Richard Treptow/Visuals Unlimited; p. 224 Dept. of Physics Imperial College, Science Photo Library; p. 226 © NASA STSCI; p. 228 CCRS/Canadian Space Agency; p. 230 top ©NASA, bottom ©NASA-JDC/Visuals Unlimited; p. 231 © NASA, Canadian Space Agency; p. 232 © NASA Canadian Space Agency; p. 233 © NASA/Science Photo Library; p. 234 Dave Starrett; p. 235 top © NASA/ Visuals Unlimited, bottom © NASA/Science Photo Library; p. 236 top © NASA STSCI/Visuals Unlimited, bottom Eyewire/Getty Images; p. 237 top © Lindsay Hebberd/Corbis, bottom © Penny Tweedie/Getty Images; p.238 top © NASA, bottom Corbis; p. 243 Terrence Dickinson; p. 244 David Nunuk/Firstlight.ca; p. 243 Terrence Dickinson; p. 244 David Nunuk/Firstlight.ca; p. 245 © NASA JPL-Cal Tech.

Skills Handbook

Unit Opener: Chris Salvo/Getty Images; p. 251 Dave Starrett; p. 252 Dave Starrett; p. 260 left Corbis/Magma, right Dave Starrett; p. 261 courtesy of Boreal; p. 264 left Dave Starrett, right Loren Winters/Visuals Unlimited; p. 265 Dave Starrett; p. 267 Custom Medical Stock Photo; p. 269 Dave Starrett; p. 270 Dave Starrett; p. 283 top to bottom PhotoDisc/Getty Images, Todd Ryoji, Dick Hemingway Photographs; Omni Photo Communications Inc.; p. 294 Dave Starrett.

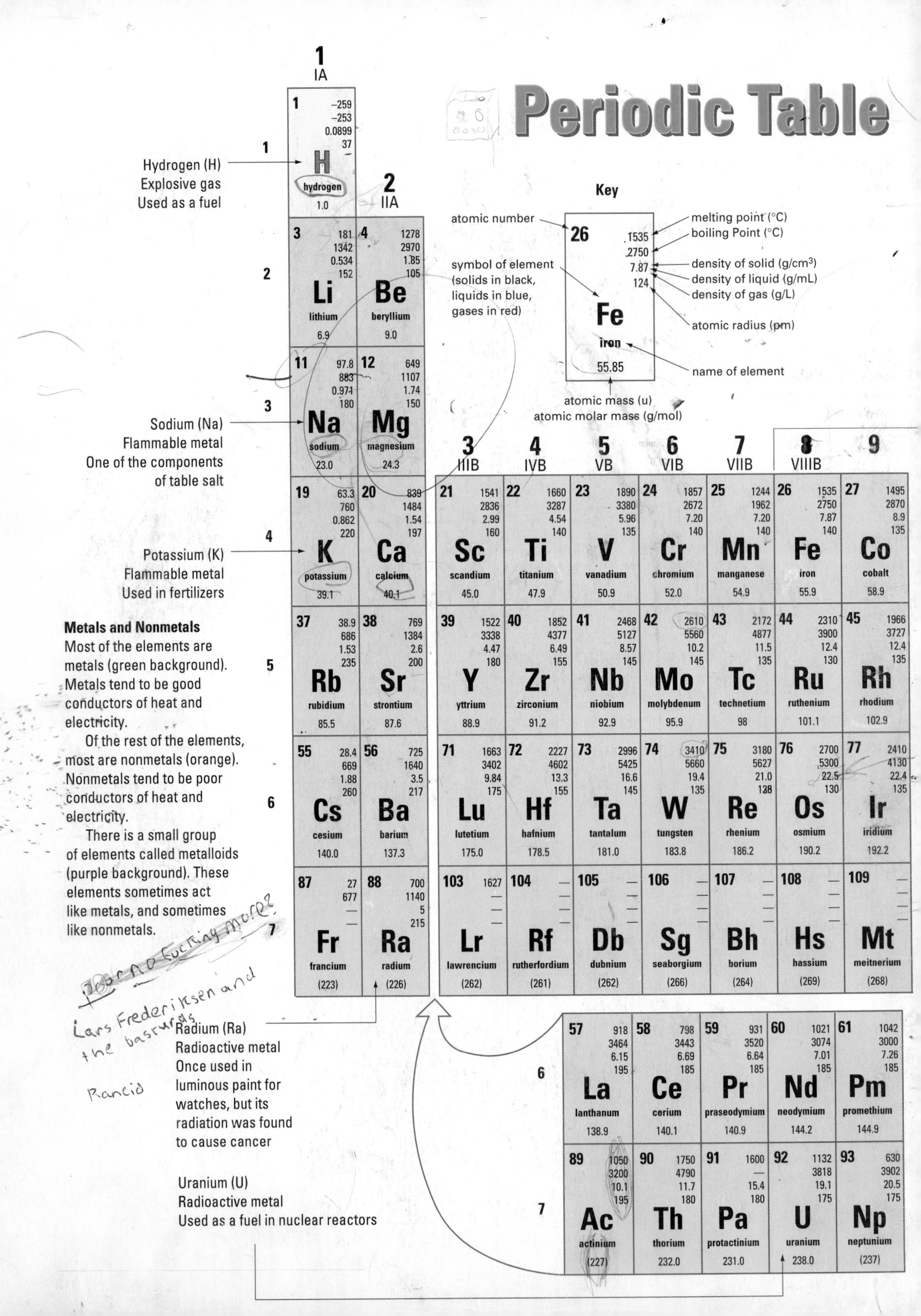
Periodic Table
1
IA
2
IIA
Key
atomic number
symbol of element (solids in black, liquids in blue, gases in red)
26
1535
2750
7.87
124
Fe
iron
55.85
melting point (°C)
boiling Point (°C)
density of solid (g/cm3)
density of liquid (g/mL)
density of gas (g/L)
atomic radius (pm)
name of element
atomic mass (u)
atomic molar mass (g/mol)
3 IIIB
4 IVB
5 VB
6 VIB
7 VIIB
8 VIIIB
9
1
2
3
4
5
6
7
1 −259 −253 0.0899 37 H hydrogen 1.0
3 181 1342 0.534 152 Li lithium 6.9
4 1278 2970 1.85 105 Be beryllium 9.0
11 97.8 883 0.971 180 Na sodium 23.0
12 649 1107 1.74 150 Mg magnesium 24.3
19 63.3 760 0.862 220 K potassium 39.1
20 839 1484 1.54 197 Ca calcium 40.1
21 1541 2836 2.99 160 Sc scandium 45.0
22 1660 3287 4.54 140 Ti titanium 47.9
23 1890 3380 5.96 135 V vanadium 50.9
24 1857 2672 7.20 140 Cr chromium 52.0
25 1244 1962 7.20 140 Mn manganese 54.9
26 1535 2750 7.87 140 Fe iron 55.9
27 1495 2870 8.9 135 Co cobalt 58.9
37 38.9 686 1.53 235 Rb rubidium 85.5
38 769 1384 2.6 200 Sr strontium 87.6
39 1522 3338 4.47 180 Y yttrium 88.9
40 1852 4377 6.49 155 Zr zirconium 91.2
41 2468 5127 8.57 145 Nb niobium 92.9
42 2610 5560 10.2 145 Mo molybdenum 95.9
43 2172 4877 11.5 135 Tc technetium 98
44 2310 3900 12.4 130 Ru ruthenium 101.1
45 1966 3727 12.4 135 Rh rhodium 102.9
55 28.4 669 1.88 260 Cs cesium 140.0
56 725 1640 3.5 217 Ba barium 137.3
71 1663 3402 9.84 175 Lu lutetium 175.0
72 2227 4602 13.3 155 Hf hafnium 178.5
73 2996 5425 16.6 145 Ta tantalum 181.0
74 3410 5660 19.4 135 W tungsten 183.8
75 3180 5627 21.0 138 Re rhenium 186.2
76 2700 5300 22.5 130 Os osmium 190.2
77 2410 4130 22.4 135 Ir iridium 192.2
87 27 677 — — Fr francium (223)
88 700 1140 5 215 Ra radium (226)
103 1627 — — — Lr lawrencium (262)
104 — — — — Rf rutherfordium (261)
105 — — — — Db dubnium (262)
106 — — — — Sg seaborgium (266)
107 — — — — Bh borium (264)
108 — — — — Hs hassium (269)
109 — — — — Mt meitnerium (268)
6
57 918 3464 6.15 195 La lanthanum 138.9
58 798 3443 6.69 185 Ce cerium 140.1
59 931 3520 6.64 185 Pr praseodymium 140.9
60 1021 3074 7.01 185 Nd neodymium 144.2
61 1042 3000 7.26 185 Pm promethium 144.9
7
89 1050 3200 10.1 195 Ac actinium (227)
90 1750 4790 11.7 180 Th thorium 232.0
91 1600 — 15.4 180 Pa protactinium 231.0
92 1132 3818 19.1 175 U uranium 238.0
93 630 3902 20.5 175 Np neptunium (237)
Hydrogen (H)
Explosive gas
Used as a fuel
Sodium (Na)
Flammable metal
One of the components of table salt
Potassium (K)
Flammable metal
Used in fertilizers
Metals and Nonmetals
Most of the elements are metals (green background). Metals tend to be good conductors of heat and electricity.
Of the rest of the elements, most are nonmetals (orange). Nonmetals tend to be poor conductors of heat and electricity.
There is a small group of elements called metalloids (purple background). These elements sometimes act like metals, and sometimes like nonmetals.
Radium (Ra)
Radioactive metal
Once used in luminous paint for watches, but its radiation was found to cause cancer
Uranium (U)
Radioactive metal
Used as a fuel in nuclear reactors

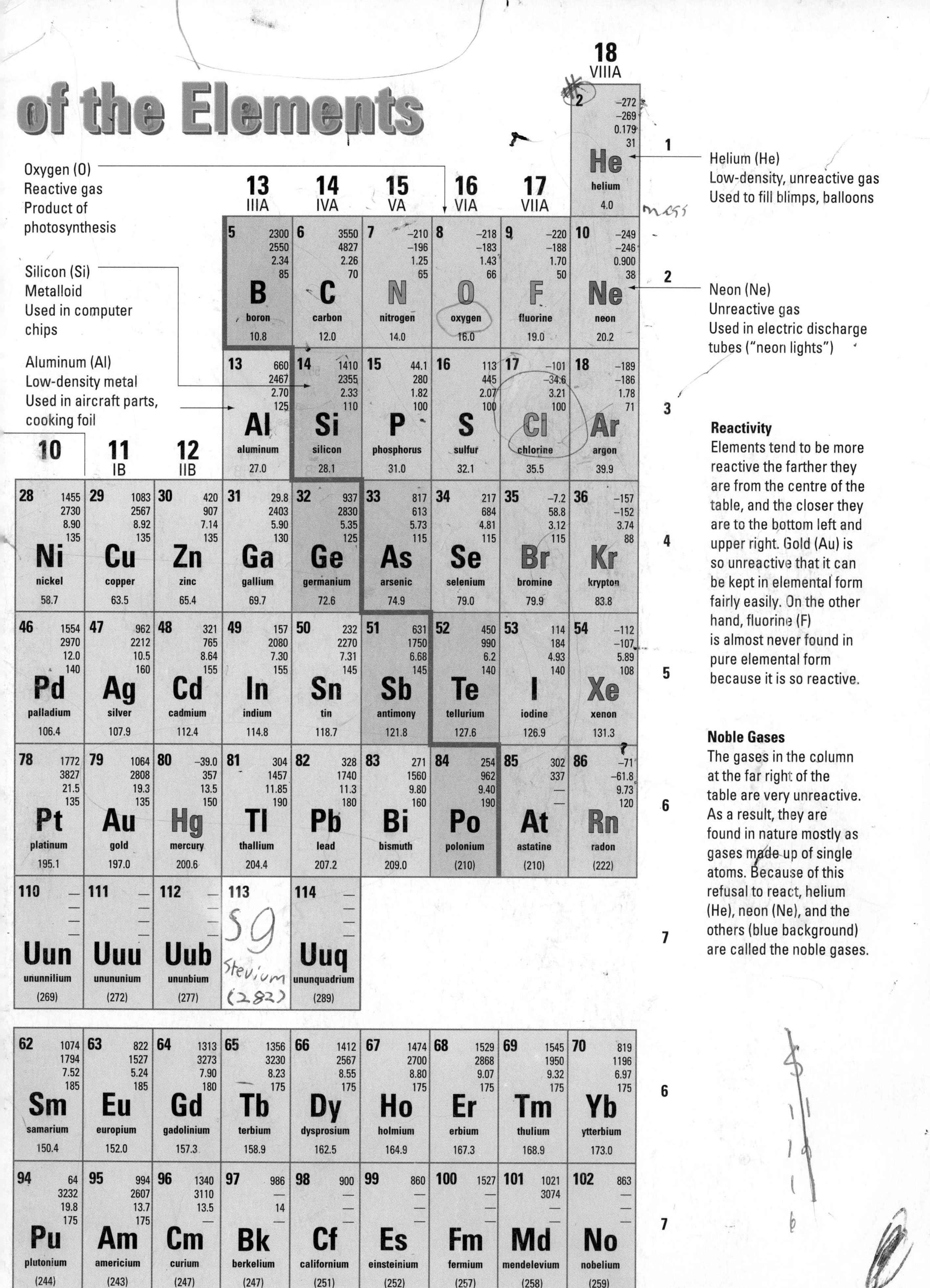

of the Elements

Oxygen (O)
Reactive gas
Product of photosynthesis

Silicon (Si)
Metalloid
Used in computer chips

Aluminum (Al)
Low-density metal
Used in aircraft parts, cooking foil

10	11 IB	12 IIB	13 IIIA	14 IVA	15 VA	16 VIA	17 VIIA	18 VIIIA	
								2 −272 −269 0.179 31 **He** helium 4.0	1
			5 2300 2550 2.34 85 **B** boron 10.8	6 3550 4827 2.26 70 **C** carbon 12.0	7 −210 −196 1.25 65 **N** nitrogen 14.0	8 −218 −183 1.43 66 **O** oxygen 16.0	9 −220 −188 1.70 50 **F** fluorine 19.0	10 −249 −246 0.900 38 **Ne** neon 20.2	2
			13 660 2467 2.70 125 **Al** aluminum 27.0	14 1410 2355 2.33 110 **Si** silicon 28.1	15 44.1 280 1.82 100 **P** phosphorus 31.0	16 113 445 2.07 100 **S** sulfur 32.1	17 −101 −34.6 3.21 100 **Cl** chlorine 35.5	18 −189 −186 1.78 71 **Ar** argon 39.9	3
28 1455 2730 8.90 135 **Ni** nickel 58.7	29 1083 2567 8.92 135 **Cu** copper 63.5	30 420 907 7.14 135 **Zn** zinc 65.4	31 29.8 2403 5.90 130 **Ga** gallium 69.7	32 937 2830 5.35 125 **Ge** germanium 72.6	33 817 613 5.73 115 **As** arsenic 74.9	34 217 684 4.81 115 **Se** selenium 79.0	35 −7.2 58.8 3.12 115 **Br** bromine 79.9	36 −157 −152 3.74 88 **Kr** krypton 83.8	4
46 1554 2970 12.0 140 **Pd** palladium 106.4	47 962 2212 10.5 160 **Ag** silver 107.9	48 321 765 8.64 155 **Cd** cadmium 112.4	49 157 2080 7.30 155 **In** indium 114.8	50 232 2270 7.31 145 **Sn** tin 118.7	51 631 1750 6.68 145 **Sb** antimony 121.8	52 450 990 6.2 140 **Te** tellurium 127.6	53 114 184 4.93 140 **I** iodine 126.9	54 −112 −107 5.89 108 **Xe** xenon 131.3	5
78 1772 3827 21.5 135 **Pt** platinum 195.1	79 1064 2808 19.3 135 **Au** gold 197.0	80 −39.0 357 13.5 150 **Hg** mercury 200.6	81 304 1457 11.85 190 **Tl** thallium 204.4	82 328 1740 11.3 180 **Pb** lead 207.2	83 271 1560 9.80 160 **Bi** bismuth 209.0	84 254 962 9.40 190 **Po** polonium (210)	85 302 337 — — **At** astatine (210)	86 −71 −61.8 9.73 120 **Rn** radon (222)	6
110 — — — — **Uun** ununnilium (269)	111 — — — — **Uuu** unununium (272)	112 — — — — **Uub** ununbium (277)	113	114 — — — — **Uuq** ununquadrium (289)					7

62 1074 1794 7.52 185 **Sm** samarium 150.4	63 822 1527 5.24 185 **Eu** europium 152.0	64 1313 3273 7.90 180 **Gd** gadolinium 157.3	65 1356 3230 8.23 175 **Tb** terbium 158.9	66 1412 2567 8.55 175 **Dy** dysprosium 162.5	67 1474 2700 8.80 175 **Ho** holmium 164.9	68 1529 2868 9.07 175 **Er** erbium 167.3	69 1545 1950 9.32 175 **Tm** thulium 168.9	70 819 1196 6.97 175 **Yb** ytterbium 173.0	6
94 64 3232 19.8 175 **Pu** plutonium (244)	95 994 2607 13.7 175 **Am** americium (243)	96 1340 3110 13.5 — **Cm** curium (247)	97 986 — 14 — **Bk** berkelium (247)	98 900 — — — **Cf** californium (251)	99 860 — — — **Es** einsteinium (252)	100 1527 — — — **Fm** fermium (257)	101 1021 3074 — — **Md** mendelevium (258)	102 863 — — — **No** nobelium (259)	7

Helium (He)
Low-density, unreactive gas
Used to fill blimps, balloons

Neon (Ne)
Unreactive gas
Used in electric discharge tubes ("neon lights")

Reactivity
Elements tend to be more reactive the farther they are from the centre of the table, and the closer they are to the bottom left and upper right. Gold (Au) is so unreactive that it can be kept in elemental form fairly easily. On the other hand, fluorine (F) is almost never found in pure elemental form because it is so reactive.

Noble Gases
The gases in the column at the far right of the table are very unreactive. As a result, they are found in nature mostly as gases made up of single atoms. Because of this refusal to react, helium (He), neon (Ne), and the others (blue background) are called the noble gases.